INTERNATIONAL SERIES OF MONOGRAPHS ON PHYSICS

SERIES EDITORS

An Introduction to Non-Perturbative Foundations of Quantum Field Theory

Franco Strocchi

OXFORD

UNIVERSITY PRESS

OXFORD
UNIVERSITY PRESS

Great Clarendon Street, Oxford, OX2 6DP,
United Kingdom

Oxford University Press is a department of the University of Oxford.
It furthers the University's objective of excellence in research, scholarship,
and education by publishing worldwide. Oxford is a registered trade mark of
Oxford University Press in the UK and in certain other countries

First published 2013
First published in paperback 2016

Published in the United States of America by Oxford University Press
198 Madison Avenue, New York, NY 10016, United States of America

British Library Cataloguing in Publication Data
Data available

Library of Congress Cataloging in Publication Data
Data available

ISBN 978–0–19–967157–1 (Hbk.)
ISBN 978–0–19–878923–9 (Pbk.)

Preface

The subject of these notes was the content of courses given at Scuola Normale Superiore (Pisa) during the last ten years and addressed to graduate students in theoretical physics.

Quantum field theory (QFT) has proved to be the most useful strategy for the description of elementary particle interactions, and as such is regarded as a fundamental part of modern theoretical physics. The textbooks on QFT are hundreds and some of them excellent, such as the recent book by S. Weinberg. In most presentations the emphasis is on the effectiveness of the theory in producing experimentally testable predictions, which at present essentially means perturbative QFT.

Clearly, it would be silly to underestimate the extraordinary success of perturbative quantum electrodynamics (QED) and of the perturbative standard model of electroweak interactions, the agreement with experiments being very impressive. However, also in view of the fact that QFT cannot be regarded as the final theory, it would be somewhat reductive to disdain questions of consistency and/or of mathematical soundness. The reasons are many.

The perturbative series is known to diverge, and actually, for the prototypic model of self-interacting scalar field (φ^4-theory), on which most of the textbooks are based, the perturbative expansion is misleading, since the theory has been proved to be trivial in $s+1$ dimensions (with $s \geq 3$). Clearly, this makes it difficult to define a QFT through its perturbative expansion and raises conceptual problems such as the mathematical consistency of such an expansion.

As a matter of fact, after more than fifty years of QFT we are still in the embarrassing situation of not knowing a single non-trivial (even non-realistic) model of QFT in 3+1 dimensions, allowing a non-perturbative control.

As a reaction to these consistency problems one may take the position that they are related to our ignorance of the physics of small distances and that QFT is only an effective theory, so that radically new ideas are needed for a consistent quantum theory of relativistic interactions (in 3+1 dimensions). Actually, proposals (like string theory, non-commutative spacetime) have recently been put forward for possible descriptions of the ultraviolet (UV) regime, leading to a theory more fundamental than QFT.

In this perspective, it may be useful to critically re-examine the *conceptual motivations for the birth of QFT*. Even if the difficulties of a relativistic quantum mechanics (RQM) based on relativistic wave equations are mentioned in any textbook on QFT, a general discussion of the conflict between quantum stability (i.e., spectral condition), relativistic covariance, and locality (i.e., localization and finite propagation speed) may be of help in understanding the transition from one-particle Schrödinger quantum mechanics, briefly relativistic wave mechanics (RWM), to QFT, with the necessary involvement of infinite degrees of freedom. No arbitrariness is involved in the free field case with mass gap, since the Fock representation is uniquely required by the existence of the Hamiltonian, but there is still no non-perturbative control of a non-trivial interaction in 3+1 dimensions, compatibly with the constraints of the positivity of the energy and locality.

In Chapter 2 the basic ideas and the *mathematical problems of the perturbative expansion* are outlined. The lack of convergence of the perturbative series is argued also on the basis of simple low-dimensional models (Dyson argument, non-convergence, asymptotic series, Borel summability). The general mathematical problems are discussed with reference to the mathematical inconsistency of the basic ingredients: namely, the interaction picture, the canonical quantization, and the choice of the Fock representation.

The *non-perturbative and mathematical foundations of QFT* are discussed in Chapter 3, in terms of general *physical* requirements, according to the principles of QM and relativistic invariance (Wightman formulation). The emerging theory provides a substitute for canonical quantization, which has been proved to be mathematically inconsistent in the presence of non-trivial interactions in 3+1 dimensions, and reduces to it in the free field case.

General non-perturbative results, with relevant experimental implications, are discussed in Chapter 4. In particular, the *spin–statistics theorem*, the *PCT theorem*, the occurrence of Schwinger terms in current commutators and the *axial anomaly*, which governs the $\pi_0 \to 2\gamma$ decay.

The analysis of the analyticity properties implied by the *physical* requirements of positivity of the energy, relativistic invariance and locality allows for a *non-perturbative definition of the functional integral* approach to QFT, which has proved useful for practical purposes and for the non-perturbative constructive approach. Since the Feynman path integral does not have the property (σ additivity), which allows computation of the integral by approximating the integrand, one needs the imaginary time version and the Wick rotation leading to it requires suitable analyticity properties. The imaginary times (Schwinger) functions define the so-called Euclidean QFT (Chapter 5), which displays remarkable analogies with *classical statistical mechanics*; the full equivalence with the real-time formulation is guaranteed by the Osterwalder–Schrader (OS) or reflection positivity, which plays a crucial role in the lattice approach to QFT.

In Chapter 6 the non-perturbative definition and existence of the S-matrix is briefly outlined through the derivation of the *LSZ asymptotic condition* from the general principles of quantum mechanics and relativity, both in the case of mass gap (Haag–Ruelle theory) and for fields describing massless particles (Buchholz theory). Needless to say, a non-perturbative approach to the S-matrix is a crucial issue for the control of the infrared problem in QED and for the discussion of color confinement in quantum chromodynamics (QCD). A brief discussion of the infrared problem in QED is given in Section 4.

The general problem of quantizing gauge theories is discussed in Chapter 7, with emphasis on the general structures and mechanisms; in particular, it is argued that the crucial property is not gauge invariance (which must be broken by a gauge fixing for the standard quantization procedures), but rather the local *Gauss law* (and the related

Ward identities), which is implied by gauge invariance (second Noether theorem) and actually survives the gauge-breaking by a gauge-fixing.

Most of the peculiar properties of gauge theories, like the non-locality of the charged states, the non-local field algebra, the *superselection of Gauss charges* associated to the gauge group, the evasion of the Goldstone theorem (Higgs mechanism), the "linearly rising quark potential" requiring a violation of the cluster property, and the infraparticle structure (i.e., the lack of a definite mass) of the charged particles in QED, are all strictly related to the local Gauss law and can be understood in general, without reference to a particular Lagrangian model. In view of the relevance of the Higgs mechanism for the standard model of elementary particle interactions, its realization in the local renormalizable gauges and in the physical Coulomb gauge is discussed in detail. As a result one obtains a *non-perturbative proof of the Higgs mechanism*: namely, that only the vector bosons corresponding to the broken generators acquire a mass, and their two-point function gives the Goldstone spectrum, so that there are no massless Goldstone bosons as well.

The last chapter is devoted to the *non-perturbative derivation of the vacuum structure and chiral symmetry breaking in QCD*, which govern the mass generation. Rather than relying on the semiclassical approximation in terms of the instanton dilute gas picture and on the topological classification of the gauge-field configurations in the functional integral, we revisit the Jackiw strategy of exploiting the topological structure of the gauge group. This avoids the problem of the zero functional measure of the regular gauge-field configurations and allows a localization of the topological structure, which is precluded for the instanton configurations by their property of minimizing the action integral. A careful formulation of the temporal gauge avoiding the mathematical inconsistencies and paradoxes, which invalidate the existing treatment, provides a *solution of the so-called $U(1)$ problem* and derives the θ vacuum structure solely from the *non-trivial topology of the gauge group, which is shown to force the breaking of chiral symmetry in each θ sector*.

The aim of these notes is to provide material complementary to the standard teaching of QFT, addressed to students interested in questions of principle and in those foundational aspects which have a non-trivial physical impact. For these reasons, an effort is made to correct widespread prejudices and misconceptions which partly invalidate the standard textbook presentation of crucial issues such as the infrared problem, the Higgs mechanism, the interplay between topology and vacuum structure in QCD, etc.

Moreover, the choice has been made of emphasizing the main ideas and problems without aiming to provide an exhaustive treatment of the subjects discussed, for which other more extended treatments, quoted in the footnotes, are available. The idea is to provide basic information for a possible wide audience, leaving to the mathematically minded reader the task of further deepening the outlined problems, according to his taste and his mathematical education, e.g., by referring to the very systematic book by J. Glimm and A. M. Jaffe.

The discussion of gauge field theories is largely based on collaborations and illuminating discussions with Gianni Morchio, to whom I am greatly indebted. For the general look to the non-perturbative foundations of QFT, the mentorship by A. S. Wightman is gratefully acknowledged.

Contents

1
Relativistic quantum mechanics

1 Quantum mechanics and relativity

Soon after the birth of quantum mechanics it became clear that in order to describe microscopic systems, such as electrons, protons, etc., at high energies one should combine quantum mechanics and relativity.

Both theories are very sound and fully under control in their domain of applications, also from a mathematical point of view; however, as we shall see, the combination of the two is a non-trivial problem. Quantum field theory (QFT) is supposed to provide a satisfactory relativistic quantum mechanics, but up to now we have only a perturbative control of QFT, and no non-trivial (even non-realistic) model in four (spacetime) dimensions is under non-perturbative control. Actually, the prototypical model of self-interacting scalar field, which is used in most textbooks for developing (non-trivial) perturbation theory, has been proved to be trivial under general conditions, when treated non-perturbatively (namely, the non-perturbatively renormalized coupling constant vanishes when the ultraviolet cutoff is removed). Such a negative result seems to apply also to quantum electrodynamics (QED) and more generally to quantum field theories which are not asymptotically free.[1] The mathematical consistency of the perturbative expansion in QED was indeed questioned soon after the setting of perturbation theory, since Dyson argued[2] that the perturbative series of QED cannot be summed and that big oscillations overwhelming the successful lowest orders are expected to arise (typically at order $n = 1/\alpha = 137$).

This means that in general the perturbative expansion is not reliable, and in general one cannot use it for defining a QFT model. The possibility of rescuing the success of the perturbative expansion by interpreting it as an asymptotic expansion in any case requires non-perturbative information, since an asymptotic series does not identify a unique function.

It should be stressed that most of our wisdom on QFT is derived from the perturbative expansion, and it would be silly to underestimate the extraordinary success of perturbative QED in yielding theoretical predictions which agree with the

---

[1] For a review of the arguments for the triviality of φ^4 theories, see R. Fernandez, J. Fröhlich, and A. D. Sokal, *Random Walks, Critical Phenomena, and Triviality in Quantum Field Theory*, Springer 1992, esp. Sects. 1.5.1.5–6, 1.5.2.

[2] F. Dyson, Phys. Rev. **85**, 631 (1952).

experiments up to the eleventh significant figure; also, the success of the perturbative treatment of the standard model is impressive.

The puzzling conflict between the successful predictions and the divergence of the (renormalized) perturbative series legitimate the need for a non-perturbative approach to the problem of combining quantum mechanics and relativity, with the aim of either validating the foundations of quantum field theory or displaying the need for radical changes and new ideas. In the history of theoretical physics there are famous examples in which the combination of different theories, with different origins, such as electromagnetism and thermodynamics or electromagnetism and mechanics, turned out to be impossible without conceptual revolutions, such as Planck energy quanta and special relativity. Furthermore, it might be shortsighted to dismiss consistency questions on the basis of an interpretation of quantum field theory as an effective low energy theory, one of the main motivations for the discovery of the standard model of elementary particle interactions having been that of curing the high energy inconsistency of the Fermi theory.

In the following sections we will start by briefly discussing the conceptual difficulties of a relativistic quantum mechanics based on relativistic wave equations, according to the ideas of one- (or few-)particle Schrödinger quantum mechanics. We shall then discuss the arguments in favor of quantum field theory.

2   Relativistic Schrödinger wave mechanics

A relativistic quantum mechanics should satisfy the following general requirements:
i) **spectral condition**, namely the energy spectrum should be bounded below, actually positive, and in the free case should satisfy the *relativistic energy–momentum dispersion law* (for simplicity we choose units such that $\hbar = c = 1$)

$$E = \sqrt{\mathbf{p}^2 + m^2}$$

(the positivity of the energy is needed for the stability of the corresponding quantum system under small external perturbations and for a reasonable physical interpretation);
ii) **hyperbolicity**, namely the evolution equation should admit solutions allowing the *localizability* of the observables, and the time evolution of localized observables should take place with a propagation speed less than the velocity of light. Such a condition, also briefly called *locality*, is necessary for the implementation of Einstein causality.

As we shall show in the following sections, such conditions cannot be satisfied within the framework of Schrödinger quantum mechanics, i.e., in terms of relativistic wave equations, and a radical change of perspective is needed.

Historically, the first attempts to combine quantum mechanics (QM) and relativity (R) went in the direction of writing a relativistic version of the Schrödinger equation (*relativistic wave equations*), and it was soon realized that serious problems emerge especially in the presence of interactions.

2.1 Relativistic Schrödinger equation

The simplest step is to replace the non-relativistic energy–momentum dispersion law $E = \mathbf{p}^2/2m$ by the relativistic one, $E = \sqrt{\mathbf{p}^2 + m^2}$, so that one obtains the *relativistic Schrödinger equation*

$$i\partial_t\, \psi = \sqrt{-\Delta + m^2}\, \psi. \tag{1.2.1}$$

On the right-hand side we have a pseudodifferential operator, whose mathematical meaning is that of acting as the multiplication operator $\sqrt{\mathbf{p}^2 + m^2}$ on the Fourier transform $\tilde{\psi}(\mathbf{p}, t)$ of ψ.

Such an operator is non-local, and actually it implies that even if the initial data is of compact support, it cannot remain so at later times.

Proposition 2.1 *Equation (1.2.1) does not have solutions of compact support.*

Proof. In fact, if $\psi(\mathbf{x}, t)$ is of compact support in $\mathbf{x}$ for any $t \in [0, \varepsilon)$, so is its time derivative and, therefore, the Fourier transforms of both, with respect to $\mathbf{x}$, are analytic functions of $\mathbf{p}$. Now, by eq. (1.2.1), $\partial_t\tilde{\psi}(\mathbf{p}, t) = -i\sqrt{\mathbf{p}^2 + m^2}\,\tilde{\psi}(\mathbf{p}, t)$, and the discontinuity of the square root across the cut in the complex $\mathbf{p}^2$ plane running from $-m^2$ to ∞ cannot be removed by multiplication by the analytic function $\tilde{\psi}(\mathbf{p}, t)$. Hence, $\tilde{\psi}(\mathbf{p}, t)$ cannot be analytic, and $\psi(\mathbf{x}, t)$ cannot have compact support.

The lack of locality of the time evolution is a serious drawback. First, localizability of the wave functions is strictly related to the possibility of implementing relativistic causality—namely, the property that observable densities can be localized—so that their support evolves in time with velocity less that $c = 1$.

For example, such a localization problem arises for the current density $j_\mu(\mathbf{x}) = (j_0(\mathbf{x}), j_i(\mathbf{x}))$, defined by

$$j_0 = (1/2m)\,[\psi^*(\mathbf{x})\,(\sqrt{-\Delta + m^2}\psi)(\mathbf{x}) + (\sqrt{-\Delta + m^2}\psi^*)(\mathbf{x})\,\psi(\mathbf{x})],$$

$$j_i(\mathbf{x}) = (i/2m)\,[\psi^*(\mathbf{x})\,\partial_i\psi(\mathbf{x}) - (\partial_i\psi^*)(\mathbf{x})\,\psi(\mathbf{x})], \tag{1.2.2}$$

equivalently by $(\partial_\mu = \partial/\partial x^\mu)$

$$j_\mu(\mathbf{x}, t) = (i/2m)[\,\psi^*\,\partial_\mu\psi - (\partial_\mu\psi^*)\,\psi\,](\mathbf{x}, t) \equiv (i/2m)(\psi^*\,\overset{\leftrightarrow}{\partial_\mu}\,\psi)(\mathbf{x}, t).$$

The interpretation of $j_0(\mathbf{x})$ as a probability density is supported by its being positive and by the fact that it reduces to the standard Schrödinger probability density in the non-relativistic limit $E \to mc^2$. Furthermore, j_μ satisfies the continuity equation

$$\partial^\mu j_\mu = 0$$

and it transforms as a four-vector under the Lorentz transformations $x \to x' = \Lambda x$, if ψ transforms as a scalar, i.e., $\psi'(x') = \psi(x)$.

However, if j_μ is of compact support at the initial time (obtained by taking $\psi(\mathbf{x}, 0)$ of compact support), it does not remain so at any later time.

Moreover, if the form of j_0 is used to define the scalar product between two wave functions, the hermiticity of operators is not the same as in Schrödinger QM; for example, the multiplication by $\mathbf{x}$ is not hermitian and cannot describe the position operator.

A hermitian position operator

$$x_i^{op} \equiv \tfrac{1}{2}\left(x_i + x_i^*\right) = x_i - \tfrac{1}{2}\left(-\Delta + m^2\right)^{-1}\partial_i$$

can be introduced, but one cannot have wave functions with localization as close as one likes to δ functions. Quite generally, the maximum localization has a tail of exponential decay with rate given by the inverse of the mass and, moreover, such a "localization" property is not stable under Lorentz transformations.[3]

Another serious problem arises if one tries to formulate the interaction case. In fact, the non-locality of the free Hamiltonian makes it very difficult to introduce the interaction, such as the minimal coupling with an electromagnetic potential. In fact, it is hard to give a meaning to the formal operator $\sqrt{(\mathbf{p} - e\mathbf{A}(\mathbf{x}))^2 + m^2}$, without *a priori* knowing the spectrum of the operator under square root, and this is difficult to control, since $\mathbf{p}$ and $\mathbf{A}(\mathbf{x})$ do not commute.

For these reasons, such an attempt of a relativistic quantum mechanics was soon abandoned in favor of an hyperbolic wave equation (see the next section). Actually, as we shall see, a satisfactory interpretation and solution of such problems is provided by quantum field theory, which however requires a strong departure from the one- (or few-)particle picture of Schrödinger quantum mechanics.

2.2 Klein–Gordon equation

A local time evolution is given by the Klein–Gordon equation

$$(\Box + m^2)\varphi(x) = 0, \tag{1.2.3}$$

which is a hyperbolic equation and therefore preserves the localization of the initial data with a finite propagation speed.

However, eq. (1.2.3) has solutions whose Fourier transform with respect to time has support unbounded below, i.e., the frequency ω may take arbitrarily large values with both signs $\omega = \pm\sqrt{\mathbf{k}^2 + m^2}$. The negative sign violates the positive energy spectral condition.

Moreover, for solutions with negative frequencies also the time component $\rho(\mathbf{x}, t) \equiv j_0(\mathbf{x}, t)$ of the conserved current

$$j_\mu = (i/2m)\,\varphi^*\,\overset{\leftrightarrow}{\partial_\mu}\,\varphi$$

becomes negative, and there is no good candidate for a probability density.

Thus, for a quantum-mechanical interpretation, the solutions with negative frequencies must be excluded by means of a supplementary condition. Since eq. (1.2.3) is of second order in the time derivative, the initial data involve both the value of φ and

---

[3] For a discussion of the localization problem, see, e.g., S. S. Schweber, *An Introduction to Relativistic Quantum Field theory*, Harper and Row 1961, Section 3c and references therein. For the conflict between localization or hyperbolicity and positive energy (Propositions 2.1 and 2.2), see F. Strocchi, Foundations of Physics, **34**, 510 (2004).

of its time derivative, and the energy spectral condition requires that the initial data must satisfy the condition

$$i\,\partial_t\varphi(\mathbf{x},0) = \sqrt{-\Delta + m^2}\,\varphi(\mathbf{x},0). \tag{1.2.4}$$

However, as discussed previously, this is a non-local condition, and therefore, by the argument of Proposition 2.1, the initial data for solutions satisfying the positive energy spectral condition cannot have compact support. One is therefore facing the same localization problems of eq. (1.2.1).

An advantage with respect to the Schrödinger equation is that the minimal coupling electromagnetic interaction is described by local terms; namely, by the following equation:

$$D_\mu D^\mu \varphi(x) + m^2\varphi(x) = 0, \quad D_\mu \equiv \partial_\mu - ieA_\mu(x).$$

There is, however, an additional serious problem (beyond the localization); namely, the positive energy condition is not stable under the interaction; in fact, quite generally, even if the initial data satisfy the positive energy condition, the Fourier transform of the corresponding solution may contain negative frequencies.

For example, an interaction term $\mathcal{U}(x)\varphi(x)$, in eq. (1.2.3), with $\mathcal{U}(x)$ a potential of compact support in space and time, induces transitions to negative frequencies, since the frequency spectrum of $\mathcal{U}$ is unbounded below. An interaction of this form is provided by a minimal coupling with an external electromagnetic field, since it gives rise to a similar term, $eA_\mu^2(x)\,\varphi(x)$, in the equations of motion. This phenomenon is known as the *Klein paradox*, and represents a serious obstacle for the interpretation of $\varphi(\mathbf{x},t)$ as the wave function of a quantum particle.

The analysis of the energy spectrum is conveniently done in the (equivalent) first-order formulation, which in the free case reads

$$i\,\partial_t u = \begin{pmatrix} 0 & 1 \\ -\Delta + m^2 & 0 \end{pmatrix} u \equiv H_0\,u, \quad u = \begin{pmatrix} u_1 \\ u_2 \end{pmatrix}. \tag{1.2.5}$$

In this way, the quantum-mechanical structure is better displayed. The Hilbert space $\mathcal{H}$ is defined by the scalar product

$$(u,\,v) = \tfrac{1}{2}\int d^3x \left[\nabla u_1 \nabla v_1 + m^2 u_1\,v_1 + u_2\,v_2\right].$$

Technically, with $H^i(\mathbf{R}^3) \equiv \{f;\,(1+\mathbf{k}^2)^{i/2}\,\tilde{f}(\mathbf{k}) \in L^2(\mathbf{R}^3)\}$, $i = 1, 2$, one has that $u_1 \in H^1(\mathbf{R}^3)$, $u_2 \in L^2(\mathbf{R}^3)$; the Hamiltonian H_0 is self-adjoint on $D(H_0) = H^2(\mathbf{R}^3) \oplus H^1(\mathbf{R}^3)$, and its spectrum is symmetric with respect to the origin.

The interaction with external (tempered) fields—e.g., the minimal coupling electromagnetic interaction—corresponds to a bounded, in general non-symmetric, perturbation. In general, the interaction with external fields having compact support in time does not commute with the projection operator P_+, which projects on the initial data satisfying eq. (1.2.4), and therefore transitions to negative energies are induced.[4]

2.3 Dirac equation

One might think that the problems of the relativistic wave equations discussed above originate because time and space derivatives do not appear in a symmetric way and that the second time derivative gives rise to both negative frequencies and a non-positive density $\rho(\mathbf{x})$. These seem to have been the motivations for the Dirac equation,[5] which is linear in time and space derivatives at the expense of a four-component wave function; in the free case it reads

$$[-i\gamma^{\mu}\partial_{\mu} + m]\psi(x) = 0, \tag{1.2.6}$$

where γ^{μ}, $\mu = 0, 1, 2, 3$ are 4×4 matrices satisfying

$$\gamma^{\mu}\gamma^{\nu} + \gamma^{\nu}\gamma^{\mu} = 2g^{\mu\nu},$$

with $g^{\mu\nu}$ the Minkowski metric ($g^{00} = 1 = -g^{ii}$, $i = 1, 2, 3$). The γ's are determined up to a similarity transformation, and therefore there exist different explicit representations of them. A widely used possibility is

$$\gamma^{0} = \begin{pmatrix} 0 & 1 \\ 1 & 0 \end{pmatrix} \equiv \tau_1 \times \mathbf{1}\,, \quad \gamma^{i} = \begin{pmatrix} 0 & \sigma^{i} \\ -\sigma^{i} & 0 \end{pmatrix} \equiv i\tau_2 \times \sigma^{i},$$

where σ_i, τ_i, $i = 1, 2, 3$ are the Pauli matrices.[6] Relativistic invariance is obtained if ψ transforms as a continuous unitary representation $U(\Lambda(A))$, $A \in SL(2, C)$, of $SL(2, C)$, the universal covering of the restricted Lorentz group $L_{+}^{\uparrow} \equiv$ the group of Lorentz transformations Λ with no space and time inversions,

$$(U(\Lambda(A))\,\psi)(x) = (S(A)\,\psi)(\Lambda(A^{-1})x), \tag{1.2.7}$$

where $S(A)$ is a 4×4 matrix representation of $SL(2, C)$.[7]

---

[4] For a discussion of the external field problem, see the Proceedings of the 1977 Erice School, *Invariant Wave Equations*, G. Velo and A. S. Wightman (eds.), Springer 1978, in particular the contribution by R. Seiler.

[5] P. A. M. Dirac, Proc. Roy. Soc. (London), **A117**, 610 (1928).

[6] The free Dirac equation can be easily derived from the conditions that i) the Dirac differential operator D be linear in time and space derivatives, i.e., of the form (1.2.6), ii) the Hamiltonian defined by $i\,\partial_t\psi = H\,\psi$ be hermitian (this requires γ_0^{-1} and $\gamma_0^{-1}\gamma^i$ hermitian), iii) the spectrum of H satisfy the mass-shell condition $H^2 = p^2 + m^2$ (this implies $\gamma_0^2 = 1$, $\{\gamma_0, \gamma_i\} = 0$ and $\{\gamma_i, \gamma_j\} = -2\delta_{ij}$), v) the conserved current density $\rho = j_0$ be free of derivatives (this implies $j_\mu = \psi^*\gamma_0\gamma_\mu\psi$). For a detailed discussion of the Dirac equation see S. S. Schweber, *An Introduction to Relativistic Quantum Field Theory*, Harper and Row 1961; E. Corinaldesi and F. Strocchi, *Relativistic Wave Mechanics*, North-Holland 1963; B. Thaller, *The Dirac Equation*, Springer 1992. A brief account is given in the Appendix below.

[7] The homomorphism between the group $SL(2, C)$ of 2×2 matrices A with $\det A = 1$ and the restricted Lorentz transformations Λ is obtained by representing the four-vector x by the 2×2 matrix $X \equiv \sum x^\mu \tau^\mu$, $\tau^0 = \mathbf{1}$, τ^i the Pauli matrices; since $x^\mu x_\mu = \det X$, AXA^* defines a Lorentz transformation $\Lambda(A) = \Lambda(-A)$. For more details see R. F. Streater and A. S. Wightman, *PCT, Spin and Statistics, and All That*, Benjamin 1964, pp. 9–16.

The form invariance of the Dirac equation is obtained provided

$$S(A)^{-1}\,\gamma^\mu\,S(A) = \Lambda(A)^\mu_\lambda\,\gamma^\lambda.$$

The Dirac conserved current is

$$j_\mu(x) = \bar\psi(x)\gamma_\mu\psi(x), \quad \bar\psi(x) \equiv \psi^*\gamma^0, \tag{1.2.8}$$

and the (charge) density $j_0(x) = \psi^*(x)\psi(x)$ is positive definite (with no condition) and can be interpreted as a probability density. This is one of the advantages with respect to the Klein–Gordon equation.

Another important feature of the Dirac equation is that its time evolution is hyperbolic and the initial data can be taken of compact support. This property persists for a large class of (local) interactions, including the electromagnetic minimal coupling.[8]

However, as seen in the case of the Klein–Gordon equation, the spectrum of the Hamiltonian

$$H_0 = -i\alpha^i\partial_i + \gamma^0 m, \quad \alpha^i \equiv \gamma^0\gamma^i = (\gamma^0\,\gamma^i)^* \tag{1.2.9}$$

is not positive and one faces the problem of eliminating the negative energy solutions as before. Again, one could use initial data satisfying a positive energy condition

$$P_+\psi = \psi, \quad P_+ = (-\Delta + m^2)^{-1/2}[\gamma^0\sqrt{-\Delta + m^2} - i\gamma^i\nabla_i + m],$$

but i) such a projection operator is non-local, so the initial data satisfying the above condition cannot have compact support, ii) time-dependent interactions, such as an electromagnetic interaction, induce transitions between positive and negative energy solutions, as for the Klein–Gordon equation discussed above (*Klein paradox*[9]).

Due to the occurrence of negative energy solutions, the interpretation of the coordinate **x** as the position operator is problematic; in fact,

$$\dot x_i(t) = i\,[\,H_0,\,x_i(t)\,] = e^{iH_0 t}\alpha_i\,e^{-iH_0 t}$$

and, since $\alpha_i^2 = 1$, $\dot x$ has eigenvalues ± 1, corresponding (in our units) to $\pm c$ (*Zitterbewegung*); moreover, even in the free case $\ddot x_i \neq 0$. As in the Klein–Gordon case, the position operator is obtained by taking projections on the positive energy subspace $\mathbf{x}^{op} = P_+\mathbf{x}\,P_+$, with the inevitable delocalization problems discussed above.

To cure the negative energy problem, Dirac proposed his "hole theory", according to which in the ground state all negative energy states are occupied and the Pauli exclusion principle precludes any transition to them. This step represents a radical departure from Schrödinger (few-particle) quantum mechanics, since the actual picture involves both the Dirac wave function and the Dirac sea of occupied negative energy states, i.e., infinite degrees of freedom. Interactions can induce a transition from a

---

[8] D. Buchholz, S. Doplicher, G. Morchio. J. Roberts and F. Strocchi, Ann. Phys. **209**, 53 (2001), Sect. 2.

[9] See, e.g., B. Thaller, 1992, Sect. 4.5 and p. 307; for a comprehensive treatment see A. S. Wightman, Invariant wave equations: general theory and applications to the external field problem, in *Invariant Wave Equations*, Erice 1977, G. Velo and A. S. Wightman (eds.), Springer 1978, pp. 76ff.

negative to a positive energy state, giving rise to a hole in the Dirac sea, which will appear as a particle of opposite charge, with the same mass m and *positive energy with respect to the ground state*, interpretable as an *antiparticle*. The net result is the creation of a pair consisting of a particle and its antiparticle.

It is not difficult to recognize in this bold Dirac idea the seeds of quantum field theory, where the field operators contain both positive and negative frequencies and the positive energy spectrum is a property of the states.[10]

2.4 The general conflict between locality and energy positivity

The roots of the difficulties discussed so far are rather deep, since one can show that the following general requirements:

   i) *positive energy spectrum*
  ii) *finite propagation speed (hyperbolicity)*
 iii) *localization of the initial data (compact support in space)*

for a time evolution described by relativistic wave equations are in conflict with a reasonable quantum-mechanical interpretation of the solutions.[11]

Proposition 2.2 *For a relativistic wave equation, if the corresponding time evolution satisfies the properties i)–iii), then it cannot induce a displacement of the compact support in space of the initial data.*

Furthermore, there is no non-trivial localized solution $\varphi(\mathbf{x}, t)$, transforming covariantly under the (restricted) Lorentz group $L_+^\uparrow$,

$$\varphi(x) \to \varphi^\Lambda(x') = S(A)\,\varphi(\Lambda(A^{-1})x'), \quad \Lambda(A) \in L_+^\uparrow, \quad x' = \Lambda x, \tag{1.2.10}$$

with $A \in SL(2,C)$ and $S(A)$ a finite dimensional representation of $SL(2,C)$, the universal covering of the restricted Lorentz group $L_+^\uparrow$.

Proof. In fact, the quantum-mechanical interpretation of a relativistic wave equation requires that positivity of the energy is equivalent to the occurrence of only positive frequencies in the Fourier transform (with respect to time) of the wave function $\varphi(\mathbf{x}, t)$. Therefore $\varphi(\mathbf{x}, t)$ has an analytic continuation to complex times

$$\varphi(\mathbf{x}, z = t - i\tau) = (2\pi)^{-1/2} \int d\omega\, \tilde{\varphi}(\mathbf{x}, \omega)\, e^{-i\omega(t-i\tau)},$$

---

[10] It is often stated that one of the main merits of the Dirac equation (as emphasized in the motivation of the Nobel prize) is the prediction of the positron as a hole of the Dirac sea. Actually, the occurrence of negative energy solutions and the related prediction of antiparticles (in the QFT interpretation) is common to all relativistic local equations (Proposition 2.2). Also, the achieved positivity of $\rho = j_0$ is not a must, since its interpretation changes in QFT. The real achievement of the Dirac equation is the *local relativistic description of spin 1/2 particles*, in a way which departs from eq. (1.2.1) with crucial predictions in agreement with experiments, when the electromagnetic interaction is switched on, in particular, the prediction of the fine structure of the hydrogen spectrum (see, e.g., E. Corinaldesi and F. Strocchi, *Relativistic Wave Mechanics*, North Holland 1963, pp. 202–6).

[11] F. Strocchi, Foundations of Physics **34**, 501 (2004).

where $\tilde{\varphi}$ denotes the Fourier transform of φ with respect to time, and $\varphi(\mathbf{x}, z)$ is analytic for $z = t - i\tau$, $t \in \mathbf{R}$, $\tau > 0$, i.e., in the lower half of the complex z plane. Furthermore, $\varphi(\mathbf{x}, t)$ is the continuous limit of $\varphi(\mathbf{x}, z)$, when $\tau = \mathrm{Im}\, z \to 0$.

Now, if R is any (compact) region in space disjoint from the compact support K of the initial data, $\varphi(\mathbf{x}, 0)$, by hyperbolicity, for a sufficiently small interval of time $I_\varepsilon = (0, \varepsilon)$ the solution remains zero in R.

This means that for any f, with $\mathrm{supp}\, f \subseteq R$, there is a time interval I_ε, such that

$$\varphi(f, t) \equiv \int d^3x\, f(\mathbf{x})\, \varphi(\mathbf{x}, t) = 0, \quad t \in I_\varepsilon.$$

This implies that the function $\varphi(f, z)$, which is analytic in the lower half plane, has a continuous limit on the real axis and vanishes on the open interval $(0, \varepsilon)$ of the real axis.

Hence, by Schwarz's reflection principle for analytic functions, $\varphi(f, z)$ vanishes everywhere.[12] Since this is true for any compact R disjoint from K, it follows that $\mathrm{supp}\, \varphi(\mathbf{x}, t) \subseteq K, \forall t$. This property of the time evolution of a wave function raises serious difficulties for a reasonable physical interpretation, since there is no room for a particle displacement in space.

For the proof of the second statement, for simplicity we consider the case of a wave function transforming as a scalar, i.e., $S(A) = \mathbf{1}$, so that eq. (1.2.10) reads

$$\varphi^\Lambda(\Lambda(\mathbf{x}, t)) = \varphi(\mathbf{x}, t).$$

Then, if $\mathrm{supp}\, \varphi(\mathbf{x}, 0) \subseteq K$, one has

$$\mathrm{supp}\, \varphi^\Lambda(\mathbf{x}', t' = 0) = \mathrm{supp}\, \varphi(\Lambda^{-1}(\mathbf{x}', 0)) \subseteq K^\Lambda \equiv \{\mathbf{x}'; \Lambda^{-1}(\mathbf{x}', 0) \in K\}$$

and by the above argument $\mathrm{supp}\, \varphi^\Lambda(\mathbf{x}', t') \subseteq \mathrm{supp}\, \varphi^\Lambda(\mathbf{x}', 0) \subseteq K^\Lambda$, for all t'. On the other hand, given a point $\mathbf{x} \notin K$, one can find a Λ and a time t such that

---

[12] We recall that the Schwarz's reflection principle is a special case of the edge of the wedge theorem (see R. F. Streater and A. S. Wightman, *PCT, Spin and Statistics, and All That*, Benjamin 1964, Theorem 2-13, for the case of one complex variable and Theorem 2-14 for several complex variables). In the case of one complex variable, it says that given two functions $F_i(z), i = 1, 2$, analytic in domains D_i contained in the upper and lower half complex plane respectively, with a real interval (a, b) as a common part of their boundaries, if the limits of F_i are continuous and coincide on (a, b), then the two functions are the analytic continuation one of the other. In fact, one can show that if F_i have continuous limits on the common boundary, the convergence is uniform. Then, let C be the union of two closed contours in D_1 and D_2, with $(a', b') \subset (a, b)$ as a common boundary, run in opposite directions and consider the function $F(z) = F_i(z)$, for $z \in D_i$, respectively, and equal to their common value on (a', b'). It is easy to see that the integral

$$(2\pi i)^{-1} \int_C dz'\, F(z')/(z' - z)$$

defines a function which is analytic inside C_i and also on (a', b'), so that it gives the required analytic continuation.

As a special case, if a function $F(z)$ analytic in a domain D of the upper complex plane, with an interval (a, b) as part of its boundary, has a continuous real limit on (a, b), then the function $\overline{F}(\bar{z})$ analytic in $\overline{D}$ provides an analytic continuation of $F(z)$; in particular, if such a limit on (a, b) vanishes, the function $F(z)$ vanishes.

$(\mathbf{x}',t') = \Lambda(\mathbf{x},t)$ satisfies $\Lambda^{-1}(\mathbf{x}',0) \in K^\Lambda$ and $\varphi^\Lambda(\mathbf{x}',t') \neq 0$. Hence, $\varphi(\mathbf{x},t) = \varphi^\Lambda(\Lambda(\mathbf{x},t)) \neq 0$, which contradicts the derived stability of the support of $\varphi(\mathbf{x},t)$. Hence, the general requirements i)–iii) plus Lorentz covariance cannot be satisfied.

In conclusion, quite generally for relativistic wave equations there is a conflict between the localization in space and the positivity of the frequency spectrum of the solutions.

Thus, it is impossible to consider the relativistic equations as the evolution equations for wave functions having the Schrödinger interpretation of state vectors. In fact, in this case the occurrence of negative frequencies implies that the spectrum of the generator of time translations is not positive, and since in the quantum-mechanical framework such a generator has the meaning of the energy operator, one has states with negative energy.

One often finds in the literature the statement that the problem of the negative frequencies in the solutions of the relativistic *wave* equations is solved by their interpretation in terms of antiparticles, but by the above remark this is not possible within a (Schrödinger) quantum-mechanical framework, because then antiparticles would have negative energy.

To overcome this crucial difficulty, Feynman suggested an interpretation in which antiparticles are running backwards in time.[13] In our opinion this recipe is problematic in the framework of Schrödinger quantum mechanics, and its effectiveness in picturing the perturbative expansion of Green's function in terms of Feynman diagrams essentially relies on a change of perspective; namely, it is effectively an expansion of the time evolution of a quantum field, rather than of a wave function.[14]

As we shall see, the possibility of interpreting the unavoidable negative frequencies as describing antiparticles requires abandonement of the one-particle Schrödinger quantum mechanics—the starting point of quantum field theory.

In conclusion, the difficulties of a one particle interpretation of Dirac equation wave equations (instability against interactions and conflict with locality) indicate that almost inevitably a relativistic quantum mechanics (RQM) must involve infinite degrees of freedom.

As we have remarked before, such a possible solution of the problems discussed above was first foreseen by Dirac in connection with the interpretation of the negative energy solutions of the Dirac equation (Dirac hole theory), by crucially exploiting the Pauli exclusion principle. The above Proposition 2.2 points out that the problem arises quite generally for the relativistic wave equations, even in cases where one cannot have recourse to the Pauli principle.

The above Proposition 2.2 may be regarded as a counterpart of the TCP theorem of quantum field theory. Already at the level of relativistic wave equations, the necessary occurrence of negative frequencies, which signal the need of antiparticles in the quantum field theory interpretation, follows from basic physical properties;

[13] R. P. Feynman, Phys. Rev. **76**, 749 (1949).

[14] For a discussion of this point see E. Corinaldesi and F. Strocchi, *Relativistic Wave Mechanics*, North-Holland 1963, Part III, Chap. VI.

namely, relativistic locality, positive energy spectrum, and Lorentz covariance. Exactly the same ingredients are responsible for the TCP theorem in quantum field theory.

Summarizing, as a general requirement for a relativistic quantum theory, the condition of positivity of the energy may be better formalized as the *relativistic spectral condition*, i.e., the spectrum of the four-momentum P_μ must be contained in the *closed forward cone*:

$$\overline{V}_+ \equiv \{p; p^2 \geq 0, p_0 \geq 0\}.$$

Furthermore, quite generally the *locality requirement* may be better formalized by appealing to Einstein causality: if two observables $A_{\mathcal{O}_1}$, $B_{\mathcal{O}_2}$ are localized in relatively spacelike regions of spacetime $\mathcal{O}_1$, $\mathcal{O}_2$, then they must be independent in the quantum-mechanical sense, i.e., they must commute:

$$[A_{\mathcal{O}_1}, B_{\mathcal{O}_2}] = 0.$$

3   Relativistic particle interactions and quantum mechanics

Another source of problems for combining Schrödinger quantum mechanics and relativity is their basically different description of particle interactions. In Schrödinger QM the treatment of interaction between particles is based on the canonical (Hamiltonian) formalism and on the Newtonian concept of *force at a distance* (typically described by an interaction potential), which makes use of simultaneity and therefore cannot be relativistically invariant. Indeed, as we shall see below, there are serious obstructions, even at the classical level, for building up a relativistic dynamics of particles interacting by forces at a distance. The natural concept of interaction compatible with relativity is that of contact interaction or more generally of local interaction with a dynamical medium or a field. One is then led to abandon the Newtonian picture of few particle interactions and to consider the infinite degrees of freedom associated with the (dynamical) field responsible for the interaction.

3.1   Problems of relativistic particle interactions

A relativistic dynamics of particles in terms of forces at a distance meets the problem that interactions cannot be instantaneous (an inevitable delay resulting from the finite propagation speed), and that simultaneity is not a relativistically invariant concept.[15] In fact, in the case of N particles of definite masses, if $x^{(i)}(\tau^{(i)})$ denotes the world line of the ith particle, $\tau^{(i)}$ the corresponding proper time, and $\dot{x}^{(i)}_\mu \equiv dx^{(i)}_\mu/d\tau^{(i)}$ the four-velocity, Lorentz invariance implies that

$$\dot{x}^{(i)}_\mu \, \dot{x}^{(i)\,\mu} = 1. \tag{1.3.1}$$

Hence, the stability of this condition under time evolution imposes the following constraint on the accelerations

---

[15] L. D. Landau and E. Lifshitz, *The Classical Theory of Fields*, Addison-Wesley 1962, Chap. III, Sect. 15.

$$\dot{x}_{\mu}^{(i)}\, \ddot{x}^{(i)\,\mu} = 0 \tag{1.3.2}$$

and therefore on the forces.

Now, a spacetime translation invariant force at a distance on a particle depends on the relative positions (and possibly on the velocities) of the other particles at the same time, i.e., on the Cauchy data at the given time, which can be assigned freely, so that in general eq. (1.3.2) will not be satisfied. For example, as discussed by Wigner[16] a spacetime reflection invariant central force between the i, j pair of particles is of the form

$$F_{\mu}^{ij} = (x_{\mu}^{(i)} - x_{\mu}^{(j)})f = -F_{\mu}^{ji},$$

where f is a function of the invariants which can be constructed in terms of the four-vectors x_{μ} and the four-velocities;[17] it is clear that in general the four-vector $x_{\mu}^{(i)} - x_{\mu}^{(j)}$ will not be orthogonal to both tangents of the two world lines (by the freedom of the Cauchy data).

To cure this problem, van Dam and Wigner[18] proposed to use *non-local* "forces at a distance" such that the force $F^{(ij)}$ on the ith particle by the jth particle depends on *all* the points of the trajectory of the jth particle, which are spacelike with respect to $x^{(i)}$.[19] In this way, however, the dynamical problem is no longer formulated in terms of a Cauchy problem for differential equations and becomes almost intractable, since it involves *a priori* knowledge of part of the particle trajectories.

No interaction theorems for relativistic particle dynamics have been proved within the canonical (Hamiltonian) formalism, on which quantum mechanics crucially relies.

A substantial step for the proof of such results[20] is the formalization of the property of relativistic invariance. A simple and natural translation of Poincaré invariance is that the ten *generators of the Poincaré group* (spacetime translations, space rotations, and Lorentz transformations) are realized by functions of the canonical variables and

---

[16] E. P. Wigner, Relativistic Interaction of Classical Particles, in *Fundamental Interactions at High Energy*, Coral Gables 1969, T. Gudehus *et al.* (eds.), Gordon and Breach 1969, p. 344.

[17] The general form of the force F_{μ}^{ij} satisfying the condition $F_{\mu}^{ij} = -F_{\mu}^{ji}$ is

$$F_{\mu}^{ij} = (x_{\mu}^{(i)} - x_{\mu}^{(j)})f + (\dot{x}_{\mu}^{i} - \dot{x}_{\mu}^{j})g + \varepsilon_{\mu}^{\nu\rho\lambda}(x_{\nu}^{(i)} - x_{\nu}^{(j)})\dot{x}_{\rho}^{i}\,\dot{x}_{\lambda}^{j}\,h,$$

where f, g, h are functions of the invariants which can be constructed in terms of the positions and the four-velocities of the ith and jth particles. For central forces $g = 0$ and space-time reflection invariance requires $h = 0$.

[18] H. van Dam and E. P. Wigner, Phys. Rev. **138**, B1576 (1965); **142**, 838 (1966).

[19] Similarly, in the Feynman and Wheeler theory of particle interactions (J. A. Wheeler and R. P. Feynman, Rev. Mod. Phys. **21**, 425 (1949)) the force $F^{(ij)}$ depends on the points of the jth trajectory which lie on the light cone centered at $x^{(i)}$; with such a choice, the conservation laws of the particle energy–momentum are not satisfied.

[20] D. G. Currie, T. F. Jordan, and E. C. G. Sudarshan, Rev. Mod. Phys. **35**, 350 (1963), for the two-particle case; H. Leutwyler, Nuovo Cim. **XXXVII**, 556 (1965) for the N-particle case; D. G. Currie and T. F. Jordan, Interactions in relativistic classical particle mechanics, in (Boulder) Lectures in Theoretical Physics Vol. X-A, *Quantum Theory and Statistical Mechanics*, A. O. Barut and W. E. Brittin eds., Gordon and Breach 1968, p. 91, for a general review; V. V. Molotkov and I. T. Todorov, Comm. Math. Phys. **79**, 111 (1981), for a proof in the constraint Hamiltonian formulation.

that their Lie algebra is satisfied with the Lie product $[\,,\,]$ given by Poisson brackets.[21] Then, if H, P_i, J_i, K_i, $i = 1, 2, 3$ denote the generators of time translations, space translations, space rotations, and pure Lorentz transformations, respectively, they must satisfy

$$[H, P_i] = 0, \quad [H, J_i] = 0, \quad [H, K_i] = -P_i,$$

$$[P_i, P_k] = 0, \quad [P_i, J_k] = \varepsilon_{ikl} P_l, \quad [P_i, K_k] = -\delta_{ik} H,$$

$$[J_i, J_k] = \varepsilon_{ikl} J_l, \quad [J_i, K_k] = \varepsilon_{ikl} K_l, \quad [K_i, K_k] = -\varepsilon_{ikl} J_l. \tag{1.3.3}$$

Proposition 3.1[22] *If the particle coordinates (on the trajectories) q_i^α, $\alpha = 1, 2, \ldots N$, $i = 1, 2, 3$ transform correctly under the Poincaré transformations, i.e.,*

$$[q_i^\alpha, P_k] = \delta_{ik}, \quad [q_i^\alpha, J_k] = \varepsilon_{ikl} q_l^\alpha, \quad [q_i^\alpha, K_j] = q_j^\alpha [q_i^\alpha, H], \tag{1.3.4}$$

(world line conditions) and the equations of motion are not degenerate, i.e.,

$$\det \frac{\partial^2 H}{\partial p_i^\alpha \, \partial p_k^\beta} \neq 0, \tag{1.3.5}$$

then the particle accelerations vanish:

$$[[q_i^\alpha, H], H] = 0. \tag{1.3.6}$$

Remark. The technical non-degeneracy condition states that the positions and velocities form a complete set of dynamical variables, so that the transition to a Lagrangian is possible in the standard way. Eqs. (1.3.4) correspond to the world-line condition of Currie *et al.*[23]

The proof proceeds through two steps. Firstly, one shows that by means of a canonical transformation, which does not affect the coordinates, the generators P_i and J_i can be written as sums of single-particle momenta and angular momenta

$$P_i = \sum_\alpha p_i^\alpha, \quad J_i = \sum_\alpha \varepsilon_{ikl} \, q_k^\alpha \, p_l^\alpha. \tag{1.3.7}$$

---

[21] P. A. M. Dirac, Rev. Mod. Phys. **21**, 392 (1949); E. C. G. Sudarshan, Structure of dynamical theories, in 1961 Brandeis Summer Institute *Lectures in Theoretical Physics*, Vol. 2, Benjamin 1962, esp. Sect. 5, p. 143; D. G. Currie, T. F. Jordan, and E. C. G. Sudarshan, Rev. Mod. Phys. **35**, 350 (1963).

[22] H. Leutwyler, 1965.

[23] We briefly sketch the argument (see the review by Currie and Jordan pp. 93–4 for more details). Let $X_i \equiv q_i(t)$, $X_i' \equiv q_i'(t')$ denote the coordinates of the particle position on the trajectory at time t and the corresponding ones in a Lorentz-transformed frame. We consider an infinitesimal transformation so that second-order terms in the Lorentz boost parameter α_j are neglected. Then, putting $v_i \equiv dq_i(t)/dt|_{t=0} = [q_i(0), H]$, and choosing $t' = 0$, one has

$$X_i \simeq X_i' - \alpha_i t' = X_i' = q_i'(0), \quad t \simeq t' - \alpha_j X_j' = -\alpha_j X_j',$$

$$q_i'(0) = X_i = q_i(t = -\alpha_j X_j') \simeq q_i(0) - \alpha_j X_j' v_i \simeq q_i(0) - \alpha_j q_j(0) v_i,$$

$$\alpha_j [q_i(0), K_j] = \delta^{\alpha j} q_i(0) = q_i(0) - q_i'(0) = \alpha_j q_j(0) v_i.$$

This means that, as expected, there is no interaction momentum or angular momentum; such a property could actually be taken as the characterization of a system consisting of N particles (with no fields). Indeed, in the non-relativistic case, the interaction shows up only at the level of the Hamiltonian, which cannot be written as a sum of single-particle Hamiltonians H^α, each depending only on the coordinate and momentum of the α-particle. The substantial part of the proof (for which we refer to Sect. 5 of Leutwyler's paper) is then to show that, by exploiting eqs. (1.3.3–5), one can find a canonical transformation, which does not affect P_i, J_i and the coordinates, such that H and K_i can be brought to their free particle form

$$H = \sum_\alpha H^\alpha = \sum_\alpha ((\mathbf{p}^\alpha)^2 + (m^\alpha)^2)^{1/2}, \quad K_i = \sum_\alpha H^\alpha\, q_i^\alpha. \qquad (1.3.8)$$

Clearly, the first of eqs. (1.3.8) implies eq. (1.3.6).

As discussed by Ekstein,[24] the point at the basis of the argument is that in the relativistic case the Hamiltonian H is given by the Lie product of $[K_i, P_i]$, and therefore, if space translations and boosts have a kinematical character—i.e., can be written as sums of single particle functions—so is also the Hamiltonian, i.e., there is no interaction.

Such a constraint does not exist in the non-relativistic case, where the non-vanishing Lie products of the generators of the Galilei group are those which state the vector character of P_k, J_k, and G_k (the generators of the Galilei boosts) and $[G_i, H] = P_i$, $[P_i, G_k] = m\,\delta_{ik}$. Such Lie products are compatible with a non-trivial particle interaction.

For a better qualification of Ekstein's no-interaction theorem we spell out the assumptions of his analysis:

i) for each spacelike hyperplane σ there is an associated algebra of observables $\mathcal{A}(\sigma)$ containing all the observables of the system which can be measured on σ;

ii) a system of N distinguishable particles is characterized by the fact that $\mathcal{A}(\sigma)$ is generated by single particle subalgebras $\mathcal{A}_i(\sigma)$, $i = 1, \ldots N$, with no element in common except zero and the identity; the subalgebra $\mathcal{A}_i(\sigma)$ can be thought as generated only by the "kinematical" variables of the ith particle on σ, e.g., its canonical variables on σ, no variable of the other particles being involved. Clearly, such an assumption does not mean that any observable $A \in \mathcal{A}(\sigma)$ can be written as a sum of single-particle observables $A = \sum A_i$, $A_i \in \mathcal{A}_i(\sigma)$;[25]

iii) for each ith particle the variables of $\mathcal{A}_i(\sigma)$ transform independently of the variables of the other particles under space translations, space rotations, and Lorentz transformations (e.g., p^i_μ transforms as a four-vector, independently of variables of $\mathcal{A}_j$, $j \neq i$). This means that the individuality of the ith particle is a concept invariant under such (kinematical) transformations, no mixing being induced between $\mathcal{A}_i$ and $\mathcal{A}_j$.

A mixing between different single-particle algebras can be induced by time evolution, and its occurrence can be taken as the characterization of a non-trivial

---

[24] H. Ekstein, Comm. Math. Phys. **1**, 6 (1965).

[25] The characterization ii) of an N-particle system codifies the property that the evolution of the system is fully determined by the knowledge of one-particle (kinematical) observables $\mathcal{A}_i(\sigma)$, $i = 1, \ldots N$ on σ, and no mediating field influencing the particle dynamics is present.

interaction. This happens for non-relativistic particle systems with interaction, for which the above assumptions hold with the pure Lorentz transformations replaced by the Galilei boosts. For the relativistic case we have the following:

Theorem 3.2 *(Ekstein no-interaction theorem) If an algebra $\mathcal{A}_i$ is stable under the subgroup of space translations and Lorentz transformations, it is also stable under time translations and therefore the N-particle systems characterized by the above assumptions do not admit interactions between particles.*

Proof. The proof follows easily from the Lie algebra relations (1.3.3).

3.2   Field interactions and quantum mechanics

The exclusion of interactions at a distance suggests contact interactions, the distinguished (if not the exclusive) case being field mediated interactions with a contact action of the field on the particles. The interaction is a result of energy–momentum exchanges between the particles through the field, which propagates energy and momentum and can transfer them to the particles by contact. Then, Lorentz covariance becomes transcribed in the Lorentz invariance of the field equations. Clearly, the prototypical example of such a way of describing relativistic particle interactions is the electromagnetic interaction, but the above arguments indicate that this is the general case.

When such a picture is confronted with quantum mechanics, interesting considerations emerge, as clearly emphasized by Heisenberg.[26] If the classical particles are promoted to Schrödinger particles, the question arises concerning the quantum-mechanical status of the interaction mediating field. The possibility of keeping a classical structure for the fields is ruled out by Heisenberg uncertainty relations. The point of the Heisenberg argument is that if the measurement of field momentum and its localization were not constrained by quantum-mechanical limitations, one could use the particle–field interaction to violate the Heisenberg uncertainty relations in the measurement of the particle position and momentum. Thus, the uncertainty relations for the position and momentum require that, e.g., for an electromagnetic field localized in a small volume $\delta v = (\Delta l)^3$, the uncertainty ΔP_1 of the first component of field momentum, is constrained by

$$\Delta P_1 \, \Delta l \geq \hbar/2, \quad P_1 = \delta v \, (4\pi)^{-1} \, (E_2 \, B_3 - E_3 \, B_2).$$

By considering, for simplicity, the uncertainty corresponding to the deviations from the point $\mathbf{E} = 0 = \mathbf{B}$, one has $\Delta P_1 = \delta v \, (4\pi)^{-1}(\Delta E_2 \Delta B_3 - \Delta E_3 \Delta B_2)$ and therefore

$$\Delta E_2 \, \Delta B_3 \geq h/(\Delta l)^4, \quad \Delta E_3 \, \Delta B_2 \geq h/(\Delta l)^4 \tag{1.3.9}$$

(and similar relations for the other components of the field momentum), so that the electromagnetic field cannot be treated as a classical field.

---

[26] W. Heisenberg, *The Physical Principles of the Quantum Theory*, Dover 1930, esp. Chap. III.

The same uncertainty relations would be reached by the canonical quantization prescription, i.e., by replacing the Poisson brackets between the electromagnetic potential and its canonically conjugated momentum by quantum commutators.[27]

In elementary quantum mechanics the interaction with a *classical external field* does not meet such problems, because by definition the field is simply a source of energy and momentum and its time dependence is preassigned; the situation changes drastically if the field mediates the interaction between particles, so that the field dynamics cannot be preassigned and is interlaced with the particle dynamics.

In conclusion, the above considerations about relativistic particle interactions, in the context of quantum mechanics, strongly suggest the introduction of interaction mediating fields and their quantization. As we shall see in the next section, such a strategy opens the way of overcoming the conflict between locality, stability, and Lorentz covariance pointed out in Section 2 above.

4   Free field equations and quantum mechanics

A relativistic description of particle interactions mediated by a local action of fields requires that the field dynamics be governed by *local relativistic equations*. We have already seen such equations in Section 2 above, *albeit* with different motivations, and the interpretation of fields as Schrödinger wave functions would cause the same problems.

In the case of the free Klein–Gordon equation, the Hamiltonian H_0 in eq. (1.2.5) governs the frequency spectral support of (the Fourier transform of) the solution u in the Hilbert space $\mathcal{H} = H^1 \oplus L^2$, and does not have a positive spectrum (by Proposition 2.2, positive spectrum would exclude localizability of u). On the other hand, the classical expression of the field energy ($\varphi = u_1, \dot{\varphi} = u_2$)

$$H(\varphi) = \tfrac{1}{2} \int d^3x \, [\nabla \varphi^* \, \nabla \varphi + m^2 \varphi^* \, \varphi + \dot{\varphi}^* \, \dot{\varphi}] \tag{1.4.1}$$

is positive definite and, in the canonical formulation, the Poisson brackets of $H(\varphi)$ give the time derivatives of the canonical field

$$\partial_t \, \varphi = -\{H(\varphi), \, \varphi\}.$$

Thus, one recovers the role of $H(\varphi)$ as the *positive* quantum generator of time translations, provided the field is considered as a quantum operator (*field quantization*) and the Poisson brackets are replaced by commutators according to the canonical quantization of classical theories. In this way, *positivity of the energy* is obtained, even if the spectral support of the field contains frequencies of both signs, since, as

---

[27] W. Heisenberg and W. Pauli, Zeitschrift f. Phys. **56**, 1 (1929); W. Heisenberg, *The Physical Principles of the Quantum Theory*, Dover 1930, Appendices 11, 12. For a textbook discussion of the canonical formalism for classical fields and for the derivation of the expression of the energy–momentum and charge via the Noether theorem, see N. N. Bogoliubov and D. V. Shirkov, *Introduction to the Theory of Quantized Fields*, Interscience 1959, Chap. I; S. S. Schweber, *An Introduction to Relativistic Quantum Field Theory*, Harper and Row 1961, pp. 186–93 and Sect. 7g and J. W. Leech, *Classical Mechanics*, Methuen 1965, Chap. IX.

we shall see, it describes the energy–momentum variations that the field operator may induce by its application on a state. A similar picture is displayed by the harmonic oscillator, where the Hamiltonian $H = \frac{1}{2}(p^2 + q^2)$ is positive, the position operator contains both positive and negative frequencies $\sqrt{2}\,q(t) = a e^{-i\omega t} + a^* e^{i\omega t}$, $\omega > 0$, and the states obtained by applying $q(t)$ to Ψ_0 have all positive energies (e.g., $H a^* e^{i\omega t}\Psi_0 = \omega a^* e^{i\omega t}\Psi_0$). Then, a *local field dynamics* is compatible with energy positivity.

Furthermore, the problems connected with the non-positive definiteness of $\rho(\mathbf{x}, t) = j_0 = -i(\varphi^*\dot{\varphi} - \dot{\varphi}^*\varphi)$ disappear, since ρ has the meaning of charge density operator, rather than of probability density.

In order to see how all this works, we consider a free Klein–Gordon quantum (complex) field. For the physical interpretation of physical quantities associated to it, it is convenient to go to momentum space ($k_0 \equiv \sqrt{\mathbf{k}^2 + m^2}$),

$$\varphi(\mathbf{x},\, t) = (2\pi)^{-3/2} \int d^3k \, (\sqrt{2}k_0)^{-1}[\, b(\mathbf{k})e^{-ikx} + a^*(\mathbf{k})\, e^{ikx}\,], \tag{1.4.2}$$

where we have split the positive and negative frequency parts (b, a^* respectively) and introduced a factor k_0^{-1} in order to obtain a Lorentz-invariant measure $d^3k/k_0 \equiv d\Omega_m(\mathbf{k})$. For a real field, $b(\mathbf{k}) = a(\mathbf{k})$.

Now, canonical quantization (for the canonical formulation of fields see e.g., Section 7 below) yields the equal-time commutators

$$[\,\varphi(\mathbf{x},t),\, \pi(\mathbf{y},\, t)\,] = i\,\delta(\mathbf{x} - \mathbf{y}), \quad \pi(\mathbf{x},\, t) = \dot{\varphi}^*(\mathbf{x},\, t), \tag{1.4.3}$$

all other equal-time commutators vanishing. Equivalently,

$$[\,a(\mathbf{k}),\, a^*(\mathbf{q})\,] = k_0\,\delta(\mathbf{k} - \mathbf{q}), \quad [\,b(\mathbf{k}),\, b^*(\mathbf{q})\,] = k_0\,\delta(\mathbf{k} - \mathbf{q}), \tag{1.4.4}$$

all other commutators vanishing.

In passing from the classical expressions of the observables to quantum operators, a problem of ordering of factors arises, to be decided on the basis of physical and mathematical considerations . By substituting eq. (1.4.2) in the classical expressions for the field four-momentum $P_0(\varphi) = H(\varphi)$, $P_k(\varphi) = \int d^3x\,[\dot{\varphi}^*\partial_k\varphi + \partial_k\varphi^*\,\dot{\varphi}]$ and charge $Q(\varphi) = \int d^3x\,\rho(\mathbf{x})$, and by using the above commutation relations, one can obtain expressions in which the operators a^*, b^* stay to the right. Thus, neglecting (actually divergent) c-number terms, one obtains[28]

$$P_\mu(\varphi) = \int d\Omega_m(\mathbf{k})\, k_\mu \,[\, a^*(\mathbf{k})\, a(\mathbf{k}) + b^*(\mathbf{k})\, b(\mathbf{k})\,], \tag{1.4.5}$$

$$Q(\varphi) = \int d\Omega_m(\mathbf{k})\,[\, a^*(\mathbf{k})\, a(\mathbf{k}) - b^*(\mathbf{k})\, b(\mathbf{k})\,]. \tag{1.4.6}$$

---

[28] Such c-number terms are irrelevant for the interpretation of P_μ and Q as generators of symmetry groups, and moreover, as we shall see below, only due to the subtraction of such terms, called *Wick ordering*, one obtains well-defined operators P_μ and Q. This is the simplest example of *renormalization* needed for obtaining well-defined operators starting from the classical field expressions.

The above formulas display the positivity of the energy operator and the indefiniteness of the charge operator; the "modes" described by the a's give a positive contribution to both, whereas the b's contribute positively to the energy but negatively to the charge. Thus, the a^*, a, b^*, b play the same role of the creation and destruction operators of the harmonic oscillator, and the corresponding excitations describe field quanta. In fact, the canonical commutation relations, eqs. (1.4.4), give

$$[\,P_\mu,\, a(\mathbf{k})\,] = -k_\mu\, a(\mathbf{k}), \quad [\,P_\mu,\, b(\mathbf{k})\,] = -k_\mu\, b(\mathbf{k}). \tag{1.4.7}$$

Hence, neglecting for simplicity distributional and domain problems, if Ψ is an eigenvector of P_μ with eigenvalue p_μ, $P_\mu\,\Psi = p_\mu\,\Psi$, the state $a(\mathbf{k})\,\Psi$ has eigenvalue $p_\mu - k_\mu$

$$P_\mu\, a(\mathbf{k})\,\Psi = [\,P_\mu,\, a(\mathbf{k})\,]\,\Psi + p_\mu\, a(\mathbf{k})\,\Psi = (p_\mu - k_\mu)a(\mathbf{k})\,\Psi. \tag{1.4.8}$$

Similarly, the action of the operator $a^*(\mathbf{k})$ leads to an increase of energy–momentum by k_μ. The same conclusions hold for $b(\mathbf{k})\,\Psi$ and $b^*(\mathbf{k})\,\Psi$.

Therefore, the field operator describes the possible addition or subtraction of energy–momentum quanta, through its action on a state. In this way, as pointed out by Dirac, one recovers and explains Einstein's interpretation of the electromagnetic field as the carrier of energy–momentum quanta (the photons) and the laws of their emission and absorption.[29] Thus, relativistic quantum fields describe particle interactions through the local absorption or emission of energy–momentum quanta carried by the field. Since the number of such quanta is unlimited, one is forced to deal with quantum systems with infinite degrees of freedom.

Quite generally, as in the case of the electromagnetic field, the field quanta may be interpreted as particles. In fact, the condition that the formally positive operator P_0 is a well- (densely) defined Hilbert operator with positive spectrum requires that the lowering of the energy, according to eq. (1.4.8), must terminate, i.e., there must be a lowest energy state Ψ_0, such that $a(\mathbf{k})\Psi_0 = 0 = b(\mathbf{k})\Psi_0$. Thus, $a^*(\mathbf{k})\Psi_0$ is a state with momentum k_μ and charge one, i.e., it has the same properties of a one-particle state of charge one. The same argument applies to $b^*(\mathbf{k})\Psi_0$, which describes a particle with the same mass but opposite charge, i.e., an *antiparticle*. The *existence of antiparticles* is therefore related to the inevitable presence of the negative-frequency part of the field *required by locality and covariance* (Proposition 2.2).

Field quantization also solves the problems of the Dirac equation. In fact, a momentum space analysis for the free Dirac field reads ($p_0 = \sqrt{\mathbf{p}^2 + m^2}$, $\sigma = \pm$ being the helicity),

$$\psi(\mathbf{x}, t) = \int d\Omega_m(\mathbf{p}) \sum_{\sigma=\pm} [\,a(\mathbf{p}, \sigma)u(\mathbf{p}, \sigma)e^{-ipx} + b^*(\mathbf{p}, \sigma)u^c(\mathbf{p}, \sigma)e^{ipx}\,], \tag{1.4.9}$$

---

[29] P. A. M. Dirac, Proc. Royal Soc. London, A **114**, 243 (1927). For a brief account, see F. Strocchi, *Elements of Quantum Mechanics of Infinite Systems*, World Scientific 1985, Part A, Sect. 2.3.

where $u(\mathbf{p}, \sigma)$, $\sigma = \pm$, are two orthonormal solutions of the momentum-space Dirac equation,

$$(-\gamma^\mu p_\mu + m)u(\mathbf{p}, \sigma) = 0,$$

u^c is the so-called charge conjugate of u (for the reason for such a name, see below) defined by $u^c(\mathbf{p}, \sigma) = C^{-1}u(\mathbf{p}, \sigma)$, where C is the charge conjugation matrix defined by $C\gamma^\mu C^{-1} = \overline{\gamma^\mu}$. Now, if one substitutes eq. (1.4.9) in the Dirac Hamiltonian, momentum, and charge,

$$P_\mu(\psi) = \int d^3x\, \psi^*(x)\, i\partial_\mu\, \psi(x), \quad Q(\psi) = \int d^3x\, \psi(x)^*\, \psi(x),$$

by (using the orthonormality of the u's and) keeping the order of factors, one obtains

$$P_\mu = \int d\Omega_m(\mathbf{p})\, p_\mu \sum_{\sigma=\pm} [\, a^*(\mathbf{p}, \sigma)\, a(\mathbf{p}, \sigma) - b(\mathbf{p}, \sigma)\, b^*(\mathbf{p}, \sigma)\,],$$

$$Q = \int d\Omega_m(\mathbf{p}) \sum_{\sigma=\pm} [\, a^*(\mathbf{p}, \sigma)\, a(\mathbf{p}, \sigma) + b(\mathbf{p}, \sigma)\, b^*(\mathbf{p}, \sigma)\,].$$

The apparent non-positivity of the energy can be cured by adopting (canonical) anticommutation relations ($[A, B]_+ \equiv AB + BA$):

$$[\, a(\mathbf{p}, \sigma),\, a^*(\mathbf{p}', \sigma')\,]_+ = \delta_{\sigma,\sigma'}\, p_0\, \delta(\mathbf{p} - \mathbf{p}') = [\, b(\mathbf{p}, \sigma),\, b^*(\mathbf{p}', \sigma')\,]_+. \tag{1.4.10}$$

In fact, in this case, by neglecting (actually divergent) c-number terms, whose subtraction may be ascribed to the ordering problem, one obtains the following expressions[30] for P_μ and Q:

$$P_\mu = \int d\Omega_m(\mathbf{p})\, p_\mu \sum_{\sigma=\pm} [\, a^*(\mathbf{p}, \sigma)\, a(\mathbf{p}, \sigma) + b^*(\mathbf{p}, \sigma)\, b(\mathbf{p}, \sigma)\,], \tag{1.4.11}$$

$$Q = \int d\Omega_m(\mathbf{p}) \sum_{\sigma=\pm} [\, a^*(\mathbf{p}, \sigma)\, a(\mathbf{p}, \sigma) - b^*(\mathbf{p}, \sigma)\, b(\mathbf{p}, \sigma)\,]. \tag{1.4.12}$$

For the same reasons discussed in the scalar case, the a^*a terms are interpreted as the contribution of particles, and the b^*b terms as the contribution of antiparticles.

The condition of positivity of the energy selects the choice between commutation and anticommutation relations. Thus, at least in the free case, a link between integer/half-integer spin and field commutator/anticommutator, namely, a field theory transcription of *Pauli principle* emerges as a consequence of the positive energy spectrum.[31]

In conclusion, in the free case the problems of local relativistic equations and positive energy spectrum are solved in the following way: i) the hyperbolic field

---

[30] The mechanism is that by an infinite subtraction (which corresponds to the filling of the Dirac sea), the negative term $-b(\mathbf{p})\, b^*(\mathbf{p})$ is turned into a positive one $b^*(\mathbf{p})\, b(\mathbf{p})$.

[31] W. Pauli, Phys. Rev. **58**, 716 (1940); this approach to field quantization is discussed in detail in N. N. Bogoliubov and D. V. Shirkov, *Introduction to the Theory of Quantized Fields*, Interscience 1959, Chap. II.

equations allow for both positive and negative frequencies and, in fact, the field has support in both the upper and lower hyperboloids $p^2 = m^2$, ii) the positive/negative frequency parts of the field φ increase/lower the energy of the state to which they are applied (they will be called the positive/negative energy parts of the field and denoted by $\varphi^{\pm}$), iii) since the local relativistic equations describe the time evolution of field operators, rather than of wave functions, the inevitable occurrence of both positive and negative frequencies is compatible with the positivity of the Hamiltonian, iv) the fields, required for describing particle interactions, also describe particles.

5 Particles as field quanta

The realization that local relativistic equations govern the dynamics of field operators, rather than that of Schrödinger wave functions, was regarded as a substantial change with respect to Schrödinger QM, and was given the name *second quantization*. Actually, the change does not involve the principles of quantum mechanics, but rather accounts for the need of dealing with an unlimited number of quanta or particles for the quantum description of relativistic systems. Thus, whereas in the Hilbert space of Schrödinger quantum mechanics all the states have the same fixed number of particles and are described by wave functions of a fixed number of variables, in the case of relativistic quantum fields this is no longer possible, and the Hilbert space decomposes as the direct sum of n-particle subspaces $\mathcal{H}^n$, $n = 0, 1, 2, \ldots$.

The particle-quanta content of a quantum field and its Hilbert space representation is particularly simple in the *free case*. Given a complete orthonormal set of single-particle wave functions $\{f_i(\mathbf{x})\}$, a complete set of n-particle states is obtained by specifying how many particles are in the state f_1, how many in f_2, etc.: $\Psi^n(n_1, n_2, \ldots)$, $\sum_k n_k = n$ (*occupation number representation*). Thus, in the simplest case of free scalar neutral (identical) particles, *creation and destruction operators* are defined by

$$a_i^* \Psi^n(n_1, \ldots, n_i, \ldots) = \sqrt{n_i + 1}\, \Psi^{n+1}(n_1, \ldots, n_i + 1, \ldots),$$

$$a_i \Psi^n(n_1, \ldots, n_i, \ldots) = \sqrt{n_i}\, \Psi^{n-1}(n_1, \ldots, n_i - 1, \ldots),$$

with the meaning of increasing, and respectively decreasing, the number of particles in ith state. One immediately obtains

$$[a_i, a_j^*] = \delta_{ij}, \tag{1.5.1}$$

all other commutators vanishing. If the particles are charged, one has also the creation and destruction operators for the corresponding antiparticles (with the same mass but opposite charge). $\Psi_0 \equiv \Psi^{n=0}$ is the state with no particles (and no antiparticles); clearly $a_i \Psi_0 = 0$, $\forall i$.

Similarly, for fermions one defines

$$c_i^* \Psi^n(n_1, \ldots, n_i, \ldots) = (-1)^{\theta_i} (1 - n_i)\, \Psi^{n+1}(n_1, \ldots, n_i + 1, \ldots),$$

$$c_i \Psi^n(n_1, \ldots, n_i, \ldots) = (-1)^{\theta_i} n_i\, \Psi^{n-1}(n_1, \ldots, n_i - 1, \ldots), \quad \theta_i \equiv \sum_{k=1}^{i-1} n_k,$$

with the suitable factors to account for the Pauli exclusion principle. In this case one obtains

$$[c_i, c_j^*]_+ = \delta_{ij}, \tag{1.5.2}$$

all other anticommutators vanishing.

A representation of a free neutral scalar field (of mass m) in the space $\mathcal{H} = \oplus_n \mathcal{H}^n$ is obtained by identifying $a_i = a(f_i)$, $a_i^* = a(f_i)^*$,

$$a(f_i) \equiv \int d\Omega_m(k) a(\mathbf{k}) \overline{f}_i(\mathbf{k}), \quad \int d\Omega_m(k) f_i(\mathbf{k}) \overline{f}_j(\mathbf{k}) = \delta_{ij}.$$

Thus, as in the case of the harmonic oscillator, in such a representation of the field, the field operators $a(\mathbf{k})$, $a^*(\mathbf{k})$ act as destruction and creation operators of particles. Such a representation, called the *Fock representation*, clearly displays the particle content and can be characterized as the representation in which the particle number operator $N = \sum_k a_k^* a_k = \sum_k N_k$ exists (as a densely defined operator).[32] In fact, we have:

Proposition 5.1 *In an irreducible representation of the Heisenberg algebra $\mathcal{A}_H$, i.e., of the polynomial algebra generated by a_i, a_i^*, in a dense domain D, the following properties are equivalent*
1) the total number operator $N = \sum_j a_j^* a_j$, exists in the sense that

$$strong - \lim_{K \to \infty} e^{i\alpha \sum_j^K a_j^* a_j} \equiv e^{i\alpha N} \equiv T(\alpha), \quad \forall \alpha \in \mathbf{R}, \tag{1.5.3}$$

exists and defines a one-parameter group of unitary operators strongly continuous in α, leaving stable the common dense domain D of the polynomials of a_i^*, a_j
2) there exists a vector Ψ_0, called the Fock vacuum vector, such that

$$a_j \Psi_0 = 0, \quad \forall j. \tag{1.5.4}$$

Proof. Property 1) and the commutation relations (1.5.1) imply that

$$T(\alpha) a_j T(\alpha)^{-1} = e^{-i\alpha} a_j, \quad [T(2\pi), \mathcal{A}_H] = 0.$$

By the irreducibility of $\mathcal{A}_H$, it follows that $T(2\pi) = \mathbf{1} \exp i\theta$. The spectral representation of $T(\alpha)$ then gives (in the improper Dirac notations $dE(\lambda) \sim |\lambda ><\lambda| d\lambda$)

$$0 = (T(2\pi) - \mathbf{1} e^{i\theta})^*(T(2\pi) - \mathbf{1} e^{i\theta}) = \int_{\sigma(N)} dE(\lambda) |e^{i(2\pi\lambda - \theta)} - 1|^2,$$

which implies $2\pi\lambda - \theta \in 2\pi \mathbf{Z}$, i.e., the spectrum of N is discrete.
Now, if $\lambda > 0$ is a point of the spectrum of N and Ψ_λ a corresponding eigenvector, then $0 < \lambda ||\Psi_\lambda||^2 = (\Psi_\lambda, N \Psi_\lambda) = \sum_j ||a_j \Psi_\lambda||^2$, so that there must be at least one j such that $a_j \Psi_\lambda \neq 0$, and one has

$$T(\alpha) a_j \Psi_\lambda = e^{i(\lambda-1)\alpha} a_j \Psi_\lambda.$$

<hr>

[32] G. F. Dell' Antonio, S. Doplicher, and D. Ruelle, Comm. Math. Phys. **2**, 223 (1966); G. F. Dell' Antonio and S. Doplicher, Jour. Math. Phys. **8**, 663 (1967); J. M. Chaiken, Comm. Math. Phys. **8**, 164 (1968).

Thus, also $\lambda - 1 \in \sigma(N)$ and, since the spectrum of N is non-negative, in order that this process of lowering the eigenvalues terminates, $\lambda = 0$ must be a point of the spectrum of N, and

$$a_j \, \Psi_0 = 0, \quad \forall j. \tag{1.5.5}$$

Conversely, if the Fock vacuum Ψ_0 exists, then $\mathcal{A}_H \, \Psi_0 = \mathcal{P}(a^*) \, \Psi_0$, where $\mathcal{P}(a^*)$ denotes the polynomial algebra generated by the a^*'s and on such a domain, which is dense by the irreducibility of $\mathcal{A}_H$, N exists as a self-adjoint operator. Moreover, the exponential series converges strongly on $\mathcal{P}(a^*)\Psi_0$ and defines a one-parameter group of unitary operators $e^{i\alpha N}$, since the monomials of a^* applied to Ψ_0 yield eigenstates of N and generate such domains.

It is clear from the above argument that the occupation number representation is characterized by the existence of a Fock vacuum, with the meaning of no-particle state; all other n-particle states are obtained by acting on it by monomials of the creation operators a_i^*.

It is worthwhile noting that whereas in the finite-dimensional case the Fock representation is the only possibility under general regularity conditions, which do not involve the dynamics, in the infinite-dimensional case there are many inequivalent representations of the Heisenberg algebra. Different representations correspond to different physical properties, and the choice of one instead of another must invoke some additional physical requirement. This problem will become particularly acute in the presence of interactions.[33] In the free field case, one has:

Proposition 5.2   *For an irreducible representation of the Heisenberg algebra $\mathcal{A}_H$, in a dense domain D, the Fock representation is selected by the condition that the free Hamiltonian (with mass gap m)*

$$H_0 = \sum_i \omega_i a_i^* a_i, \quad \omega_i \geq m > 0$$

exists in the sense that

$$strong - \lim_{K \to \infty} e^{i\alpha \sum_j^K \omega_j a_j^* a_j} \equiv e^{i\alpha H_0}, \quad \forall \alpha \in \mathbf{R} \tag{1.5.6}$$

exists and defines a one-parameter group of unitary operators strongly continuous in α, leaving stable the common dense domain D of the polynomials of a_i^*, a_j.

Proof.   In fact, since

$$\sum_{j=1}^K \omega_j a_j^* a_j \geq m \sum_{j=1}^K a_j^* a_j,$$

---

[33] K. Friedrichs, *Mathematical Aspects of the Quantum Theory of Fields*, Interscience, 1953; A. S. Wightman and S. Schweber, Phys. Rev. **98**, 812 (1955). See in particular the beautiful lectures by A. S. Wightman, Introduction to some aspects of the relativistic dynamics of quantized fields, in *High Energy Electromagnetic Interactions and Field Theory*, M. Lèvy (ed.), Gordon and Breach 1967, esp. Part II, Sect. 6.

the convergence of the above exponentials implies the convergence of the exponentials of eq. (1.5.3), i.e., the existence of N. Then, Proposition 5.1 applies.

It is not difficult to adapt this argument to the free field expression, eq. (1.4.5), in the case of mass gap.[34]

Therefore if H_0, with a mass gap, is well defined, N is well defined, and the representation is the Fock representation.

The selection of the Fock representation by the existence of the free Hamiltonian no longer holds if there is no mass gap, typically in the case of free massless fields. Such a freedom plays a crucial role in quantum field theory models, where interaction involving massless particles gives rise to the so-called infrared problem (see, e.g., Chapter 2, Section 3.2).

If we denote by $\Psi^n(\mathbf{k}_1, \mathbf{k}_2, \ldots \mathbf{k}_n)$ the momentum space wave functions of the n-particle states Ψ^n, the representation of the field operators can be written in the form

$$(\varphi(\mathbf{k}, 0)\Psi)^n(\mathbf{k}_1, \mathbf{k}_2, \ldots \mathbf{k}_n) = \sqrt{n+1}\,\Psi^{n+1}(\mathbf{k}, \mathbf{k}_1, \mathbf{k}_2, \ldots \mathbf{k}_n)$$

$$+ n^{-1/2} \sum_i \delta(\mathbf{k} - \mathbf{k}_i)\,\Psi^{n-1}(\mathbf{k}_1, \ldots \hat{\mathbf{k}}_i, \ldots \mathbf{k}_n),$$

where $\hat{\mathbf{k}}_i$ means that the ith variable has to be omitted.[35]

It is easy to see that the Wick ordering yields well-defined operators for the energy, momentum, and charge, eqs. (1.4.5) and (1.4.6). In fact, by the Fock condition for the annihilation operators, eq. (1.5.4),

$$P_\mu \Psi_0 = 0, \quad Q \Psi_0 = 0,$$

and on the n-particle states one has

$$P_\mu a^*(f_1)a^*(f_2)\ldots a^*(f_n)\Psi_0 = [P_\mu, a^*(f_1)a^*(f_2)\ldots a^*(f_n)]\Psi_0.$$

Since $[a^*(\mathbf{k})a(\mathbf{k}), a^*(f_i)] = a^*(\mathbf{k})\,f_i(\mathbf{k})$, the commutator on the right-hand side is well defined and therefore so is P_μ.

Now, if the restriction of the energy–momentum to the one-particle space is self-adjoint on a suitable dense domain, then also the restriction to the n-particle states

---

[34] In fact, one has

$$P_0 = H = \int d\Omega(\mathbf{k})\, d\Omega(\mathbf{q}) k_0(\mathbf{k}) a^*(\mathbf{q})\, a(\mathbf{k}) \sum_i \bar{f}_i(\mathbf{k})\, f_i(\mathbf{q}) \geq$$

$$\geq m \int d\Omega(\mathbf{k})\, d\Omega(\mathbf{q})\, a^*(\mathbf{q})\, a(\mathbf{k}) \sum_i \bar{f}_i(\mathbf{k})\, f_i(\mathbf{q}) = m \sum_i a_i^* a_i = m\, N.$$

For details, see H. J. Borchers, R. Haag, and B. Schroer, Nuovo Cim. **29**, 148 (1963).

[35] In the Dirac notation the above representation reads

$$a^*(\mathbf{k})\,|\mathbf{k}_1, \mathbf{k}_2, \ldots \mathbf{k}_n> = \sqrt{n+1}\,|\mathbf{k}, \mathbf{k}_1, \mathbf{k}_2, \ldots \mathbf{k}_n>,$$

$$a(\mathbf{k})\,|\mathbf{k}_1, \mathbf{k}_2, \ldots \mathbf{k}_n> = (\sqrt{n})^{-1} \sum_i k_0\, \delta(\mathbf{k} - \mathbf{k}_i)|\mathbf{k}_1, \ldots \hat{\mathbf{k}}_i, \ldots \mathbf{k}_n>.$$

is self-adjoint on the corresponding domain, and the energy–momentum is essentially self-adjoint on the direct sum of n-particle domains.

The same argument applies to the angular momentum and obviously to the charge. In this way one obtains well-defined self-adjoint operators in the Fock space.[36]

The energy–momentum P_μ and the angular momentum are the generators of strongly continuous unitary groups $U(a) = e^{iPa}$, $a \in \mathbf{R}^4$, $U(R)$, $R \in SU(2)$, respectively, which describe spacetime translations and rotations; e.g.,

$$U(\mathbf{a})\, a(f)\, U(\mathbf{a})^{-1} = a(f_\mathbf{a}), \quad f_\mathbf{a}(\mathbf{k}) \equiv e^{i\mathbf{k}\cdot\mathbf{a}}\, f(\mathbf{k}). \tag{1.5.7}$$

Thus, one has

$$U(a)\, \varphi(x)\, U(a)^* = \varphi(x + a), \quad a \in \mathbf{R}^4. \tag{1.5.8}$$

The following properties of the Fock representation constructed above, where the Fock no-particle vector is cyclic for $\mathcal{A}_H$, are easy to prove.

Proposition 5.3   *i) the Fock representation of $\mathcal{A}_H$ is irreducible,*
ii) all (irreducible) Fock representations are unitarily equivalent,
iii) the no-particle state is unique (i.e., the zero eingenvalue of N has multiplicity 1),
iv) the Fock vacuum is the only state invariant under space translations in the Fock representation space

$$U(\mathbf{a})\Psi_0 = \Psi_0. \tag{1.5.9}$$

Proof. In fact, i) if $[\,C, \mathcal{A}_H\,] = 0$, then $C\mathcal{H}^n \subseteq \mathcal{H}^n$ and one is reduced to the irreducibility of the Schrödinger representation for finite degrees of freedom; then C is a multiple of the identity in each $\mathcal{H}^n$ and actually in the whole $\mathcal{H}$, since C commutes with a, a^*;

ii) the mapping $U : \pi(\mathcal{P}(a, a^*))\,\Psi_0 \to \pi'(\mathcal{P}(a, a^*))\,\Psi_0'$ is densely defined, together with its inverse, and both preserve the scalar products, which are exclusively determined by the canonical commutation relations and the Fock condition;

iii) $N\,\Psi = 0$ implies $a\,\Psi = 0$ and therefore $(\Psi, \mathcal{P}(a^*)\,\Psi_0) = 0$, so that, by the cyclicity of Ψ_0, $\Psi = 0$;

iv) since the space translations commute with N, $U(\mathbf{a})\,\Psi = \Psi$ implies $U(\mathbf{a})\,\Psi^n = \Psi^n$, with Ψ^n the component of Ψ in $\mathcal{H}^n$, and $\mathcal{H}^n$, $n \neq 0$, does not contain translationally invariant states.

It is not difficult to check that in the Fock space also the Lorentz transformations Λ are described by unitary operators $U(\Lambda)$.

For a scalar field, one has

$$U(\Lambda)\, \varphi(x)\, U(\Lambda)^* = \varphi(\Lambda x), \tag{1.5.10}$$

and for the Dirac field

$$U(\Lambda(A))\, \psi_\alpha(x)\, U(\Lambda(A))^* = \sum_\beta S(A^{-1})_{\alpha\,\beta}\psi_\beta(\Lambda(A)x), \quad A \in SL(2, C), \tag{1.5.11}$$

---

[36] For a detailed discussion, see, e.g., O. Bratteli and D. W. Robinson, *Operator Algebras and Quantum Statistical Mechanics, Vol. 2*, 2nd ed., Springer 1996, Sect. 5.2.1.

where $S(A)$ is the 4×4 representation of $SL(2, C)$, the universal covering of the restricted Lorentz group.

From eqs. (1.4.2) and (1.4.4), and the Fock condition $a(\mathbf{k}) \Psi_0 = 0$, one may derive the so-called two-point function of the free scalar neutral field

$$i \Delta^+(x - y) \equiv < \varphi(x) \varphi(y) > \equiv (\Psi_0, \varphi(x) \varphi(y) \Psi_0) =$$

$$= (2\pi)^{-3} \int d^4 k \, \delta(k^2 - m^2) \, \theta(k_0) \, e^{-ik(x-y)}, \tag{1.5.12}$$

where θ is the Heaviside function $\theta(x) = 1$, for $x > 0$, $\theta(x) = 0$ for $x < 0$.

Similarly, for the free Dirac field one obtains

$$< \psi(x) \overline{\psi}(y) > = (i\gamma^\mu \, \partial_\mu + m) \, i \Delta^+(x - y) \equiv -i \, S^+(x - y). \tag{1.5.13}$$

It is important to remark that the (anti)commutation relations of eqs. (1.4.4) and (1.4.10) imply that the fields defined by eqs. (1.4.2) and (1.4.9) satisfy relativistic *locality*, also called *microscopic causality*, namely, at spacelike separated points, x, y, $(x - y)^2 < 0$, the scalar fields commute,

$$[\varphi(x), \, \varphi^*(y)] = 0 = [\varphi(x), \, \varphi(y)], \tag{1.5.14}$$

and the Dirac fields anticommute,

$$[\psi(x), \, \overline{\psi}(y)]_+ = 0 = [\psi(x), \, \psi(y)]_+. \tag{1.5.15}$$

This assures that the observables, described by polynomial functions of the fields, satisfy Einstein causality—one of the basic requirements stated in Section 2 above. In fact, if A, B are polynomial functions of fields localized in the spacetime regions $\mathcal{O}_1, \mathcal{O}_2$, respectively, eqs. (1.5.14) and (1.5.15) imply that $[A, \, B] = 0$, if $\mathcal{O}_1$ is relatively spacelike with respect to $\mathcal{O}_2$.

Furthermore, the free field equations and the (anti)commutation relations imply

$$[\varphi(x), \, \varphi(y)] = i\Delta(x - y); \quad \Delta(x) = \Delta^+(x) - \Delta^+(-x), \tag{1.5.16}$$

$$[\psi(x), \overline{\psi}(y)]_+ = -i \, S(x - y), \quad S(x) = S^+(x) - S^+(-x), \quad [\psi(x), \psi(y)]_+ = 0. \tag{1.5.17}$$

In conclusion, the preceding discussion shows that for free fields one has a satisfactory solution of the problems of a relativistic quantum mechanics. Actually, in this case one can obtain as much mathematical control and rigor as one likes.

In the following chapter we will discuss the problems which arise in the case of interactions, starting with the standard perturbative approach, which has been so successful in predicting physical results in quantum electrodynamics. Clearly, in the interacting case the equations (1.5.16) and (1.5.17) will not hold, but one may reasonably expect that the locality property, eqs. (1.5.14) and (1.5.15), keeps being satisfied.

6   Appendix: The Dirac equation

It is instructive to derive the Dirac equation following the original Dirac motivations for the problems of relativistic wave equations (see Section 2 above).

The first requirement by Dirac is that the evolution equation for the wave function be of *first order in time*, so that an Hamiltonian interpretation in terms of the generator of time translations is obtained, in agreement with the basic structure of quantum mechanics (in Dirac language the transformation theory). Then, Lorentz invariance demands that also the space derivatives occur only at first order. Another requirement is that the conserved current j_μ associated to the evolution equation should have a positive definite density $\rho = j_0$, free of derivatives so that a probability density interpretation is possible as in Schrödinger quantum mechanics, i.e., one asks for a $\rho(x)$ *of the form* $\psi^*(x)\psi(x)$. Hence, quite generally the equation must be of the form

$$(-i\gamma^\mu \partial_\mu + \Gamma M)\psi(x) = 0, \qquad (1.6.1)$$

where γ^μ, $\mu = 0, 1, 2, 3$, and Γ are operators independent of space and time, which will be taken as non-singular.[37] Then, one can take $\Gamma = 1$, with the redefinition $\Gamma^{-1}\gamma^\mu \to \gamma^\mu$, and by an overall rescaling M can be identified with the particle mass m. Thus, one has the equation

$$i\partial_0\, \psi = [-i(\gamma^0)^{-1}\gamma^i \partial_i + (\gamma^0)^{-1}m)]\psi \equiv H\psi. \qquad (1.6.2)$$

The Hamiltonian form further requires that H be hermitian with respect to the scalar product defined by $\rho(x)$; thus $(\gamma^0)^{-1}$ and $(\gamma^0)^{-1}\gamma^i$ must be hermitian. The relativistic energy–momentum relation requires that $H^2 = \mathbf{p}^2 + m^2$. This implies:

i) $(\gamma^0)^{-2} = 1$, i.e., $\gamma^0 = (\gamma^0)^{-1}$, equivalently $(\gamma^0)^2 = 1$,

ii) $\gamma^0\gamma^i + \gamma^i\gamma^0 = 0$, which implies that the γ^i are antihermitian, since $\gamma^0\gamma^i = (\gamma^0\gamma^i)^* = (-\gamma^i\gamma^0)^* = -\gamma^0(\gamma^i)^*$,

iii) $\gamma^0\gamma^i\gamma^0\gamma^j \partial_i\, \partial_j = -\Delta$, namely, $\{\gamma^i, \gamma^j\} = -2\delta_{ij}$.

The relations between the γ^μ may be summarized by the equation

$$\gamma^\mu\, \gamma^\nu + \gamma^\nu\, \gamma^\mu = 2g^{\mu\nu}, \qquad (1.6.3)$$

where $g^{\mu\nu}$ denotes the Minkowski metric. Such an algebraic structure is well known in the mathematical literature. Eq. (1.6.3) says that the γ^μ define a Clifford algebra.[38] It is easy to show that eq. (1.6.1), with ΓM replaced by m, leads to the following conserved current:

---

[37] For an excellent treatment of the general case, see A. S. Wightman, Invariant wave equations; General theory and applications to the external field problem, in *Invariant Wave Equations*, G. Velo and A. S. Wightman (eds.), Springer 1978, Chap. 2, Sects. 1–4.

[38] For the definition and properties of Clifford algebras see Y. Choquet-Bruhat and C. DeWitt-Morette, *Analysis, Manifolds and Physics*, Part I, North-Holland 1991, p. 64; Part II, p. 6. A Clifford algebra exists for any real n-dimensional vector space V, $n = n_+ + n_-$, equipped with a metric g, $g_{il} = 0$ if $i \neq l$, $g_{ii} = 1$ for $i \leq n_+$, $g_{ll} = -1$, for $l > n_+$. The Clifford algebra $\mathcal{C}(n_+, n_-)$ associated with V is the real vector space generated by the symbols γ_j and a unit $\mathbf{1}$, equipped with an associative product, distributive with respect to the addition, and satisfying

$$\gamma_j\gamma_i + \gamma_i\gamma_j = 2g_{ij}\mathbf{1}\,.$$

If n is even, say $n = 2m$, there is only one (up to equivalence) irreducible representation of the algebra by $2^m \times 2^m$ matrices. The explicit case of eq. (1.6.3) has been worked out by Pauli; for a review see R. H. Good, Rev. Mod. Phys. **27**, 187 (1955); see also S. Schweber, *An Introduction to Relativistic Quantum Field Theory*, Harper and Row 1961, pp. 69–74.

$$j_\mu(x) \equiv \overline{\psi}(x)\gamma_\mu\psi(x), \quad \overline{\psi}(x) \equiv \psi^*(x)\gamma^0, \quad \rho(x) \equiv j_0(x) = \psi^*(x)\,\psi(x).$$

For the Lorentz invariance, one notes that $\gamma'^\mu \equiv \Lambda^\mu_\nu\gamma^\nu$ satisfies eq. (1.6.3) and therefore define the same Clifford algebra. By a general result, all irreducible representation of the Clifford algebra defined by eq. (1.6.3) are equivalent, i.e., they are related by a similarity transformation and given by 4×4 matrices (see the previous footnote). Hence, for any $A \in SL(2, C)$, which defines a Lorentz transformation $\Lambda(A)$, there exists a non-singular matrix $S(A)$ such that

$$S(A)^{-1}\gamma^\mu S(A) = \Lambda(A)^\mu_\nu\,\gamma^\nu. \tag{1.6.4}$$

Lorentz covariance of the Dirac equation is obtained by requiring the following transformation for ψ:

$$\psi^\Lambda(x) = S(A)\psi(\Lambda(A^{-1})x). \tag{1.6.5}$$

For infinitesimal transformations $\Lambda^\nu_\mu = g^\nu_\mu + \varepsilon\lambda^\nu_\mu$, $S(A) \simeq 1 + \varepsilon T$, and eq. (1.6.4) gives $[\gamma_\mu, T] = \lambda^\nu_\mu\gamma_\nu$. The condition $1 = \det S = 1 + \mathrm{Tr}\,T$ (where Tr denotes the trace) and irreducibility give

$$S(A) = \mathbf{1} + (1/8)\lambda^{\mu\,\nu}[\gamma_\mu,\,\gamma_\nu].$$

The content of the transformation (1.6.5) becomes very clear by choosing the following representation of the γ's : $\gamma^0 = \tau_1 \times \mathbf{1}$, $\gamma^i = i\tau_2 \times \sigma^i$ (as in Section 2.3 above). Then, $S(A)$ takes a block diagonal form

$$S(A) = \begin{pmatrix} S_1(A) & 0 \\ 0 & S_2(A) \end{pmatrix},$$

i.e., one has a reducible representation of $SL(2, C)$ and of the Lorentz group. The reason for such a reducibility is the invariance under space inversion I_P represented by $S(I_P) = \gamma^0$, i.e., parity interchanges the first two components of ψ with the second ones.

For the convenience of the reader we briefly recall that the finite dimensional irreducible representations of $SL(2, C)$ are provided by vectors, called *spinors*, labeled by two sets (α), $(\dot\beta)$, $\alpha = \alpha_1\ldots\alpha_j$, $\dot\beta = \dot\beta_1\ldots\dot\beta_k$, $\alpha_i = 1, 2$, $\dot\beta_i = 1, 2$, of (symmetric) undotted and dotted indices such that $\forall A \in SL(2,\,C)$

$$\psi_{\alpha_1\ldots\alpha_j\,\dot\beta_1\ldots\dot\beta_k} \to A_{\alpha_1\,\gamma_1}\cdots A_{\alpha_j\,\gamma_j}\,\bar{A}_{\dot\beta_1\,\dot\delta_1}\cdots\bar{A}_{\dot\beta_k\,\dot\delta_k}\,\psi_{\gamma_1\ldots\gamma_j\,\dot\delta_1\ldots\dot\delta_k}, \tag{1.6.6}$$

(the sum over repeated indices being understood).[39] The conjugate spinor is given by $\psi^*_{\alpha,\dot\beta} = \psi_{\dot\alpha,\beta}$. Such representations are usually denoted by $\mathcal{D}^{(j/2,k/2)}$; any such representation is a (one-valued) representation of $L^\uparrow_+$ iff $A = -1$ is represented by the identity, i.e., iff $j + k$ is even, corresponding to integer spin representations ($j + k$

---

[39] For a synthetic and simple account of the representations of $SL(2, C)$ see S. Weinberg, *The Quantum Theory of Fields*, Vol. I, Cambridge University Press 1995, pp. 229–32; see also B. L. van der Waerden, *Group Theory and Quantum Mechanics*, Springer 1980; S. Schweber, *An Introduction to Relativistic Quantum Field Theory*, Harper and Row 1961, Chap. 2.

is odd for half-integer spin). The numbers j, k corresponds to the standard labels of the representations of the $SU(2)$ groups generated by $\tilde{J}_i$, $\tilde{K}_i$, $i = 1, 2, 3$,

$$\tilde{J}_j \equiv \tfrac{1}{2}(J_j + iK_j), \quad \tilde{K}_j \equiv \tfrac{1}{2}(J_j - iK_j),$$

where J_j, K_j are the generators of the space rotations and of the Lorentz boosts, respectively.

Such representations become unitary if restricted to the $SU(2)$ subgroup of $SL(2, C)$ generated by unitary A's; in fact, the representations $SU(2) \ni A \to \mathcal{D}^{(j/2,k/2)}(A)$ are the direct product, $\mathcal{D}^{(j/2)} \otimes \mathcal{D}^{(k/2)}$, of irreducible representations of $SU(2)$, of angular momentum $j/2$ and $k/2$, respectively ($\mathcal{D}^{(m/2)}$ is equivalent to $\mathcal{D}^{(m/2,0)}$ and to $\mathcal{D}^{(0,m/2)}$).

Another way of obtaining the Dirac equation is via the unitary representations of the Poincaré group. Thanks to the fundamental paper by Wigner,[40] an elementary particle of spin s and mass m is described by a unitary irreducible representation of the Poincaré group given by

$$\mathbf{P} = \mathbf{p}, \quad \mathbf{J} = \mathbf{x} \wedge \mathbf{p} + \mathbf{s}, \quad x_i = i\partial/\partial p_i,$$

$$P_0 = \sqrt{\mathbf{p}^2 + m^2} \equiv \omega, \quad \mathbf{K} = \omega\,\mathbf{x} - (\omega + m)^{-1}\mathbf{s} \wedge \mathbf{p}.$$

Thus, the evolution equation is the relativistic Schrödinger equation discussed in Section 2.1 above, which is non-local and does not allow the introduction of local interactions; furthermore, the transformation law under Lorentz boosts is highly non-local. By Proposition 2.2 we know that in order to obtain a local time evolution we have to give up positivity of the energy; thus the minimal change (amounting to work with reducible representations) is to consider a representation in which P_0 is represented by

$$P_0 = \beta \sqrt{\mathbf{p}^2 + m^2}, \quad \beta = \tau_3. \tag{1.6.7}$$

In order to preserve the group commutation relations $[P_i, \, K_i] = P_0$, $[K_i, \, K_j] = i\varepsilon_{ijk}\,J_k$, one also has to multiply K_i by β. Thus, for spin $1/2$ one has four-component wave functions. It is now easy to see that there exists a unitary U such that

$$U^* P_0 U = (\alpha_i\, p_i + \beta\, m), \quad \alpha_i \equiv \beta\,\gamma^i,$$

so that if ψ_S is a (four-component) solution of the relativistic Schroedinger-like equation, $i\partial_t \psi_S = \beta\sqrt{\mathbf{p}^2 + m^2}\psi_S$, then $U\psi_S$ is a solution of the Dirac equation.[41]

For the detailed properties of the plane wave solutions of the Dirac equation, we refer the reader to the textbooks quoted in footnote 6.

---

[40] E. P. Wigner, Ann. Math. **40**, 149 (1939); A. S. Wightman, Nuovo Cim. (Suppl.) **1**, 81 (1959); E. P. Wigner, Unitary Representations of the Inhomogeneous Lorentz Group Including Reflections, in *Group Theoretical Concepts and Methods in Elementary Particle Physics*, F. Gursey (ed.), Gordon and Breach 1964.

[41] Actually, $U = (2\omega(\omega + m))^{-1/2}[\beta(\omega + m) + \alpha_i p_i]$ does the job.

7 Appendix: Canonical field theory

For the convenience of the reader we briefly review the main features of the canonical formulation of classical field theory. For simplicity, we discuss the case of a real scalar field φ.

As in the case of finite degrees of freedom, the dynamics of φ is governed by the Euler–Lagrange equations which can be derived from a (local) Lagrangian density $\mathcal{L}(x) = \mathcal{L}(\varphi(x), \partial_\mu \varphi(x))$, by looking for the stationary points of the action

$$A_{\mathcal{V}} = \int_{\mathcal{V}} d^4 x \, \mathcal{L}(\varphi(x), \partial_\mu \varphi(x)),$$

for any finite four-dimensional volume $\mathcal{V}$. The stationarity condition $\delta A_{\mathcal{V}} = 0$, $\forall \mathcal{V}$, with respect to arbitrary variations $\delta \varphi$ of the field, with the only restriction that $\delta \varphi(x)$ vanishes on the boundary $\partial \mathcal{V} \equiv \Sigma_{\mathcal{V}}$, gives the Euler equations

$$\frac{\partial \mathcal{L}(x)}{\partial \varphi(x)} - \frac{\partial}{\partial x^\mu} \frac{\partial \mathcal{L}(x)}{\partial(\partial_\mu \varphi(x))} = 0. \tag{1.6.9}$$

By introducing a complete set $\{f_n(\mathbf{x})\}$, $n \in \mathbf{N}$, of real orthonormal functions, a real scalar field is completely identified by a denumerable set of Lagrangian variables

$$q_n(t) \equiv \varphi(f_n, t) \equiv \int d^3 x \, f_n(\mathbf{x}) \varphi(\mathbf{x}, t), \quad \dot{q}_n(t) \equiv \dot{\varphi}(f_n, t). \tag{1.7.1}$$

Clearly

$$\varphi(x) = \sum_n q_n(t) \, f_n(\mathbf{x}), \quad \dot{\varphi}(x) = \sum_n \dot{q}_n(t) \, f_n(\mathbf{x}).$$

The above Euler–Lagrange field equations are equivalent to the following denumerable set of equations $(L(t) \equiv \int d^3 x \, \mathcal{L}(x))$:

$$\frac{\partial L(t)}{\partial \dot{q}_n(t)} = \int d^3 x \, \frac{\partial \mathcal{L}(x)}{\partial \dot{\varphi}(x)} \frac{\partial \dot{\varphi}(x)}{\partial \dot{q}_n(t)} = \int d^3 x \frac{\partial \mathcal{L}(x)}{\partial \dot{\varphi}(x)} \, f_n(\mathbf{x}),$$

$$\frac{\partial L(t)}{\partial q_n(t)} = \int d^3 x \, \Big(\frac{\partial \mathcal{L}(x)}{\partial \varphi(x)} - \sum_{i=1}^{3} \frac{\partial}{\partial x^i} \frac{\partial \mathcal{L}(x)}{\partial(\partial_i \varphi(x))} \Big) f_n(\mathbf{x}).$$

A canonical formalism may be introduced by defining the canonical momenta $p_i(t)$ by

$$p_n(t) = \partial L(t)/\partial \dot{q}_n(t) = \int d^3 x \, \pi(x) \, f_n(\mathbf{x}), \quad \pi(x) \equiv \partial \mathcal{L}/\partial \dot{\varphi}(x).$$

Then, the Hamiltonian is defined by

$$H = \sum_n p_n(t) \, \dot{q}_n(t) - L(t) = \int d^3 x \, (\partial \mathcal{L}(x)/\partial \dot{\varphi}(x) \, \dot{\varphi}(x) - \mathcal{L}(x)).$$

The canonical (classical) Poisson brackets read

$$\{q_i(t),\, p_j(t)\} = \delta_{ij},\tag{1.7.2}$$

all other Poisson brackets vanishing. They are equivalent to the following Poisson brackets for the fields:

$$\{\varphi(\mathbf{x}, t),\, \pi(\mathbf{y}, t)\} = \delta^3(\mathbf{x} - \mathbf{y}),\tag{1.7.3}$$

all other Poisson brackets vanishing.

It is instructive to check that the Hamilton equations derived in terms of the canonical variables q_i, p_i are equivalent to the Euler–Lagrange equations for the fields.

2

Mathematical problems of the perturbative expansion

1   Dyson's perturbative expansion

Dyson's perturbative expansion of the S-matrix makes use of the *interaction picture* which is defined by subtracting the free (or more generally the asymptotic) time evolution $U_0(t)$ from the Schrödinger time evolution of the states. If the Hamiltonian H is split as a sum (of well-defined self-adjoint operators)

$$H = H_0 + gH_{int},$$

where g denotes the coupling constant, the time evolution of the states in the interaction and in the Schrödinger picture, $\Psi_I(t)$, $\Psi_S(t)$, respectively, is given by $(U(t) = e^{-iHt}, U_0(t) \equiv e^{-iH_0t})$:

$$\Psi_I(t) = V(t)\,\Psi_S(0), \qquad V(t) \equiv U_0(t)^*\,U(t).$$

The time evolution of the operators $O_I(t)$ is defined by

$$(\Psi_I(t), O_I(t)\,\Psi_I(t)) = (\Psi_S(t), O_S\,\Psi_S(t)) \quad \text{i.e.,} \quad O_I(t) = U_0^*(t)\,O_S\,U_0(t); \qquad (2.1.1)$$

namely, is governed by the action of the free Hamiltonian on the Schrödinger operators O_S. The relation to the time evolution of the operators in the Heisenberg picture $O_H(t)$ is

$$O_H(t) = V(t)^*O_I(t)V(t), \quad V(t)^* = U(t)^*U_0(t). \qquad (2.1.2)$$

The main advantage of the interaction picture is to yield the solution of the time evolution in terms of a power series in the coupling constant. In fact, the interaction picture propagator $V(t,t')$, which governs the time evolution of the states,

$$\Psi_I(t) = V(t,t')\Psi_I(t'),$$

$$V(t,t') = V(t)\,V(t')^{-1} = U_0(t)^*\,U(t,t')\,U_0(t'), \quad U(t,t') = U(t)\,U(t')^*,$$

satisfies

$$V(t,t')\,V(t',t'') = V(t,t''), \quad V(t,t) = 1,$$

$$i\partial_t V(t,t') = H_I(t)\,V(t,t'), \quad H_I(t) = U_0(t)^*g\,H_{int}U_0(t).$$

Then, $V(t, t')$ is the solution of the integral equation

$$V(t, t') = 1 - i \int_{t'}^{t} ds\, H_I(s)\, V(s, t'). \tag{2.1.3}$$

The iterative solution of eq. (2.1.3) provides a series in powers of g, whereas the expansion of the exponential of the Schrödinger time evolution $U(t, t_0) = e^{-iH(t-t_0)}$ does not have this property, and its computation requires a control of the Hamiltonian spectrum.

Furthermore, the operator $V(t, t')$ gives direct information on the S-matrix, since the Møller operators $\Omega^{\pm}$ (see Section 6 Appendix below) are the asymptotic limits of $V(t)^{-1}$

$$\Omega^{\pm} = \lim_{t \to \mp \infty} V(t)^{-1},$$

and therefore

$$S = \Omega^{-*} \Omega^{+} = \lim_{t \to \infty,\ t' \to -\infty} V(t, t').$$

Another useful feature of the interaction picture is that eq. (2.1.3) applies equally well to time-dependent interactions[1] $H_S(t)$, and a time-dependent Hamiltonian arises as a consequence of the adiabatic switching $H_I(t) \to H_I^{\varepsilon}(t) = e^{-\varepsilon |t|}\, H_I(t)$, the limit $\varepsilon \to 0$ being taken at the very end (see Section 6.3 in the Appendix below).

The perturbative expansion of the S-matrix follows from the expansion of $V_{\varepsilon}(t, t_0)$ (where the subscript ε denotes the introduction of an adiabatic switching), in a power series of the coupling constant g

$$V_{\varepsilon}(t, t_0) =$$

$$= \sum_{n=0}^{\infty} (-i)^n \int_{t_0}^{t} dt_1 \int_{t_0}^{t_1} dt_2 \ldots \int_{t_0}^{t_{n-1}} dt_n\, e^{-\varepsilon(|t_1| + \ldots |t_n|)} H_I(t_1) \ldots H_I(t_n)$$

$$= \sum_{n=0}^{\infty} \frac{(-i)^n}{n!} \int_{t_0}^{t} dt_1 \ldots \int_{t_0}^{t} dt_n\, e^{-\varepsilon(|t_1| + \ldots |t_n|)} T(H_I(t_1) \ldots H_I(t_n)), \tag{2.1.4}$$

where $T(\ )$ means that the product of the operators within the round brackets must be chronologically ordered, with time increasing from right to left (*chronologically ordered product*). For example,

$$T(H_I(t_1)\, H_I(t_2)) = \theta(t_1 - t_2)\, H_I(t_1)\, H_I(t_2) + \theta(t_2 - t_1) H_I(t_2)\, H_I(t_1), \tag{2.1.5}$$

where θ denotes the Heaviside step function $\theta(t) = 1$ for $t > 0$, $\theta(t) = 0$ for $t < 0$; for $t_1 = t_2$, θ is not defined, but this is irrelevant, since then $[H_I(t_1), H_I(t_2 = t_1)] = 0$.

As clearly displayed by the $n = 2$ term, the last equality in eq. (2.1.4) is obtained by a change of variables and orders of integrations and by realizing that the $n!$ θ terms appearing at the nth order give all the same contribution. The series corresponds to

---

[1] This was in fact one of Dirac motivations: P. A. M. Dirac, Proc. Roy. Soc (London) **A 112**, 661 (1926); **A 114**, 243 (1927).

the Neumann–Liouville expansion of the integral in eq. (2.1.3), which converges if H_I is a bounded operator; in quantum field theory such a convergence fails (see below), and the expansion has only a formal meaning.

Thanks to the adiabatic switching, the limit $t \to \infty$, $t_0 \to -\infty$ is term-by-term well defined, since the $n!$ ways of performing it (corresponding to the $n!$ θ functions appearing at order n) give all the same result. Thus, the expansion of V yields the following S-matrix expansion

$$S = \lim_{\varepsilon \to 0} \sum_{n=0}^{\infty} \frac{(-i)^n}{n!} \int_{-\infty}^{\infty} dt_1 \ldots \int_{-\infty}^{\infty} dt_n \, e^{-\varepsilon(|t_1|+\ldots|t_n|)} \, T(H_I(t_1) \ldots H_I(t_n)) \quad (2.1.6)$$

as a power series in g (*perturbative expansion*), and the explicit computation is reduced to matrix elements of operators with free (or asymptotic) time evolution, eq. (2.1.1), on free (or asymptotic) states.

In (most) quantum field theory models, the interaction Hamiltonian is the integral over a Lorentz scalar density $h_I(x)$

$$H_I(t) = \int d^3x \, h_I(\mathbf{x}, t), \qquad (2.1.7)$$

and the S-matrix takes the Lorentz-invariant form

$$S = \sum_{n=0}^{\infty} \frac{(-i)^n}{n!} \int dx_1^4 \ldots \int dx_n^4 \, T(h_I(x_1) \ldots h_I(x_n)), \qquad (2.1.8)$$

with the adiabatic limit understood.

Actually, the integration in eq. (2.1.7) is also ill defined (in general divergent), and requires a convergence factor, e.g., a function $f(|\mathbf{x}|/R)$, $f(x) = 1$, for $|x| \le 1$, $f(x) = 0$, for $|x| \ge 1 + \varepsilon$. The limit $R \to \infty$ should be taken at the very end, together with the removal of the adiabatic switching (after a possible regularization of local singularities of $h_I(x)$). Thus, in the expression for the S-matrix one should introduce an adiabatic four-dimensional switching of the interaction, e.g., by multiplying $h_I(x)$ by a smooth function $g(x) \in \mathcal{D}(\mathbf{R}^4)$, $0 < g(x) < 1$, $g(x) = 1$, for $x \in K$, and take the limit $K \to \mathbf{R}^4$ at the very end. The existence of such limits, at least for matrix elements on a dense set of states, in general requires a modification or regularization of h_I, typically by the addition of counter terms.[2]

---

[2] In the case of free fields, as discussed in Chapter 1, Section 5, the counter-term corresponding to Wick ordering makes the integral convergent when applied to states of the Fock representation containing a finite number of particles. However, as we shall see, the Fock representation is not allowed in the interaction case, and therefore one must develop a strategy of infinite volume regularization. The four-dimensional switching, needed for the mathematical control of the terms of the expansion, has been advocated by N. N. Bogoliubov and D. V. Shirkov, *Introduction to the Theory of Quantized Fields*, Interscience 1959, Chap. III, and it is at the basis of the mathematical control of the renormalized perturbation theory discussed by H. Epstein and V. Glaser, Le rôle de la localité dans la renormalisation en théorie quantique des champs, in *Statistical Mechanics and Quantum Field Theory*, R. Stora and C. DeWitt (eds.), Gordon and Breach 1971; Adiabatic limit in perturbation theory, in *Renormalization Theory*, G. Velo and A. S. Wightman (eds.), Reidel 1976.

The locality of the Hamiltonian densities, i.e., the property of commuting at spacelike points, plays a crucial rôle for the Lorentz invariance of the T products and, therefore, of the S-matrix.

The computation of the S-matrix elements becomes greatly simplified due to the Wick theorem,[3] by which the time-ordered products of fields, occurring in the nth order expansion of S, can be written as a sum of normal products times vacuum expectations (or contractions) of pairs of time-ordered free fields, called *Feynman propagators* (for a sketchy account see Section 7 below).

In this way, since the matrix elements of normal products on free asymptotic states are simply computed, one easily obtains the Dyson–Feynman–Schwinger expansion and its pictorial representation in terms of Feynman diagrams.

The field propagators are singular at coincident points and give rise to ultraviolet (UV) divergences in the integrations of eq. (2.1.8); the great success of perturbative renormalization is the proof that there is a finite number of local (UV-divergent) counter-terms to be added to the original Hamiltonian density, such that the resulting expansion of the S-matrix is order-by-order finite. Such counter-terms may be calculated order by order in perturbation theory, e.g., by introducing an UV cutoff. The counter-terms are then cutoff-dependent and determined by the condition that the S-matrix elements, corresponding to the Hamiltonian density with counter-terms (*renormalized Hamiltonian*), have finite limits when the UV cutoff is removed.

The need of UV divergent counter-terms implies that the renormalized Hamiltonian (or Lagrangian) at the basis of the renormalized perturbative expansion cannot be a UV-finite polynomial of the interacting fields. Inevitably, the term-by-term finiteness of the perturbative expansion is obtained from the cancellation of the divergences of the Hamiltonian counter-terms and the UV divergences of the integrations in eq. (2.1.8). A well-defined Hamiltonian can be recovered, once the solution is known, as the generator of time translations.

A detailed discussion of renormalized perturbation theory is outside the scope of these notes, and we refer to any textbook on quantum field theory.[4]

2 Dyson argument against convergence

Soon after the theoretical and experimental success of perturbative renormalization, the question arose about the convergence of the renormalized perturbative series. An argument against the convergence was given by Dyson,[5] which we briefly sketch. Let

$$F(e^2) = F(0) + e^2 F_2(0) + e^4 F_4(0) + \dots$$

be the renormalized perturbative expansion (at the origin) for a physical quantity $F(e^2)$. The convergence of the series means that $F(e^2)$ is analytic in a neighborhood

---

[3] G. C. Wick, Phys. Rev. **80**, 268 (1950); for the convenience of the reader a brief account is given in Section 7 below.

[4] For a beautiful updated discussion, see S. Weinberg, *The Quantum Theory of Fields*, Vol. I, Cambridge University Press 1996, esp. Sect. 12. For a mathematically-minded presentation, see M. Salmhofer, *Renormalization: An Introduction*, Springer 1999.

[5] F. Dyson, Phys. Rev. **85**, 631 (1952).

of the origin and therefore also for small negative values of the expansion parameter. As argued by Dyson,[6] $F(-e^2)$ corresponds to the value of the given quantity in a fictitious world in which particles of equal charge attract and those of opposite charge repel each other. Now, in the ordinary case, by grouping together pairs of opposite charges one cannot lower indefinitely the energy, since in this way also charges of equal sign will become close, and their repulsion will set a lower bound for the energy, since the potential energy cannot indefinitely overcome the total rest energy (stability of matter). On the other hand, in the fictitious world, by creating pairs of opposite charges, e.g., under the action of an external perturbation, and by bringing together, in two largely separated regions, charges of equal sign, one can indefinitely lower the energy; thus, the vacuum is no longer the lowest energy state, and it is unstable against collapse. Such drastically different stability properties cannot be reconciled with a smooth, actually analytic, behavior of the physical quantities in the transition from positive to negative values of the expansion parameter, and this indicates lack of analyticity and failure of convergence of the renormalized perturbative expansion. Dyson further argues that large-order terms (starting from $n \geq 137$) will be more and more relevant and overcome the lowest-order terms.

The correctness of Dyson's conclusion has been rigorously established in low-dimensional quantum field theory models, as discussed below.

2.1 φ^4 model in zero dimensions

A scalar field with a $g\,\varphi^4$ interaction, briefly called the φ^4 model, is the prototypical model for the perturbative expansion in QFT, and from it one can learn about coupling constant analyticity.

The simplest case is in zero space and zero time dimension, where the following partition function corresponds to the Euclidean functional integral:

$$Z(\mu, g) = \int_{-\infty}^{\infty} d\varphi \, e^{-\mu\varphi^2 - g\varphi^4}, \quad \mu > 0, \quad g > 0. \tag{2.2.1}$$

For the expansion in powers of g

$$Z(\mu, g) = \sum_n g^n \, Z_n(\mu)$$

one has

$$Z_n(\mu) = (n!)^{-1}(-1)^n \int_{-\infty}^{\infty} d\varphi \, e^{-\mu\varphi^2} \, \varphi^{4n} =$$

$$= (n!)^{-1}(-1)^n \, \mu^{-2n-\frac{1}{2}} \int dy^2 \, e^{-y^2} \, (y^2)^{2n-\frac{1}{2}}.$$

Since the integral on the right-hand side is the Γ function:

$$\Gamma(2n + \tfrac{1}{2}) = 2\sqrt{\pi}\,2^{-4n}(4n)!/(2n)! \sim_{n\to\infty} (2n)!,$$

---

[6] The argument partly relies on Feynman equivalence of QED interaction to an action at a distance given by $e^2 \delta((x_1 - x_2)^2)$, with x_1, x_2 the particle positions; R. P. Feynman, Phys. Rev. **76**, 769 (1949); **80**, 440 (1950), Appendix B.

$Z_n(\mu)$ diverges as $(2n)!/n! \sim 2^n\, n!$, so that it violates the analyticity condition $|Z_n(\mu)| \le c^n$, for some c, and the perturbative series cannot converge.

However, by estimating the rest of the Taylor series, one obtains

$$|Z(\mu, g) - \sum_{n=0}^{k} g^n Z_n(\mu)| \le C_k g^{k+1} |Z_{k+1}(\mu)|,$$

i.e., the perturbative series is asymptotic to the exact solution. We recall that a series $\sum a_n x^n$ is asymptotic at the origin to a function $f(x)$, which is C^∞ in $(0, \epsilon)$, if, $\forall k$, $f(x) - \sum_{n=0}^{k} a_n x^n = o(x^k)$, for $x \to 0^+$. This implies that $a_n = (n!)^{-1} d^n f(0^+)/dx^n$; as shown by Hadamard, there are no growth restrictions on the $d^n f(0^+)/dx^n$, and therefore on the a_n, even if $f \in C^\infty([-\varepsilon, \varepsilon])$.

It is important to note that two functions, which significantly differ (e.g., by $Ce^{-1/(Cx^2)}$) in a neighborhood of the origin, may have the same (right) derivatives at the origin, and therefore a series may be asymptotic to significantly different functions.

The asymptotic property of the series may explain why the sum of the lowest orders provides a good approximation for the exact solution near the origin, even if the series diverges and does not define a function.

For $\mu < 0$ the model mimics the double-well potential; to discuss its analyticity properties it is convenient to complete the square and consider the modified "action" $g(\varphi^2 + \mu/2g)^2$. One can show that the corresponding partition function Z_{mod} has a perturbative asymptotic series in powers of $4g/\mu^2$, but Z differs from Z_{mod} by the factor $\exp(\mu^2/4g)$ which has an essential singularity in g.[7]

2.2 φ^4 model in $0+1$ dimensions

In zero space and one time dimension, the φ^4 model coincides with the anharmonic oscillator with an Hamiltonian of the form

$$H = \pi^2 + m^2\varphi^2 + g\varphi^4 \equiv H(\mu, g), \quad \mu \equiv m^2, \ g > 0.$$

The lack of analyticity in g could be argued by a Dyson-like argument, since for $g < 0$ the Hamiltonian is not bounded below (is not even essentially self-adjoint), and instability occurs.

The model has been discussed in detail with full mathematical control by B. Simon.[8] In $L^2(\mathbf{R}, dx)$, where φ acts as the multiplication by x, the Hamiltonian is a well-defined self-adjoint operator on $D(H) \equiv D(\pi^2) \cap D(x^4)$, with discrete spectrum, since so is $\pi^2 + x^4$ and x^2 is a Kato small perturbation.[9]

[7] For a detailed analysis of the perturbative series for this very simple but instructive model, see A. S. Wightman, in *Mathematical Quantum Field Theory and Related Topics*, Montreal 1977, J. S. Feldman and L. M. Rosen (eds.), Am. Math. Soc. 1988, p. 1.

[8] B. Simon, Ann. Phys. **58**, 76 (1970); see also the excellent review by B. Simon, Int. Jour. Quantum Chemistry, **XXI**, 3 (1982).

[9] Since x^4 is positive and locally L^2, $\pi^2 + x^4$ is essentially self-adjoint on C_0^∞, and so is its extension to $D(\pi^2) \cap D(x^4)$, which is closed there. The discreteness of the spectrum of H follows from $x^4 \ge x^2 - 1$ and the min–max principle (see, e.g., the version in D. Ruelle, *Statistical Mechanics*, Benjamin 1969, p. 25). For a detailed proof of these simple facts, see B. Simon, Ann. Phys. **58**, 76 (1970).

The lack of analyticity is checked on the spectrum of H, by exploiting the fact that the scaling transformations ($\lambda > 0$)

$$x \to \lambda^{-1/2} x, \quad \pi \to \lambda^{1/2} \pi,$$

are canonical transformations described by the unitary operator $U(\lambda)$, defined by

$$(U(\lambda)\psi)(x) = \lambda^{1/4} \psi(\lambda^{1/2} x), \quad \forall \psi(x) \in L^2(dx).$$

Since $U(\lambda)D(H) \subseteq D(H)$, $U(\lambda)^* x U(\lambda) = \lambda^{-1/2} x$, $U(\lambda)^* \pi(\lambda) U(\lambda) = \lambda^{1/2} \pi$, one has

$$U(\lambda)^* H(\mu, g) U(\lambda) = \lambda H(\mu \lambda^{-2}, g \lambda^{-3}).$$

Since the spectrum is discrete and λ varies continuously, it follows that the nth eigenvalue satisfies

$$E_n(\mu, g) = \lambda E_n(\mu \lambda^{-2}, g \lambda^{-3}), \tag{2.2.2}$$

in particular, by choosing $\lambda = g^{1/3}$,

$$E_n(\mu, g) = g^{1/3} E_n(\mu g^{-2/3}, 1), \tag{2.2.3}$$

$$E_n(0, g) = g^{1/3} E_n(0, 1). \tag{2.2.4}$$

Since the quadratic term $\mu \varphi^2$ is a Kato small perturbation with respect to the rest, $E_n(\mu, 1)$ has a convergent expansion in μ for μ small and therefore, by eq. (2.2.3), $E_n(\mu g^{-2/3}, 1)$ has a convergent expansion in $\mu g^{-2/3}$. However, $E_n(\mu, g)$ is not analytic near $g = 0$, as indicated by the third-order branch point associated to the factor $g^{1/3}$ in eq. (2.2.3), and even more evidently by eq. (2.2.4), in the $\mu = 0$ case.[10]

In this simple model, $E_n(m^2, g)$ has all the right derivatives at $g = 0$, and therefore the formal expansion in powers of g is asymptotic to it; actually, one can prove that $E_n(m^2, g)$ satisfies a strong asymptotic condition, so that $E_n(m^2, g)$ can be obtained as the Borel sum of its Taylor series.

We recall that a series $\sum_n a_n z^n$ is *Borel summable* iff
i) $|a_n| \leq C^{n+1} n!$,
ii) its Borel transform $B(x) \equiv \sum_n a_n x^n / n!$, defined for $|x| < C^{-1}$, has an analytic continuation to a neighborhood of $[0, \infty]$, and for $x > 0$, $|B(x)| \leq e^{Dx}$ for some D,
iii) the integral

$$f(x) \equiv x^{-1} \int_0^\infty dy \, e^{-y/x} B(y)$$

converges absolutely.

Then, $f(x)$ is called the Borel sum of the series $\sum_n a_n x^n$, which is asymptotic to $f(x)$.

By interchanging the integral with the sum in the above integral (as it would be justified if the series is convergent) and by using that $\int_0^\infty dx \, x^n e^{-x} = n!$, one gets that formally $f(x)$ is the sum of the series.

---

[10] The non-analyticity in g of $E_n(\mu, g)$ is displayed by expanding $E_n(\mu g^{-2/3}, 1)$ for μ small in eq. (2.2.3). For a more detailed discussion and proofs, see B. Simon, 1970, Theor. II.4.1.

On the other hand, a function $f(z)$ satisfies a *strong asymptotic condition* if
i) f is analytic in a region $R_{\varepsilon,\,B}$ of the complex z plane of the form $|\arg z| < \frac{1}{2}\pi + \varepsilon;\ 0 < |z| < B$, for some ε and $B > 0$,
ii) for some A, C and for all $N \in \mathbf{N}$, and all $z \in R_{\varepsilon,\,B}$,

$$|f(z) - \sum_{n=0}^{N} a_n\, z^n| \leq A\, C^{N+1}\, (N+1)!\, |g|^{N+1}, \quad a_n = (n!)^{-1} d^n f(0^+)/dx^n.$$

Then, f is the Borel sum of its Taylor series.

The power series expansion in g is not Borel summable if the coefficient of the quadratic term is negative $(m^2 \to -m^2)$, giving rise to a double well potential; in this case the model mimics the Goldstone model of symmetry breaking with degenerate minima.[11]

2.3 φ^4 model in 1+1 and 2+1 dimensions

The non-perturbative control of φ^4 model in $1+1$ and $2+1$ spacetime dimensions has been one of the big achievements of constructive quantum field theory, but it requires non-trivial techniques.

For the present discussion, it may be relevant to mention that in both cases the Schwinger functions and the scattering matrix elements are not analytic in the coupling constant, i.e., their perturbative series diverge. In both cases, with positive mass terms, Borel summability of the perturbative series for the Schwinger functions has been proved.[12]

The recent proof of triviality of φ^4 in $3+1$ spacetime dimensions indicates that the situation becomes worse in the real world, and in particular the renormalized perturbative series of the φ^4 model seems to have little to do with the non-perturbative solution.

It may be worthwhile to mention that the φ^4 model in $3+1$ dimensions, with a negative coupling, is non-trivial, and the perturbative series is asymptotic to the solution,[13] which clearly does not satisfy basic requirements of quantum field theory, like the positive energy spectrum and the existence of a stable vacuum. Since the perturbative expansion does not distinguish between different signs of the coupling constant, the non-triviality of the negative coupling model may explain why the standard perturbative series for the positive coupling model is not trivial.

The lack of convergence of the perturbative expansion, indicated by Dyson argument and by the rigorous control of φ^4 models, precludes the possibility of defining a quantum field theory by its perturbative series, and raises the question of a non-perturbative and mathematical foundation of quantum field theory.

---

[11] For a discussion of the singularities in the complex g plane, see A. S. Wightman, 1988.
[12] For a review of the results, see J. Glimm and A. Jaffe, *Quantum Physics: A Functional Integral Point of view*, Springer 2nd ed. 1987, esp. Chap. 23. Borel summability has also been proved for Yukawa theory in 1+1 dimensions.
[13] K. Gawedzki and A. Kupiainen, Nucl. Phys. **B257**, 474 (1985).

3   Haag theorem; non-Fock representations

Apart from the convergence problems mentioned above, it is worthwhile to critically revisit the structural properties or assumptions on which the perturbative expansion is based. As discussed in Chapter 2, Section 1, a crucial role is played by the interaction picture, namely, by the existence of a unitary (time-dependent) operator which relates the fields in the Heisenberg representation to free (interaction picture) fields. The latter obey equal-time canonical (anti)commutation relations, and one can argue on general grounds (see Chapter 4, Theorem 5.1) that their representation is Fock. Thus, according to the interaction picture, the Heisenberg fields obey equal-time canonical (anti)commutation relations, and their representation is unitarily equivalent to a Fock representation.

This innocent looking picture can be shown to be in conflict with a non-trivial interaction, essentially because it implies that the ground state Ψ_0 of the total Hamiltonian $H = H_0 + gH_{int}$, hereafter called the vacuum, must coincide with the ground state Ψ_{0F} of the free Hamiltonian H_0, and this in turn implies a free field theory.[14] Here, we give only the first part of the argument,[15] which is by itself an indication of an ineffectiveness of the interaction, since the ground state is expected to depend on the interaction.

Theorem 3.1   *(Haag) In a quantum field theory in which*
i) the three-dimensional Euclidean group is implemented by unitary operators $U(\mathbf{a}, R)$,
under which the fields transform covariantly, e.g., for a scalar field

$$U(\mathbf{a}, R)\varphi(\mathbf{x}, t)U(\mathbf{a}, R)^* = \varphi(R\mathbf{x} + \mathbf{a}, t),$$

ii) the vacuum is the only Euclidean invariant state,
iii) the fields obey equal-time canonical (anti)commutation relations and their representation is a Fock irreducible representation,
the vacuum state Ψ_0 and the Fock no-particle state Ψ_{0F} coincide.

Proof.   For simplicity, we consider the case of a scalar field. Let $a(f)$, $a^*(g)$ be the canonical variables defined by the field at zero time, then the Fock no-particle state Ψ_{0F} is the unique state satisfying the Fock condition $a(f)\,\Psi_{0F} = 0$, for any (smooth) f. Since, by the transformation properties of the field $(f_{\mathbf{a}, R}(\mathbf{k}) \equiv e^{i\mathbf{k}\cdot\mathbf{a}} f(R^{-1}\mathbf{k}))$,

$$a(f)U(\mathbf{a}, R)^*\Psi_{0F} = U(\mathbf{a}, R)^*a(f_{\mathbf{a}, R})\Psi_{0F} = 0,$$

also $U(\mathbf{a}, R)^*\Psi_{0F}$ satisfies the Fock condition and therefore must coincide with Ψ_{0F} apart from a phase factor $\omega(\mathbf{a}, R)$.
By definition, $\omega(\mathbf{a}, R)$ provides a continuous one-dimensional representation of the Euclidean group and it is a general mathematical result that there is no such a representation apart from the trivial one; then, one may take $\omega = 1$. Thus, Ψ_{0F} is

---

[14] R. F. Streater and A. S. Wightman, *PCT, Spin and Statistics, and All That*, Benjamin 1964, Theors. 4-14, 4-15, 4-16.
[15] R. Haag, Dan. Mat. Fys. Medd., **29**, 12 (1955); R. Haag, *Local Quantum Physics*, Springer 1996, p. 55; the complete argument for the implication of a free theory is given in Chapter 4, Theorem 5.1.

invariant under the Euclidean group and, since the vacuum Ψ_0 is the only invariant state, Ψ_{0F} must coincide with it.

In relativistic quantum field theories, the coincidence of the vacuum state with the Fock state, and therefore a Fock representation for the interacting (Heisenberg) fields, implies a trivial interaction (see Chapter 4). The incompatibility of the Fock representation with a non-trivial interaction in turn implies that, quite differently from the case of finite degrees of freedom, the free Hamiltonian cannot be well defined and the total Hamiltonian cannot be split into a sum of well-defined self-adjoint operators corresponding to the free Hamiltonian and to the interaction Hamiltonian (in particular the interaction can in no way be a small perturbation of the free Hamiltonian); only the sum of the two may give rise to a well-defined self-adjoint operator. As it can be explicitly checked in quantum field theory models with non-trivial interaction, the representation of the fields is never a Fock representation and therefore, by Proposition 5.2 in Chapter 1, the free Hamiltonian cannot be well defined there. This implies that the interaction picture, which requires that both H_0 and H_{int} be well-defined self-adjoint operators, cannot be defined. Such features may look as mathematical subtleties with little physical interest; actually, they are related to the infinite degrees of freedom, and the strategy to overcome these difficulties, in order to produce (physically) sensible results, is linked to the problem of giving a (non-perturbative) meaning to quantum field theory models.

The following examples illustrate the above problems and the (non-perturbative) strategy of solving them.

3.1 Quantum field interacting with a classical source

We consider a quantum scalar field interacting with a classical or external (time-independent) source $j(x) = j(\mathbf{x}) = \bar{j}(\mathbf{x})$, e.g., a classical "charge" density of an extended particle, playing the role of an external field. The evolution equations are

$$(\Box + m^2)\varphi(x) = gj(x), \tag{2.3.1}$$

and quantization is obtained by requiring the canonical commutation relations, eq. (1.4.3), for φ. The formal Hamiltonian in terms of the operators $a(\mathbf{k})$, $a^*(\mathbf{k})$ (eq. (1.4.2)), is ($\omega(k) \equiv (\mathbf{k}^2 + m^2)^{1/2}$):

$$H = \int d\Omega_m(\mathbf{k})\omega(k)a^*(\mathbf{k})\,a(\mathbf{k}) + g \int d\Omega_m(\mathbf{k})[\,a(-\mathbf{k}) + a^*(\mathbf{k})]\,\tilde{j}(\mathbf{k})/\sqrt{2}$$

$$= H_0(a) + gH_{int}. \tag{2.3.2}$$

The model is easily solved by the ("normal mode") operators

$$A(\mathbf{k}) = a(\mathbf{k}) + g\,\omega(k)^{-1}\,\tilde{j}(\mathbf{k})/\sqrt{2} \equiv a(\mathbf{k}) + J(\mathbf{k}),$$

which satisfy the same canonical commutation relations as the a, a^*, and diagonalize the Hamiltonian

$$H = \int d\Omega_m(\mathbf{k})\, \omega(k)\, A^*(\mathbf{k})\, A(\mathbf{k}) + E_0 \equiv H_0(A) + E_0,$$

$$E_0 = -\tfrac{1}{2}g^2 \int d^3k\, \omega(k)^{-2}\, |\tilde{j}(\mathbf{k})|^2.$$

Everything is well defined as long as $\tilde{j}(\mathbf{k})$ decreases sufficiently fast; e.g., if a UV cutoff is introduced $\tilde{j}(\mathbf{k}) \to \tilde{j}(\mathbf{k})\theta(\Lambda - |\mathbf{k}|)$. In fact, by the discussion of Chapter 1, Section 5, the requirement that the Hamiltonian be well defined selects the Fock representation for the canonical operators $A(\mathbf{k})$, $A(\mathbf{k})^*$, and the ground state is the Fock no-particle state Ψ_0 for the A's, i.e., a *Fock coherent state*[16] for the a's:

$$A(\mathbf{k})\, \Psi_0 = 0, \quad a(\mathbf{k})\, \Psi_0 = -J(\mathbf{k})\Psi_0, \tag{2.3.3}$$

$$\Psi_0 = U(J)\Psi_{0F}, \quad U(J) = e^{[a(J) - a(J)^*]}, \quad a(J) = \int d\Omega(k)\, a(\mathbf{k})\, \bar{J}(\mathbf{k}),$$

where Ψ_{0F} is the Fock no-particle state for the a's.

When one takes the pointlike (i.e., local) limit of the interaction, $j(\mathbf{x}) \to \delta(\mathbf{x})$, which corresponds to the removal of the UV cutoff, the Hamiltonian H becomes meaningless, since E_0 is (linearly) divergent, and a counter-term δE has to be added in order to obtain a well-defined renormalized Hamiltonian,

$$H_{ren} = H + \delta E, \quad \delta E = -E_0 + M,$$

where M is an arbitrary (positive) constant. Such a counter-term, which is the only one needed for such an oversimplified model, corresponds to the mass counter-term arising from particle interaction with a field. In such a pointlike limit, all the ground state expectations of the a, a^* have a finite limit, and one has a well-defined quantum field theory model. However, the representation of the a, a^* is a *non-Fock coherent state* representation, and $H_{ren} = H_0 + gH_{int} + \delta E$ is a densely defined (self-adjoint) operator, but H_0 is not.

The ground state Ψ_0 of the total Hamiltonian, and the states of the representation defined by it, cannot be described in terms of the number of excitations which are eigenstates of the free Hamiltonian, since

$$N = \int d\Omega_m(\mathbf{k})\, a^*(\mathbf{k})\, a(\mathbf{k})$$

does not exist as a well-defined self-adjoint operator, its ground state expectation being logarithmically divergent. As a consequence of the non-existence of H_0, the Rayleigh–Schrödinger perturbative expansion is affected by divergences. For example, the expansion of Ψ_0 in terms of eigenstates of the free Hamiltonian would be

---

[16] Coherent states have been extensively studied in quantum optics, R. J. Glauber, Phys. Rev. Lett. **10**, 84 (1963); Phys. Rev. **131**, 2766 (1963); for an elementary account see, e.g., F. Strocchi, *Elements of Quantum Mechanics of Infinite Systems*, World Scientific 1985, Part A, Sects. 1.5–1.7; *An Introduction to the Mathematical Structure of Quantum Mechanics*, 2nd ed., World Scientific 2008, Sect. 3.5.

$$\Psi_0 = Z^{1/2} \sum_{n=0}^{\infty} \frac{1}{n!} \left(-\frac{g}{\sqrt{2}} \int d^3k \, \frac{\tilde{j}(\mathbf{k})}{\omega^2} a^*(\mathbf{k}) \right)^n \Psi_{0F}, \tag{2.3.4}$$

$$Z = \exp\left[-\tfrac{1}{2} g^2 \int d^3k \, |\tilde{j}(\mathbf{k})|^2 / \omega(k)^3 \right]. \tag{2.3.5}$$

If $\tilde{j} \notin L^2(d^3k/\omega^3)$, e.g., when the UV cutoff is removed, the integral in the exponent is divergent, and therefore Z vanishes.

In this case, for each value of the coupling constant g, one has an inequivalent representation, i.e., the Fock representation for A_g cannot be so also for $A_{g'}$, $g \neq g'$.

3.2   Bloch–Nordsieck model; the infrared problem

The Bloch–Nordsieck (BN) model describes the (quantum) radiation field associated to a (classical) charged particle which moves with constant velocity $\mathbf{v}$ for $t < 0$ and with velocity $\mathbf{v}'$ for $t > 0$ (idealized scattering process). The equations of motion are

$$\Box \mathbf{A}(\mathbf{x}, t) = \mathbf{j}(\mathbf{x}, t), \tag{2.3.6}$$

which are equivalent to

$$i \, d\mathbf{a}(\mathbf{k}, t)/dt = \omega(k) \, \mathbf{a}(\mathbf{k}, t) + \tilde{\mathbf{j}}(\mathbf{k}, t)/\sqrt{2}, \quad \omega(k) = |\mathbf{k}|, \tag{2.3.7}$$

where

$$\mathbf{a}(\mathbf{k}, t) \equiv (2\omega)^{-1/2} \left[\omega(k) \mathbf{A}(\mathbf{k}, t) + i \dot{\mathbf{A}}(\mathbf{k}, t) \right],$$

$$\mathbf{j}(\mathbf{x}, t) = e \, \mathbf{v}' \, \theta(t) \, \delta(\mathbf{x} - \mathbf{v}'t) + e \, \mathbf{v} \, \theta(-t) \delta(\mathbf{x} - \mathbf{v}t).$$

The solution is ($\tilde{e} \equiv e/\sqrt{2(2\pi)^3}$),

$$\mathbf{a}(\mathbf{k}, t) = e^{-i\omega(t-t_0)} \mathbf{a}(\mathbf{k}, t_0) - i e^{-i\omega t} \int_{t_0}^{t} dt' \, e^{i\omega t'} \tilde{\mathbf{j}}(\mathbf{k}, t')/\sqrt{2}. \tag{2.3.8}$$

$$\tilde{\mathbf{j}}(\mathbf{k}, t) = \sqrt{2} \, \tilde{e} \left[\mathbf{v}' \theta(t) e^{-i\mathbf{k} \cdot \mathbf{v}' \, t} + \mathbf{v} \, \theta(-t) \, e^{-i\mathbf{k} \cdot \mathbf{v} \, t} \right].$$

It is instructive to discuss the choice of the representation in terms of the asymptotic fields defined by the limits[17]

$$\mathbf{a}_{in}(h, t) = \lim_{\tau \to -\infty} e^{i\omega\tau} \, \mathbf{a}(h, t + \tau), \quad h \in \mathcal{S}(\mathbf{R}^3), \tag{2.3.9}$$

and similarly for $\mathbf{a}_{out}$.

---

[17] The existence of the limit follows from ($\omega_v(k) \equiv \omega - \mathbf{k} \cdot \mathbf{v}$, $\tau, \tau' < 0,$)

$$e^{i\omega\tau} \mathbf{a}(h, t + \tau) - e^{i\omega\tau'} \mathbf{a}(h, t + \tau') = -i\tilde{e}\mathbf{v}/\omega_v(k) \int d^3k \, h(\mathbf{k}) \, [e^{i\omega_v(k)\tau} - e^{i\omega_v(k)\tau'}],$$

which converges to zero, when $\tau, \tau' \to -\infty$, by the Riemann–Lebesgue lemma.

Thus, from eq. (2.3.8) with $t_0 = 0$, one obtains

$$\mathbf{a}_{in}(\mathbf{k}, t) = e^{-i\omega t}[\,\mathbf{a}(\mathbf{k}, 0) - i\tilde{e}\mathbf{v}/\omega_v(k)\,], \tag{2.3.10}$$

$$\mathbf{a}_{out}(\mathbf{k}, t) = e^{-i\omega t}[\,\mathbf{a}(\mathbf{k}, 0) - i\tilde{e}\mathbf{v}'/\omega_{v'}(k)\,], \tag{2.3.11}$$

$(\omega_v(k) \equiv \omega - \mathbf{k} \cdot \mathbf{v})$. Since the external current has different behaviors in the asymptotic past and future one has different asymptotic time evolutions, which can be easily read off from the free evolution of the asymptotic fields; namely $U_{in/out}(t) = e^{-it\,H_{in/out}}$,

$$H_{in/out} = \int d\Omega(k)\,\omega(\mathbf{k})\,\mathbf{a}^*_{in/out}(\mathbf{k}) \cdot \mathbf{a}_{in/out}(\mathbf{k}), \tag{2.3.12}$$

the time dependence of the external current being responsible for the different *in/out* Hamiltonians, as functions of the time zero fields $a(\mathbf{k}, 0)$, $a^*(\mathbf{k}, 0)$. The propagators $U_{in/out}(t)$ describe the asymptotic behavior of the finite-time propagator $U(t, 0)$, and exhibit analogies with the case of long-range potential (see Section 6.4).[18]

Since $\omega_v(k)$, $\omega_{v'}(k) \notin L^2(d^3k/|\mathbf{k}|)$, a Fock representation for $\mathbf{a}, \mathbf{a}^*$ cannot be a Fock representation for $\mathbf{a}_{in}, \mathbf{a}^*_{in}$, and conversely, by the same reason, a Fock representation for $\mathbf{a}_{in}$ cannot be a Fock representation for $\mathbf{a}_{out}$. In this (massless) case, both possibilities are allowed, since the existence of the free Hamiltonian for the asymptotic fields does not require a Fock representation for them.

The physical meaning of the above result is rather basic; in a scattering process of a charged particle, the emitted radiation has a finite energy but an infinite number of "soft" photons, in the sense that for any finite ε the number of emitted photons with momentum greater than ε is finite, but the total number of emitted (asymptotic) photons is infinite:

$$\lim_{\varepsilon \to 0} \int_{|k| \geq \varepsilon} d\Omega(k) < \mathbf{a}^*_{as}(\mathbf{k}) \cdot \mathbf{a}_{as}(\mathbf{k}) > = \infty.$$

Such states with an infinite number of soft photons cannot be described in terms of an occupation number representation, but rather in terms of a classical radiation field $\mathbf{f}$ (which accounts for the low-energy electromagnetic field) and hard (asymptotic) photons.

The corresponding non-Fock representation $\pi_{\mathbf{f}}$ of the creation and annihilation operators $\mathbf{a}^*, \mathbf{a}$, can be obtained from the Fock representation π_F by means of the following coherent transformation (technically *morphism*):

$$\rho(\mathbf{a}(\mathbf{k})) = \mathbf{a}(\mathbf{k}) + \mathbf{f}(\mathbf{k}), \quad \pi_{\mathbf{f}}(\mathbf{a}(\mathbf{k})) = \pi_F(\rho(\mathbf{a}(\mathbf{k}))),$$

where $\mathbf{f}$ is the classical radiation field (*non-Fock coherent state representation*).

In the Bloch–Nordsieck model, since in the asymptotic past and future one has a moving charged particle, it is reasonable to choose for the asymptotic fields the representations corresponding to non-Fock coherent states defined by the classical Lienard–Wiechert radiation fields, i.e., those corresponding to $\mathbf{f}(\mathbf{k}) = i\tilde{e}\mathbf{v}_{as}/\omega_{v_{as}}(k)$,

---

[18] Such analogies have been exploited by P. P. Kulish and L. D. Faddeev, Theor. Math. Phys., **4**, 153 (1970), for the discussion of the infrared problem in QED.

$\mathbf{v}_{in} = \mathbf{v}$, $\mathbf{v}_{out} = \mathbf{v}'$. The realization of the above basic (physical) mechanism, well displayed by the BN model, has led to the (non-perturbative) solution of the infrared problem in quantum electrodynamics. The charged (scattering) states define non-Fock coherent representations of the asymptotic electromagnetic algebra.[19] When these types of state are used to define the scattering amplitudes one obtains finite results, when the infrared cutoff is removed, also in (the correspondingly adapted) perturbation theory.[20]

It is instructive to compare the above non-perturbative analysis of the BN model with the standard perturbative expansion based on the interaction picture in a Fock representation. As in the realistic QED case, the S-matrix is affected by infrared divergences, which can be cured by the pragmatic prescription of summing over the photons with energy less than ΔE, the energy resolution of the experimental apparatus (the so-called *soft photons*).

3.3 Yukawa model; non-perturbative renormalization

The need of a non-Fock representation and of a (non-perturbative) renormalization, achieved by (divergent) subtractions or counter-terms, is well displayed by the Yukawa model of nuclear forces. As discussed in Chapter 1, the description of particle inter-actions by forces at a distance is incompatible with relativity, and one should rather introduce field-mediated interactions. This crucial idea was realized by Yukawa, who proposed a theory of nuclear forces in analogy with the electromagnetic forces.[21] The experimental data of 1947 later confirmed Yukawa's theory, with a (pseudoscalar) particle, called pion, playing, in this analogy, the role of the photon.

The free Hamiltonian H_0 is the sum of the piece $H_0(\varphi)$ for the pion field φ, see eq. (1.4.5) with m denoting the pion mass, and the free Hamiltonian $H_0(\psi)$ for the nucleon field ψ. In analogy with the electromagnetic case the interaction density is $\varphi(x)(\bar{\psi}\gamma_5\psi)(x)$. In order to simplify the discussion and actually obtain the exact solubility of the model, we shall consider the extreme non-relativistic approximation for the description of the (heavy) nucleon; namely, we replace the (free) nucleon energy $E(\mathbf{p}) = M_0 c^2 + O(\mathbf{p}^2/M_0^2 c^4)$ by the nucleon mass term $M_0 c^2$ (we shall later take $c = 1$) and the pseudoscalar nucleon density by $\psi^*(x)\psi(x)$. One can therefore neglect the nucleon spin and use a non-relativistic spinor field. In conclusion, the model is defined by the formal Hamiltonian $H = H_0 + gH_1$, with

$$H_0 = \int d^3k\,\omega_k\,a^*(\mathbf{k})a(\mathbf{k}) + M_0 \int d^3p\,\psi^*(\mathbf{p})\psi(\mathbf{p}),$$

$$H_1 = g \int d^3p\,d^3k\,(2\omega_k)^{-1/2}\,\psi^*(\mathbf{p}+\mathbf{k})\psi(\mathbf{p})[a(\mathbf{k}) + a^*(-\mathbf{k})], \tag{2.3.13}$$

[19] V. Chung, Phys. Rev. **140B**, 1110 (1965); J. Fröhlich, G. Morchio and F. Strocchi, Ann. Phys. **119**, 241 (1979); G. Morchio and F. Strocchi, Nucl. Phys. **B211**, 471 (1984); for a review, see G. Morchio and F. Strocchi, Infrared problem, Higgs phenomenon and long-range interactions, Erice Lectures, in *Fundamental Problems of Gauge Field Theory*, G. Velo and A. S. Wightman (eds.), Plenum 1986.

[20] T. W. Kibble, Phys. Rev. **173**, 1527; **174**, 1882; **175**, 1624 (1968) and references therein.

[21] H. Yukawa, Proc. Math. Soc. Japan, **17**, 48 (1935); reprinted in D. M. Brink, *Nuclear Forces: Selected Readings in Physics*, Pergamon Press 1965, pp. 214–24.

where $\omega_k \equiv \sqrt{\mathbf{k}^2 + m^2}$, and the creation and annihilation operators obey the canonical (anti)commutation relations

$$\{\psi(\mathbf{p}), \psi^*(\mathbf{p}')\} = \delta(\mathbf{p} - \mathbf{p}'), \quad [a(\mathbf{k}), a^*(\mathbf{k}')] = \delta(\mathbf{k} - \mathbf{k}'),$$

all other (anti)commutators vanishing. As already indicated by the classical source model, H is not a well- (densely) defined operator, and a renormalization is necessary.

The non-perturbative constructive strategy consists of introducing a regularization by cutting the high momenta in the interaction; e.g., by introducing a form factor in H_I: $d^3k \to \sqrt{2}\, F_\Lambda(k)d^3k$, $F_\Lambda(k) = 0$, for $k \geq \Lambda$, $F_\Lambda(k) \to_{\Lambda\to\infty} 2^{-1/2}$, $k \equiv |\mathbf{k}|$; H_Λ will denote the so-obtained Hamiltonian. Then, one must identify the cutoff dependent counter-terms δH_Λ to be added to the Hamiltonian: $H_\Lambda \to H_\Lambda + \delta H_\Lambda \equiv H_{ren\,\Lambda}$, in order to obtain a well-defined (field) dynamics when the UV cutoff Λ is removed (see below). For simplicity, the suffix Λ shall be omitted (and understood) until the discussion of the cutoff removal.

The nucleon number, $N = \int d^3p\,\psi^*(\mathbf{p})\,\psi(\mathbf{p})$, commutes with H, the representation may be chosen to be Fock for the nucleon operators, and the Hilbert space of the states of the model decomposes into a direct sum of sectors $\mathcal{H}^{(n)}$, labeled by the eigenvalues of N, $n = 0, 1 \ldots$, so that the analysis may be reduced to each sector. Due to the extreme non-relativistic approximation, the Fock vector Ψ_0 is an eigenstate (actually the ground state) of both H_0 and H, and it is reasonable to use a Fock representation also for a, a^*, at least as long as the UV cutoff is present.

In the $N = 0$ sector the Hamiltonian reduces to the free Hamiltonian for the pions, and the representation remains Fock also in the limit $\Lambda \to \infty$ (see below). For the analysis of the $N = 1$ sector (*one nucleon sector*), we note that the total momentum

$$\mathbf{P} = \int d^3p\,\mathbf{p}\,\psi^*(\mathbf{p})\psi(\mathbf{p}) + \int d^3k\,\mathbf{k}\,a^*(\mathbf{k})a(\mathbf{k})$$

commutes with H, and to simplify the discussion we shall consider the improper vectors $|\mathbf{p}>$, in the Dirac notation, labeled by the (improper) eigenvalue $\mathbf{p}$ of the total momentum. In order to discuss the definition of the Hamiltonian and its spectrum on such states, we have to solve the (improper) eigenvalue equation $H|\mathbf{p}> = M'|\mathbf{p}>$. For this purpose we expand $|\mathbf{p}>$ in terms of the eigenvectors of H_0, with $N = 1$,

$$|\mathbf{p}> = \sum_{n=0}^{\infty} \int d^3q\, d^3k_1 \ldots d^3k_n \,(n!)^{-1/2}\, c_p^n(\mathbf{q}, \mathbf{k}_1, \ldots \mathbf{k}_n) \times$$

$$\times a^*(\mathbf{k}_1) \ldots a^*(\mathbf{k}_n)\psi^*(\mathbf{q})|0>;$$

the conservation of the momentum implies that

$$c_p^n(\mathbf{q}, \mathbf{k}_1, \ldots \mathbf{k}_n) = \delta\Big(\mathbf{p} - (\mathbf{q} + \sum_i \mathbf{k}_i)\Big)\, c_p'^n(\mathbf{q}, \mathbf{k}_1, \ldots \mathbf{k}_n).$$

The matrix elements $< 0|[\,a(\mathbf{k}_1) \ldots a(\mathbf{k}_n)\psi(\mathbf{q}), H\,]|\mathbf{p}>$ yield a recursive relation between the $c_p'^n(\mathbf{q}, \mathbf{k}_1, \ldots \mathbf{k}_n)$

$$g\,(n+1)^{1/2} \int d\Omega_m(\mathbf{k})\,\omega(\mathbf{k})F(k)\,c_p'^{n+1}(\mathbf{q}-\mathbf{k},\mathbf{k},\mathbf{k}_1,\ldots\mathbf{k}_n)+$$

$$+g\,(n)^{-1/2}\sum_i \omega(\mathbf{k}_i)^{-1/2}F(k_i)\,c_p'^{n-1}(\mathbf{q}+\mathbf{k}_i,\mathbf{k}_1,\ldots\mathbf{k}_{i-1},\mathbf{k}_{i+1},\ldots\mathbf{k}_n)=$$

$$= (M'-M_0-\sum_i \omega(\mathbf{k}_i))\,c_p'^{n}(\mathbf{q},\mathbf{k}_1,\ldots\mathbf{k}_n),$$

with the following solution:

$$c_p'^{n}(\mathbf{q},\mathbf{k}_1,\ldots\mathbf{k}_n) = \sqrt{Z}\frac{(-g)^n}{\sqrt{n!}}\prod_{i=0}^{n} F(k_i)\omega_{k_i}^{-3/2}, \tag{2.3.14}$$

$$Z = \exp\left[-g^2 \int d^3k\,\omega_k^{-3}\,|F(k)|^2\right], \tag{2.3.15}$$

$$M' = M_0 - g^2 \int d^3k\,\omega_k^{-2}\,|F(k)|^2 \equiv M_0 - \delta M. \tag{2.3.16}$$

By introducing the unitary operator $\exp iS$, with

$$S \equiv ig \int d^3p\,d^3k\,\omega_k^{-3/2}F(k)\,\psi^*(\mathbf{p})\psi(\mathbf{p}+\mathbf{k})[\,a^*(\mathbf{k})-a(-\mathbf{k})\,] = S^*,$$

one obtains that, in terms of the field operators

$$A^*(\mathbf{k}) \equiv e^{iS}a^*(\mathbf{k})e^{-iS} = a(\mathbf{k})^* + g\int d^3q\,F(k)\,\omega_k^{-3/2}\psi^*(\mathbf{q}+\mathbf{k})\psi(\mathbf{q}),$$

$$\Psi^*(\mathbf{p}) \equiv e^{iS}\psi^*(\mathbf{p})e^{-iS} = (2\pi)^{-3}\int d^3q \int d^3x\,e^{-i(\mathbf{q}-\mathbf{p})\cdot\mathbf{x}}\psi^*(\mathbf{q})\times$$

$$\exp\left[-g\int d^3k\omega_k^{-3/2}F(k)\,e^{-i\mathbf{k}\cdot\mathbf{x}}(a^*(\mathbf{k})-a(-\mathbf{k}))\,\right],$$

the Hamiltonian (with the suffix Λ of the UV cutoff omitted for simplicity in the fields, but otherwise spelled out) becomes

$$H_\Lambda = \int d^3k\,\omega_k A^*(\mathbf{k})A(\mathbf{k}) + M'_\Lambda \int d^3p\,\Psi(\mathbf{p})^*\Psi(\mathbf{p})+$$

$$+\int d^3x\,d^3y\,\Psi^*(\mathbf{x})\Psi^*(\mathbf{y})V_\Lambda(\mathbf{x}-\mathbf{y})\,\Psi(\mathbf{y})\Psi(\mathbf{x}), \tag{2.3.17}$$

$$V_\Lambda(\mathbf{x}-\mathbf{y}) \equiv g^2 \int d^3k\,\omega_k^{-2}|F_\Lambda(k)|^2 e^{i\mathbf{k}\cdot(\mathbf{x}-\mathbf{y})}. \tag{2.3.18}$$

The fields $A(\mathbf{k}), \Psi(\mathbf{p})$ obey canonical (anti)commutation relations, being related to the original ones by a unitary transformation, and furthermore $A^*(\mathbf{k})$, $\Psi^*(\mathbf{p})$ create eigenstates of the (total) Hamiltonian when applied to the vacuum state $|0>$: $A^*(\mathbf{k})|0> = a^*(\mathbf{k})|0>$, $\Psi^*(\mathbf{p})|0> = |\mathbf{p}>$.

Since the Hamiltonian is the sum $H_0(A,A^*)+H_\Lambda(\Psi,\Psi^*)$ of two commuting Hamiltonians, the requirement that H_Λ is a well- (densely) defined operator implies that so is $H_0(A, A^*)$. Then, in the $N = 0, 1$ sectors the representation must be the Fock representation for the operators A, A^*. In conclusion, as long as the UV cutoff Λ is finite, one has a well-defined quantum-mechanical model showing that in the static limit, in which the nucleon energy is independent of the momentum, the direct pion–nucleon interaction may be eliminated, with the only residual effect of an effective nucleon–nucleon two-body potential $V_\Lambda(\mathbf{x})$.

The removal of the UV cutoff sheds some light on the strategy of non-perturbative renormalization. In fact, in this limit $M' \to -\infty$, and the Hamiltonian does not define an acceptable dynamics; however, the renormalized Hamiltonian $H_{ren\,\Lambda} \equiv H_\Lambda + \delta H_\Lambda$, obtained with the addition of the mass counter-term $\delta H_\Lambda = (\delta M_\Lambda + M) \int d^3p\, \psi^*(\mathbf{p})\psi(\mathbf{p})$, with δM_Λ defined in eq. (2.3.16)), and M a finite *free* parameter, defines a dynamics of the fields A, A^*, Ψ, Ψ^*, which remains well defined also in the limit $\Lambda \to \infty$. In particular, the two-body potential V_Λ has a well defined limit when $\Lambda \to \infty$, given by the *Yukawa potential*

$$V(\mathbf{x}) = \pi^2\, g^2\, \frac{e^{-m|\mathbf{x}|}}{|\mathbf{x}|}.$$

and, since $\rho(\mathbf{x}) \equiv \Psi^*(\mathbf{x})\Psi(\mathbf{x})$ is constant in time, one has

$$\Psi(\mathbf{x}, t) = e^{-i(M_0+M)t}\, e^{-it \int d^3y\, V(\mathbf{x}-\mathbf{y})\rho(\mathbf{y},0)}\, \Psi(\mathbf{x}, 0).$$

The addition of the counter-term for the renormalization of the Hamiltonian is clearly equivalent to consider a cutoff-dependent "bare mass" $M_0 \to M_0(\Lambda) = M_0 + M + \delta M_\Lambda$, with the physical constraint that $M_0(\Lambda) - \delta M_\Lambda$ converges to a finite limit, which can be fixed to coincide with the observed nucleon mass M_{phys}. This amounts to the *mass renormalization condition*

$$\lim_{\Lambda \to \infty} < \Psi(\mathbf{p})H_{ren\,\Lambda}\, \Psi^*(\mathbf{p}') >_0 = M_{phys}\delta(\mathbf{p} - \mathbf{p}').$$

As in the classical source model of Section 3.1, in the limit of cutoff removal, the "renormalized" fields A, Ψ are no longer unitarily related to the "unrenormalized fields" a, ψ. The field ψ has divergent correlation functions in the limit $\Lambda \to \infty$, e.g.,

$$< \psi(\mathbf{p})[\, H_{ren\,\Lambda},\, \psi(\mathbf{p}')^*\,] >_0 = M_0(\Lambda)\, \delta(\mathbf{p} - \mathbf{p}') \to +\infty.$$

This means that the unrenormalized field ψ does not exist as an operator in the Hilbert space of states of the model, as displayed also by the correlation function $< \psi(\mathbf{p})\, \Psi^*(\mathbf{p}') >_0 = Z^{1/2}\delta(\mathbf{p} - \mathbf{p}')$. The expression of $a^*(\mathbf{k})$ in terms of $A^*(\mathbf{k})$

$$a^*(\mathbf{k}) = A^*(\mathbf{k}) - g\, F_\Lambda(k)\omega_k^{-3/2} \int d^3x\, e^{i\mathbf{k}\cdot\mathbf{x}}\psi^*(\mathbf{x})\, \psi(\mathbf{x})$$

makes sense also in the limit $\Lambda \to \infty$; it implies that in that limit the representation is Fock for a, a^* in the $N = 0$ sector and is non-Fock in the $N = 1$ sector, where $a^*(\mathbf{k}) - A^*(\mathbf{k}) \notin L^2(d^3k)$ (e.g., the expectation of $\psi^*(\mathbf{x})\, \psi(\mathbf{x})$ on a nucleon localized at the origin is $\sim \delta(\mathbf{x})$), as in the classical source model.

It is worthwhile noting that in the limit $\Lambda \to \infty$ the renormalized Hamiltonian is not a finite polynomial of the unrenormalized fields.

In conclusion, the regularization of the original Hamiltonian by the (non-perturbative) renormalization has led to a physically acceptable field theory model, which accounts for the observed nucleon–nucleon *Yukawa potential with range given by the inverse pion mass*.

The need of a non-Fock representation for the fields a, a^* in the one-nucleon sector also arises for infrared reasons (in the presence of a UV cutoff) if the scalar field is massless. In this case, the integral in eq. (2.3.15) is logarithmically divergent (the mass renormalization is finite for finite Λ), in strong analogy with the Bloch–Nordsieck model.

4 Ultraviolet singularities and canonical quantization

The canonical quantization rule for relativistic fields, by which the equal-time classical Poisson brackets are replaced by (equal-time) commutators, takes for granted that the equal-time commutators have a purely kinematical nature, i.e., they hold independently of the Hamiltonian, as it happens in classical theories and for quantum mechanical systems with a finite number of degrees of freedom.

As we shall see in this section, relativistic quantum field theories in $3 + 1$ spacetime dimensions are afflicted by ultraviolet singularities, which depend on the dynamics, and in the presence of interactions may prevent even the existence of equal-time commutators.

In the elementary introductions to quantum field theory, fields are viewed as operators associated to each spacetime point x, with the constraint of canonical (anti)commutation relations. One may then be led to believe that the x dependence is not worse than in the classical case (for quantum-mechanical systems with finite degrees of freedom the t dependence of the canonical variables is actually C^∞).

Such a reasonably looking regularity of quantum fields is actually wrong, as shown by the following theorem,[22] by which there is no point x such that $\varphi(x)$ is a well-defined operator, since the vacuum expectation of $\varphi(x)^2$ is divergent; hence, short distance (or ultraviolet) singularities are unavoidable for relativistic quantum fields.

The divergence of the vacuum expectation of $\varphi(x)^2$ already appears in the case of free fields quantized with canonical quantization (see eq. (1.5.12) and Section 7, below), but the following theorem shows that it is an inevitable consequence solely of either Poincaré covariance or relativistic locality. The theorem also indicates that the fields may fail to have a restriction to sharp times, with the consequent appearance of a divergent renormalization constant in the equal-time (anti)commutators. Furthermore, the theorem clarifies the assumptions at the basis of the so-called Källen–Lehmann representation for the two-point function.[23]

We start by remarking that translational symmetry implies that spacetime translations are implemented by unitary operators $U(a)$, $a \in \mathbf{R}^4$, and cogent physical

[22] A. S. Wightman, Annales Inst. H. Poincaré, **I**, 403 (1964); Z. Wizimirski, Bull. Acad. Polon. Sci. (Math., Astr. et Phys.), **14**, 91 (1966).
[23] G. Källen, Helv. Phys. Acta, **25**, 417 (1952); H. Lehmann, Nuovo Cim. **11**, 342 (1954).

considerations[24] require strong continuity in the parameter a. Thus, if a quantum field is a well-defined operator at a given spacetime point x_1, with a domain stable under spacetime translations, it is also well defined at any other point x, and it is an (operator-valued) continuous function of x:

$$\varphi(x) = U(x - x_1)\,\varphi(x_1)\,U(x - x_1)^*. \tag{2.4.1}$$

Relativistic symmetry requires that the energy–momentum operator P_μ satisfies the *relativistic spectral condition*: $P_0 \geq 0$, $P_\mu P^\mu \geq 0$, i.e., spectrum$(P_\mu) \subseteq \overline{V}_+ \equiv \{p; p^2 \geq 0,\, p_0 \geq 0\}$. Moreover, Lorentz transformations must be described by (strongly continuous) unitary operators $U(\Lambda)$. For free fields, the implementation of Poincaré transformations by unitary operators $U(a, \Lambda)$ and the corresponding transformations of the fields has been checked in Chapter 1, Section 5; their validity in the interacting case is the mathematical transcription of *Poincaré symmetry*.[25] For simplicity, we discuss the case of a hermitian scalar field.

Theorem 4.1 *(Short-distance singularities) In a quantum field theory with Poincaré symmetry, if a hermitian quantum scalar field $\varphi(x)$ satisfying Poincaré covariance, eqs. (1.5.8) and (1.5.10), is a well-defined operator at a point $\bar{x}$, with the vacuum Ψ_0 in its domain, then the two-point function*

$$F(x, y) = (\varphi(x)\,\Psi_0,\, \varphi(y)\,\Psi_0)$$

is a constant; actually, if the relativistic spectrum condition holds with the vacuum as the only translationally invariant state, all the vacuum expectation values are constants
$< \varphi(x_1) \ldots \varphi(x_n) >_0 = < \varphi >_0^n$.
The same conclusion holds if Lorentz invariance is replaced by relativistic locality.

Proof. We prove the theorem under the assumption of Poincaré covariance. By the translational invariance of the vacuum one obtains, $\forall a \in \mathbf{R}^4$,

$$F(x + a, y + a) = (U(a)\varphi(x)\,\Psi_0,\, U(a)\varphi(y)\,\Psi_0) = F(x, y),$$

so that $F(x, y) = F(x - y)$, and by eq. (2.4.1) $F(x)$ is a continuous bounded function of x (and therefore a tempered distribution). Similarly, by Lorentz covariance of the field, one obtains $F(\Lambda x) = F(x)$ for all Lorentz transformations Λ.
We now analyze the Fourier transform $\tilde{F}(p)$; clearly $\tilde{F}(p) = \tilde{F}(\Lambda p)$. Furthermore, $\forall f \in \mathcal{S}(\mathbf{R}^4)$, putting $\varphi(f) = \int d^4x\, \varphi(x)\, f(x)$, one has

$$0 \leq (\varphi(f)\Psi_0,\, \varphi(f)\Psi_0) = \int |\tilde{f}(p)|^2\, \tilde{F}(p)d^4p, \tag{2.4.2}$$

---

[24] Strong continuity is equivalent to the existence of the corresponding generators—a property which cannot be dispensed with in the case of energy and momentum. In separable Hilbert spaces, strong continuity of a unitary one-parameter group is actually equivalent to the property of weak (Lebesgue) measurability, as a function of the parameter; see M. Reed and B. Simon, *Methods of Modern Mathematical Physics*, Vol. 1, Academic Press 1972, Sect. VIII.4.

[25] E. P. Wigner, *Group Theory and its Applications to the Quantum Mechanics of Atomic Spectra*, Academic Press, 1959.

i.e., F is a continuous function of positive type, and by the Bochner–Schwartz theorem $\tilde{F}d^4p$ defines a positive tempered measure.[26] The relativistic spectral condition implies that supp $\tilde{F} \subseteq \overline{V}_+$, since

$$F(x) = (\Psi_0, \varphi(0)\, U(x)^{-1}\varphi(0)\Psi_0) = \int e^{-ipx}\, d(\varphi(0)\,\Psi_0,\, E(p)\varphi(0)\,\Psi_0), \qquad (2.4.3)$$

where $dE(p)$ denotes the energy–momentum (projection-valued) spectral measure. The Lorentz-invariant positive measures are of the form[27]

$$\tilde{F}(p)d^4p = a\,\delta^4(p)\,d^4p + b\,\rho(p^2)dp^2\,d\Omega_{p^2}(p), \quad a,\, b \in \mathbf{R}^+,$$

where $d\Omega_{p^2}(p) = d^3p/\sqrt{p^2 + \mathbf{p}^2}$ is the Lorentz-invariant measure on the hyperboloid with fixed value of p^2, and $\rho(p^2)dp^2$, is a positive tempered measure, with support in $\overline{V}^+$.[28]

The continuity of $F(x)$ at $x = 0$ requires that $b = 0$, since

$$F(0) = a + b\int \rho(p^2)dp^2 \int d\Omega_{p^2}(p)$$

and the last integral is (ultravioletly) divergent.

Then, by using the spectral representation of the spacetime translations, eq. (2.4.3), one gets that the support of the spectral measure $d(\varphi(0)\Psi_0, E(p)\varphi(0)\Psi_0)$ is the point $p = 0$, and, since the vacuum is the unique translationally invariant state, $dE(0)$ is the projection on Ψ_0. Then, $a = |(\varphi(0)\Psi_0, \Psi_0)|^2 =\; <\varphi>_0^2$ and

$$\|(\varphi(x) - <\varphi>)\,\Psi_0\|^2 = 0, \quad \text{i.e.,} \quad \varphi(x)\,\Psi_0 =\; <\varphi>_0\,\Psi_0. \qquad (2.4.4)$$

By using the general form of the two-point function following from eq. (2.4.3), one can discuss the validity of the canonical equal-time (anti)commutators. Since

$$i\partial_0\Delta(x; m^2)_{x_0=0} = -i\,\delta(\mathbf{x}), \quad \Delta(x) \equiv \Delta^+(x) - \Delta^+(-x),$$

if the fields φ, $\partial_0\varphi$ have sharp time restrictions (possibly as distributions in $\mathbf{x}$) one has

$$<[\varphi(x), \partial_0\varphi(y)]_{x_0=y_0}> = i\,\delta(\mathbf{x} - \mathbf{y})\int d\rho(m^2) = i\,Z^{-1}\,\delta(\mathbf{x} - \mathbf{y}). \qquad (2.4.5)$$

---

[26] See, e.g., I. M. Gel'fand and N. Ya. Vilenkin, *Generalized Functions*, Vol. 4, Academic Press 1964, Chap. 2, Theor. 3.

[27] L. Gårding and J. L. Lions, Nuovo Cim. Supp. **14**, 45 (1959).

[28] This form corresponds to the Källen–Lehmann representation of the two-point function (including the interacting case) in terms of an integral over the mass $m^2 = p^2$ of the two-point function $i\,\Delta^+(x; m^2)$ of a free field of mass m:

$$F(x) = a + b\int d\rho(m^2)\, i\,\Delta^+(x; m^2).$$

Now, both in perturbation theory as well as a consequence of general non-perturbative arguments,[29] in the interacting case $\int d\rho(m^2)$ diverges in $3 + 1$ dimensions, and therefore the (renormalized) fields, which have well-defined vacuum correlation function, do not allow a restriction at sharp time, and the canonical (anti)commutation relation cannot be required to hold.

In conclusion, canonical quantization cannot be used as a rigorous method for quantizing relativistic interacting fields in $3 + 1$ dimensions and a non-perturbative approach to field quantization will be discussed in the next chapter.

The impossibility of defining a quantum field operator at a point requires that, from a mathematical point of view, a field operator must be regarded as an operator-valued distribution, i.e., one needs a smearing with test functions in order to obtain well- (densely) defined operators

$$\varphi(f) = \int d^4x \, f(x) \, \varphi(x)$$

with, e.g., $f \in \mathcal{S}(\mathbf{R}^4)$. In contrast with the free case, a smearing in the space variable is not enough in the interacting case, the restriction at sharp times being excluded by the divergence of Z^{-1} in eq. (2.4.5).

It is worthwhile noting that, as a consequence of the invariance under spacetime translations (i.e., the homogeneity of the spacetime), a field $\varphi(x)$ is nowhere defined, so that the distributional singularities of a field are much more severe than those of the usual (classical) distributions occurring in functional analysis, which exhibit singularities concentrated in lower-dimensional submanifolds.

As in classical field theory, with time evolution determined by non-linear partial differential equations, the class of distributions to which a quantum field belongs may depend on the field theory model in question. The possibility of smearing with C^∞ test functions of compact support is strongly suggested for the formulation of the condition of relativistic locality. On the other hand, the momentum space analysis, which enters in the relativistic (energy–momentum) spectral condition, makes use of test functions with compact support in momentum space. The class of test functions of fast decrease is stable under Fourier transform and contains enough functions of compact support (either in configuration or in momentum space variables). Thus, the class of tempered distribution appears suitable for describing quantum field singularities. The polynomial growth at infinity in momentum space agrees with the finite number of subtractions needed for defining the short distance singularities, i.e., with the finite number of renomalization constants which characterizes the so-called renormalizable quantum field theories.

---

[29] For (irreducible) fermion fields, R. T. Powers (Comm. Math. Phys. **4**, 145 (1967)) has shown that canonical anticommutation relations are compatible only with a free theory in $d + 1$ dimensions with $d > 1$, under very general conditions. No interaction theorems for the bosonic case have been proved by K. Baumann (Jour. Math. Phys. **28**, 697 (1987); Lecture at the Schladming School 1987, in *Recent Developments in Mathematical Physics*, H. Mitter and L. Pittner (eds.), Springer 1987).

5   Problems of the interaction picture

The above discussion points out the mathematical problems which affect the interaction picture.

The existence of the interaction picture field, which has a free time evolution and is unitarily related to the (interacting) Heisenberg field, is forbidden by Haag theorem; in the more complete version of Hall and Wightman,[30] the theorem says that if a field is related to a free field by a (time-dependent) unitary operator, the field is a free field.

The reasons why the formal construction of the interaction picture is mathematically inconsistent are many. First, the construction of the formal unitary operator $V(t)$, eq. (2.1.2), requires the separate existence of the free Hamiltonian (as a well-defined self-adjoint operator) and, by Proposition 5.2, this requires a Fock representation, which is incompatible with a non-trivial interaction.

Finally, and more drastically, the divergence of Z^{-1} in eq. (2.4.5) precludes the canonical (anti)commutation relations for the (interacting) Heisenberg field, which, therefore, cannot be related to a free field by a time-dependent unitary operator.

The above mathematical problems were well understood by the end of the sixties, and stimulated a non-perturbative approach and a constructive strategy.[31] The main motivation is that of possibly providing a support for the amazing success of the renormalized perturbative series. The problems outlined in this chapter do not mean that QFT is inconsistent, nor that the renormalized perturbative series cannot be rescued in some way on a non-perturbative basis (e.g., as a Borel summable series). Actually, for QFT in low dimensions the constructive (non-perturbative) strategy has proved the existence of a non-trivial solution and the Borel summability of the perturbative series.

The idea at the basis of constructive quantum field theory is to start by introducing volume and ultraviolet cutoffs, so that one has a system with a finite number of degrees of freedom, and therefore the above mathematical difficulties do not arise. One can use quantum fields (the so-called *unrenormalized fields*), which obey canonical (anti)commutation relations and use the interaction picture, which is mathematically well defined in the presence of cutoffs. The properties of Poincaré invariance and locality at the basis of the above-mentioned no-interaction theorems are destroyed by the introduction of the cutoffs. Then, the constructive strategy is to solve or control the corresponding dynamical problem for such a butchered (regularized) model and identify the necessary counterterms (to be added in the Lagrangian or in the Hamiltonian) and the field renormalization leading to the so-called renormalized fields, so that the corresponding correlation functions have well-defined limits, when the volume and UV cutoffs are removed. In this way, the role of the interaction picture and of canonical quantization is only instrumental for such a computation in the cutoff theory, which hopefully should converge to a physically acceptable theory in

[30] R. F. Streater and A. S. Wightman, *PCT, Spin and Statistics, and All That*, Benjamin 1964, Theor. 4.16; see Chapter 4, Section 5 below.

[31] R. F. Streater and A. S. Wightman, 1964; A. S. Wightman, Introduction to some aspects of the relativistic dynamics of quantized fields, in *High Energy Electromagnetic Interactions and Field Theory*, M. Lévy (ed.), Gordon and Breach 1967, esp. Part II.

the limit of cutoff removal, even if in that limit the interaction picture and canonical quantization cease to make sense. In particular, the renormalized fields—those with well-defined correlation functions in the limit of cutoff removal—will be related to the unrenormalized fields by renormalization constants Z^{-1}, which in the limit of cutoff removal are divergent and prevent canonical (anti)commutation relations for the renormalized fields. We have seen this strategy at work in the simple case of the Yukawa model, Section 3.3.

The constructive strategy has also led to the proof of the triviality of φ^4 in $3+1$ dimensions and, honestly, at the moment it is not known whether there are (non-trivial) relativistic QFT in $3 + 1$ dimensions, the only proviso for QFT being a limitation for its applicability to energy scales of the order of the Planck mass, where quantum gravity effects are expected to enter.

An important issue, inevitably arising in a constructive strategy, is the characterization of those basic properties which must be satisfied in the no cutoff limit, in order that one has a (physically acceptable) relativistic quantum field theory. This issue is also related to the question of defining a relativistic quantum field theory, independently of the canonical quantization and of the perturbative expansion. An answer to this question will be discussed in the next chapter, along the lines of the Wightman approach.

6   Appendix: Locality and scattering

6.1   Locality and asymptotic states

Scattering theory is at the basis of the S-matrix theory, and the discussion of the non-relativistic case is an almost necessary prerequisite for the full relativistic theory.

The quantum-mechanical scattering theory is a robust and well developed branch of mathematical physics, but the approach usually adopted in physics textbook somewhat hides important points.

The standard use of the interaction picture identifies the asymptotic time evolution with the free one, and this gives rise to problems for the scattering by a long-range potential, where the Coulomb-like tail leads to an asymptotic time evolution different from the free one.

Moreover, the formal theory of scattering, based on the Lippmann–Schwinger equations for plane waves, precludes the exploitation of the physically crucial localization of the asymptotic states in far-separated regions, where the interaction vanishes. Actually, localization provides a more faithful description of the experimental realization of a scattering process, and its interplay with the range of the potential allows to correctly pose and solve basic theoretical and mathematical problems.

The strict analogs of such physical features in quantum field theory are the locality property and the cluster property, whose rate of decay is the relativistic counterpart of the potential fall-off.

The localization of the incoming and outgoing states is indeed basic for the very definition of a scattering process, e.g., of two systems A and B. In fact, in the standard setting, at an initial time t_i in the distant past, the two systems are prepared and *localized* in far-separated regions, in such a way that there is no mutual interaction; then, in the *initial configuration* φ_i of the two systems, the description of each of them does not involve the other. In the case of scattering of a particle by a potential V localized in a region $\mathcal{O}$, the particle and the (external) potential play the rôle of A and B. The two systems A and B are then allowed to evolve in time and, at a final time t_f, one looks for the probability of having again localized far-separated systems, each with its own individuality described by the *final configurations* φ_f, with no mutual interaction. Roughly, for such configurations, $H \sim H_{as}^A + H_{as}^B = H_{as}$.

In the Schrödinger picture, the transition matrix is then given by

$$T(\varphi_i \to \varphi_f) = (\varphi_f, \, e^{-iH(t_f - t_i)} \, \varphi_i).$$

It is convenient to consider the asymptotic limits $t_{i/f} \to \mp\infty$, introduce a complete set of (asymptotic) configurations $\{\varphi_k^{\pm}\}$, and denote by $\mathcal{H}^{as}$ the corresponding Hilbert space, assuming, for simplicity, that the spaces of initial and final asymptotic configurations, $\mathcal{H}^-$, $\mathcal{H}^+$, coincide.

The localization of the asymptotic configurations in far separated regions, where there is no mutual interaction, means that for large times the time evolution reduces to $U_{as}(t) = U_{as}^A(t) \, U_{as}^B(t)$, each system evolving as if the other were not present. $U_{as}(t)$ describes the *asymptotic dynamics*. In the standard case, in which $U_{as}(t) = e^{-iH_{as}t}$, one expects that

$$\lim_{t \to \pm\infty} ||(H - H_{as})e^{-iH_{as}t} \varphi_k|| = 0,$$

and the rate of fall-off in t of the above norm is governed by the interplay between localization properties of the asymptotic states and the fall-off of the interaction for large distances. As we shall see in the following, the correct identification of the asymptotic dynamics is a crucial issue.

In order to appreciate the relevance of such properties for the mathematical definition of scattering, let $\Psi^{\pm}(t)$ be (smooth) solutions of the Schrödinger equation (called *scattering solutions*), such that for $t \to \pm\infty$ they approach asymptotic configurations $\varphi_{\pm}$ localized in far-separated regions and therefore with time evolution given by $U_{as}(t)$. Thus,

$$\lim_{t \to \mp\infty} ||\Psi^{\pm}(t) - U_{as}(t)\varphi_{\mp}|| = \lim_{t \to \mp\infty} ||\Psi^{\pm}(0) - e^{iHt} U_{as}(t)\varphi_{\mp}|| =$$

$$\lim_{t \to \mp\infty} ||U_{as}(t)^{-1} e^{-iHt} \Psi^{\pm}(0) - \varphi_{\mp}|| = 0.$$

Such limits define the asymptotic behavior of $\Psi^{\pm}(t)$ and conversely, for given $\varphi_{\mp}$, they identify the solutions $\Psi^{\pm}(t) = e^{-iHt} \, \Psi^{\pm}$,

$$\Psi^{\pm} = s - \lim_{t \to \mp\infty} e^{iHt} U_{as}(t)\varphi_{\mp} \equiv \Omega^{\pm}\varphi_{\mp}, \qquad (2.6.1)$$

which correspond to such asymptotic configurations. The operators $\Omega^{\pm}$ are called the *Møller operators*.

The *scattering matrix* S gives the probability amplitude that a solution $\Psi(t)$ corresponding to the asymptotic configuration φ_- in the far past, i.e., $\Psi^+(t) = e^{-iHt} \Omega^+ \varphi_-$, will look like $U_{as}(t) \, \varphi_+$ in the asymptotic future:

$$\lim_{t\to\infty} (U_{as}(t)\varphi_+,\, e^{-iHt}\,\Omega^+\varphi_-) = (\Omega^-\varphi_+, \Omega^+\varphi_-) = (\varphi_+,\, \Omega^{-\,*}\Omega^+\,\varphi_-).$$

Then, putting $U(t) \equiv e^{-iHt}$, $V(t)^* \equiv U^*(t)\,U_{as}(t)$, one has

$$S = \Omega^{-\,*}\Omega^+ = \lim_{t\to\infty,\, t'\to-\infty} V(t)\,V(t')^*. \tag{2.6.2}$$

The existence of the S-matrix is therefore related to the existence of the Møller operators. In the standard case $U_{as}(t) = e^{-iH_{as}t}$, a sufficient condition is that for large $|t|$,

$$||(H - H_{as})\,e^{-iH_{as}\,t}\varphi|| = o(1/|t|). \tag{2.6.3}$$

In fact, the equation

$$dV(t)^*/dt = i\,U^*(t)\,(H - H_{as})\,U_{as}(t), \quad V(0)^* = 1,$$

is equivalent to the following integral equation:

$$V(t)^*\,\varphi = \varphi + i\int_0^t ds\,U^*(s)\,(H - H_{as})\,U_{as}(s)\,\varphi, \tag{2.6.4}$$

and the existence of the (strong) asymptotic limit of $V(t)^*\varphi$ is equivalent to the existence of the above integral when $t \to \mp\infty$. This is guaranteed by eq. (2.6.3), since

$$\left|\left|\int_t^{t'} ds\,(\dots)\right|\right| \le \int_t^{t'} ds\,||(\dots)||.$$

The following properties of $\Omega^\pm$ are worth mentioning.
1) Eq. (2.6.1) implies

$$(\varphi, \Omega^{\pm\,*}\Omega^\pm\,\varphi) = ||\Omega^\pm\,\varphi|| = \lim_{t\to\mp\infty}||V(t)^*\,\varphi|| = ||\varphi||.$$

Hence,

$$\Omega^{\pm\,*}\Omega^\pm\varphi = \varphi,$$

i.e., $\Omega^{\pm\,*}$ are the inverse of $\Omega^\pm$ on $\mathrm{Ran}\,\Omega^\pm$ and $\Omega^{\pm\,*}(\mathrm{Ran}\Omega^\pm)^\perp = 0$. In particular, $||\Omega^{-\,*}\Omega^+\varphi|| = ||\varphi|| = ||\Omega^+\varphi||$, iff $\Omega^+\varphi \in \mathrm{Ran}\,\Omega^-$.
2) Clearly, the group properties of $U_{as}(t)$ imply

$$\lim_{t\to\mp\infty} V(t)^* = \lim_{t\to\mp\infty} V(t+s)^* = \lim_{t\to\mp\infty} U(s)^*V(t)^*U_{as}(s),$$

so that one has $\Omega^\pm = U(s)^*\,\Omega^\pm\,U_{as}(s)$ or

$$U(s)\,\Omega^\pm = \Omega^\pm U_{as}(s), \tag{2.6.5}$$

i.e., the Møller operators intertwine between the asymptotic Hamiltonian H_{as} and the total Hamiltonian H $(H\Omega^\pm = \Omega^\pm H_{as})$. Moreover, since $\Omega^{\pm\,*}$ are the inverse of $\Omega^\pm$ on $\mathrm{Ran}\,\Omega^\pm$, eq. (2.6.5) implies

$$\Omega^{\pm\,*}U(t)\,\Omega^\pm = U_{as}(t).$$

Hence, $\Omega^\pm\varphi$ has the same spectral measure for H as φ has for H_{as},

$$(\Omega^\pm\varphi, e^{-itH}\,\Omega^\pm\varphi) = (\varphi,\, e^{-itH_{as}}\,\varphi).$$

In particular, if H_{as} has a purely absolutely continuous spectrum, as is the case of $H_{as} = H_0 = -\Delta$, and $\mathcal{H} = \mathcal{H}^{as}$, H has the same continuous spectrum.

3) The isometry of the S-matrix is equivalent to the property

$$\mathrm{Ran}\,\Omega^+ = \mathrm{Ran}\,\Omega^-, \tag{2.6.6}$$

called *weak asymptotic completeness*. In fact, S preserves the norms iff $\mathrm{Ran}\ \Omega^+ \subseteq \mathrm{Ran}\,\Omega^-$ and S^* preserves the norms iff $\mathrm{Ran}\ \Omega^- \subseteq \mathrm{Ran}\ \Omega^+$.

The condition $\mathrm{Ran}\,\Omega^+ = \mathrm{Ran}\,\Omega^-$ means that the solutions of the Schrödinger equation which converge to asymptotic configurations in the far past also converge to asymptotic configurations in the asymptotic future, and vice versa (eq. (2.6.1)).[32]

A strong form of *asymptotic completeness* is

$$\mathrm{Ran}\,\Omega^+ = \mathrm{Ran}\,\Omega^- = (\mathcal{H}_{pp})^\perp, \tag{2.6.7}$$

where $\mathcal{H}_{pp}$ denotes the subspace of eigenvectors, equivalently of bound states, of H. If there are no bound states, this condition implies the unitarity of the Møller operators and of the S-matrix.

A comprehensive discussion of scattering theory and in particular of the conditions which assure the validity of eq. (2.6.6), or even the unitarity of the Møller operators, may be found in M. Reed and B. Simon, *Methods of Modern Mathematical Physics, Vol. 3, Scattering Theory*, Academic Press 1978, Sect. XI.3, Theors. XI.30, XI.31, and Vol. IV, Theors. XIII. 32, XIII.33.

In the case of scattering by a **short-range potential** with Hamiltonian

$$H = -\Delta/2m + V(\mathbf{x}), \quad V(\mathbf{x}) \sim_{|\mathbf{x}|\to\infty} o(|\mathbf{x}|^{-1-\varepsilon}), \quad \varepsilon > 0,$$

V regular everywhere except possibly at the origin, one can show that $H_{as} = -\Delta/2m$.

In fact, for smooth solutions φ of the free Schrödinger equation, one can estimate the large-time behavior by using the following Dollard estimate[33]

$$\varphi(\mathbf{x}, t) \sim_{t\to\infty} (m/it)^{3/2} e^{imx^2/2t} \tilde{\varphi}(m\mathbf{x}/t), \tag{2.6.8}$$

where $\tilde{\varphi}$ denotes the Fourier transform of φ. In fact, it is easy to show that

$$U_0(t) = e^{-itH_0} = S(t)\,T(t), \qquad (T(t)\varphi)(x) \equiv e^{imx^2/2t}\varphi(x),$$

$$(S(t)\,\varphi)(x) \equiv (m/it)^{3/2} e^{imx^2/2t}\,\tilde{\varphi}(mx/t),$$

---

[32] Such a property is often taken for granted in the discussion of scattering, but there are examples in which it fails, and therefore one must discuss conditions which assure it. For central potentials, a sufficient condition is that $V(r)$ satisfies $\int_0^1 dr\, r|V(r)| + \int_1^\infty dr\, |V(r)| < \infty$, and a behavior like $r^{-1-\varepsilon}$ at infinity is allowed. For an excellent and very clear review, see B. Simon, An Overview of Rigorous Scattering Theory, in *Atomic Scattering Theory, Mathematical and Computational Aspects*, J. Nuttal (ed.), University of Western Ontario, 1978.

[33] J. D. Dollard, Jour. Math. Phys. **5**, 729 (1964). For a complete rigorous discussion of the quantum scattering problem, see M. Reed and B. Simon, *Methods of Modern Mathematical Physics*, Vol. III, Academic Press 1979, Sects. XI. 3–4 and Appendix 1, and P. A. Perry, *Scattering Theory by the Enss Method*, Harwood 1983.

and that both $S(t)$ and $T(t)$ are unitary operators. Hence, for $t \to \infty$ one has

$$\|U_0(t)\varphi - S(t)\varphi\| = \|T(t)\varphi - \varphi\| \to 0,$$

since

$$((T(t) - 1)\varphi)(x) = (e^{imx^2/2t} - 1)\varphi(x) \to 0$$

pointwise, and one may apply the dominated convergence theorem. Thus, one obtains eq. (2.6.8). Such a large-time asymptotic behavior can also be obtained by the stationary phase method applied to the integral which expresses $\varphi(\mathbf{x}, t)$ in terms of the Fourier transform $\tilde{\varphi}$.

The above estimate implies that condition (2.6.3) holds: in fact, for $|t|$ large, for smooth solutions, one has $(\mathbf{y} \equiv m\mathbf{x}/t)$

$$\|V\varphi_t\|^2 \sim (m/t)^3 \int d^3x |V(\mathbf{x})|^2 \tilde{\varphi}(m\mathbf{x}/t)|^3 = \int d^3y \, |V(t\mathbf{y}/m)|^2 |\tilde{\varphi}(\mathbf{y})|^2$$

$$\sim (m/t)^{2+\varepsilon} \int d^3y \, |\mathbf{y}|^{-2-\varepsilon} |\tilde{\varphi}(\mathbf{y})|^2 = C|t|^{-2-\varepsilon}.$$

It is worthwhile to remark that for the validity of eq. (2.6.3), two important physical properties are involved; namely, the localizability of the asymptotic states in far-separated regions and the sufficiently fast vanishing of $H - H_{as}$ on asymptotic states. The localizability of the asymptotic states also implies that their time evolution, as given by the *total* Hamiltonian H, is also described by H_{as}.

The identification of $U_{as}(t)$ with the free evolution may not be correct, as displayed by Coulomb scattering or more generally by quantum field theory models. In both cases, the interaction term gives rise to the so-called persistent effects, which affect the asymptotic time evolution. As we shall see, in relativistic quantum field theory the localizability property needed for the existence of the S-matrix is guaranteed by the cluster property.[34]

6.2 Scattering by a long-range potential

The scattering by a long-range potential $V(\mathbf{x}) \sim e/|\mathbf{x}|$, for $|\mathbf{x}| \to \infty$, provides an example of infrared persistent effects induced by the interaction, leading to a time evolution for large $|t|$ which differs from the free dynamics defined by $H_0 = -\Delta/2m$. In fact, putting $U_0(t) \equiv e^{-iH_0 t}$, we have

$$U_0^*(t) \, \mathbf{x} \, U_0(t) = \mathbf{p} \, t/m + \mathbf{x}, \quad U_0^*(t) \, \mathbf{p} \, U_0(t) = \mathbf{p},$$

so that

$$\|V(\mathbf{x}) \, U_0(t)\varphi\| = \|U_0^*(t) \, V(\mathbf{x}) \, U_0(t)\varphi\| \sim_{|t| \to \infty} \|(e \, m/p|t| + O(t^{-2})) \, \varphi\|.$$

---

[34] The role of localizability for the definition of the asymptotic dynamics, in contrast with the usual *a priori* splitting of the Hamiltonian $H = H_0 + gH'$, has been emphasized by Haag, especially in connection with scattering in quantum field theory: see R. Haag, Quantum theory of collision processes, in *Lectures in Theoretical Physics*, Vol. III, W.E. Brittin *et al.* (eds.), Interscience 1961, pp. 326–52; Phys. Rev. **112**, 669 (1958); Nuovo Cim. **14**, 131 (1959). For a brief account see Chapter 6 below.

Because of the term $e\,m/p|t|$, the above norm is not an integrable function of t, and one obtains a logarithmic (infrared) divergence, which is responsible for the non-existence of the corresponding Møller operators. The physical meaning of this result is that the Coulomb tail contributes in an essential way to the definition of the dynamics $U_{as}(t)$ for large $|t|$, which must incorporate the "singular part V_{as} of the long-range tail e/r".[35]

$$i\,dU_{as}(t)/dt = H_{as}(t)\,U_{as}(t), \quad H_{as} = H_0 + V_{as} = H_0 + e\,m/p|t|. \tag{2.6.9}$$

The physical meaning of this phenomenon is clearly displayed by the classical case, where for large t the trajectories are straight lines, but the temporal law is not that of a free particle; this is where the effect of the long-range tail shows up for asymptotic times. The Hamiltonian H_{as} includes the asymptotic effect of the long-range potential, and in fact $H - H_{as}$ satisfies condition (2.6.3) (see below). Clearly, the operator V_{as} which must be added to H_0 in order to obtain the asymptotic dynamics is not unique—only its asymptotic form, namely the infrared singular behavior $me/p|t|$, for $|t| \to \infty$, is uniquely determined.

Since $H_{as}(t)$ depends on time, the time evolution is not described by $\exp\left(-iH_{as}(t)t\right)$. A solution for the asymptotic propagator, which incorporates the initial condition

$$\lim_{t \to t_0} U_{as}(t, t_0) = 1,$$

(where the rather delicate limit must be understood in the distributional sense), can be written in the following form for $t_0 > 0$:

$$U_{as}(t, t_0) = \exp\left(-i\frac{p^2}{2m}(t - t_0) - i\frac{e\,m}{p}\,\text{sign}\,t \,\log\frac{|t|}{t_0}\right).$$

The corresponding equations of motion of $\mathbf{x}$ and $\mathbf{p}$ induced by the asymptotic dynamics are

$$U_{as}(t, t_0)^* \,\mathbf{x}\, U_{as}(t, t_0) = \mathbf{x} + \frac{\mathbf{p}}{m}(t - t_0) - em\frac{\mathbf{p}}{p^3}\,\text{sign}\,t \,\log\frac{|t|}{t_0}, \tag{2.6.10}$$

$$U_{as}(t, t_0)^* \,\mathbf{p}\, U_{as}(t, t_0) = \mathbf{p}. \tag{2.6.11}$$

The unitary operators $U_{as}(t, t_0)$ do not define a group for finite times, but only for asymptotic times; they satisfy the physical requirement that for large t, the asymptotic dynamics of $\mathbf{x}$ and $\mathbf{p}$ is the same as the asymptotic classical law of motion, with non-constant asymptotic velocity; e.g., for $t \to \infty$, $\dot{\mathbf{x}}(t) = \mathbf{p}/m - em\mathbf{p}/(p^3\,t)$. By using eqs. (2.6.10) and (2.6.11) one can check that condition (2.6.3) is satisfied by the asymptotic dynamics. In fact, considering for simplicity the Coulomb case $V_c(\mathbf{x}) = e/|\mathbf{x}|$, and putting $U_{as}(t) \equiv U_{as}(t, t_0)$, one has

$$||(H - H_{as}(t))\,U_{as}(t)\,\varphi|| = ||U_{as}^*(t)\left(\frac{e}{|\mathbf{x}|} - \frac{em}{p|t|}\right)U_{as}(t)\,\varphi||,$$

[35] J. Dollard, Jour. Math. Phys. **5**, 729 (1964).

and, by eqs. (2.6.10) and (2.6.11), the r.h.s. decreases as t^{-2} and is an integrable function of t. Therefore, the Møller operators defined in terms of the asymptotic dynamics $U_{as}(t)$ exist:

$$\Omega^{\pm} = \lim_{t \to \mp\infty} e^{iHt} U_{as}(t) \tag{2.6.12}$$

and so does the S-matrix $S = \Omega^{-\,*}\Omega^{+}$.

The corrections to the free dynamics have the role of describing the subleading contributions for large $|t|$, and vanish in the extreme limit $|t| \to \infty$; in fact, $\Omega^{\pm\,*}U(t)\,\Omega^{\pm} = e^{iH_0 t}$.

It is worthwhile remarking that the use of H_0 instead of H_{as} would lead to a perturbative expansion of the S-matrix with the usual logarithmic infrared divergences.

Clearly, the conclusions apply to the generic case of a long-range potential, but the effectiveness of the strategy discussed above can be explicitly checked in the simplest case of the Coulomb scattering, which is exactly solvable. In fact, the energy (improper) eigenfunctions $\psi(\mathbf{x})$ for the Coulomb problem

$$(-\Delta + 2me/r)\,\psi(\mathbf{x}) = k^2\,\psi(\mathbf{x})$$

are

$$(2\pi)^{3/2}\,\psi(\mathbf{x}) = e^{-\frac{1}{2}\pi n}\Gamma(1 + in)e^{i\mathbf{k}\cdot\mathbf{x}}\,F(-in, 1, i\rho),$$

where $r = |\mathbf{x}|$, $\rho \equiv kr - \mathbf{k}\cdot\mathbf{x}$, $n \equiv em/k$, Γ is the function, and F is a function of hypergeometric type,[36] which has the following asymptotic behavior for large r ($\theta =$ the angle between $\mathbf{k}$ and $\mathbf{x}$)

$$(2\pi)^{3/2}\psi(\mathbf{x}) \sim \left(1 - i\frac{n^2}{(kr - \mathbf{k}\cdot\mathbf{x})}\right) e^{i(\mathbf{k}\cdot\mathbf{x} + n\log(kr - \mathbf{k}\cdot\mathbf{x}))} + f(\theta)\frac{e^{i(kr - n\log(kn))}}{r}.$$

This shows that the asymptotic configurations are not (improper) eigenstates of H_0; a distortion $\exp\left(in\log(kr)\right)$ occurs both in the incoming and in the outgoing wave, due to the Coulomb tail. Without such a proper identification of the asymptotic dynamics one would obtain logarithmically divergent phase shifts. The correct (finite) phase shifts η_l are given by

$$\psi(\mathbf{x}) = (2\pi)^{3/2}\sum_{l=0}^{\infty}(2l + 1)\,e^{i\eta_l}\,L_l(r)\mathcal{P}_l(\cos\theta),$$

$$L_l(r) \sim_{r \to \infty} (1/kr)\left[\sin kr - \tfrac{1}{2}l\pi + \eta_l - n\log 2kr\right].$$

---

[36] F has the following power series expansion

$$F(in, 1, i\rho) = 1 + (1!)^{-2}n\,\rho + (2!)^{-2}n(n + i)\rho^2 + (3!)^{-2}n(n + i)(n + 2i)\rho^3 + \cdots$$

For details, see L. D. Landau and E. M. Lifshitz, *Quantum Mechanics*, Pergamon Press 1958, Sect. 133, or E. Corinaldesi and F. Strocchi, *Relativistic Wave Mechanics*, North-Holland 1963, Part III, Chap. I, Sect. 9.

The asymptotic behavior differs by the logarithmic term from that corresponding to short-range potential.

6.3   Adiabatic switching

A technical tool for justifying manipulations, which would otherwise require a careful handling, is the so-called *adiabatic switching*. It consists in exponentially switching off the interaction at asymptotic times, $V \equiv H - H_{as} \to e^{-\varepsilon|t|} (H - H_{as}) = e^{-\varepsilon|t|} V$, $\varepsilon > 0$, and taking the limit $\varepsilon \to 0$ at the end of the calculations. With such a prescription, eqs. (2.6.1) and (2.6.4) give

$$\Psi^{\pm} = \varphi + \lim_{\varepsilon \to 0} i \int_0^{\mp\infty} ds\, e^{iHs} e^{-\varepsilon|s|} V\, e^{-iH_{as}s}\, \varphi.$$

This trick clearly improves the convergence of the asymptotic limits, and allows the extension of such limits to idealized asymptotic states which are not localized, such as energy (improper) eigenstates. For simplicity, in the following we assume that H_{as} is time-independent and has only a continuous spectrum. Then, by formally expanding φ in terms of a complete set of "improper" eigenstates φ_α of H_{as}, with eigenvalues E_α, and using the Dirac notations, $\varphi = \int d\alpha\, c_\alpha\, \varphi_\alpha$, $c_\alpha = <\varphi_\alpha|\varphi>$, one obtains

$$\Psi^{\pm} = \int d\alpha\, c_\alpha (1 + i \int_0^{\mp\infty} ds\, e^{-i(-H\pm i\varepsilon + E_\alpha)s}\, \varphi_\alpha),$$

i.e., $\Psi^{\pm} = \int d\alpha\, \Psi_\alpha^{\pm}$, where

$$\Psi_\alpha^{\pm} = \varphi_\alpha + \frac{1}{E_\alpha \pm i\varepsilon - H} V\, \varphi_\alpha, \tag{2.6.13}$$

the limit $\varepsilon \to 0$ being understood. From eq. (2.6.13) one has

$$(H - E_\alpha)\, \Psi_\alpha^{\pm} = 0,$$

so that the coefficients of the expansion of $\Psi^{\pm}$ in terms of eigenfunctions of H are the same as the coefficients of the expansion of φ in terms of the eigenfunction of H_{as}.

By using the operator identity $A^{-1} = B^{-1} + B^{-1}(B - A) A^{-1}$ (which easily follows from $BA^{-1} = 1 + (B - A)A^{-1}$) for $A = E_\alpha \pm i\varepsilon - H$, $B = E_\alpha \pm i\varepsilon - H_{as}$, eq. (2.6.13) becomes

$$\Psi_\alpha^{\pm} = \varphi_\alpha + \frac{1}{E_\alpha \pm i\varepsilon - H_{as}} V\, \Psi_\alpha^{\pm}. \tag{2.6.14}$$

Such equations are the (integral) equations which determine $\Psi_\alpha^{\pm}$ given φ_α, and are known as the *Lippmann–Schwinger equations.*[37] The advantage with respect to eq. (2.6.13) is that it involves the inversion of a function of the simpler operator H_{as}, rather than of H. Moreover, if the potential is written in the from gV, with g a

---

[37] B. A. Lippmann and J. Schwinger, Phys. Rev. **79**, 469 (1950); M. Gell-Mann and M. L. Goldberger, Phys. Rev. **91**, 398 (1953).

(small) coupling constant, the iterative solution of eq. (2.6.14) is given by a power series expansion in powers of g (*perturbative expansion*).

By using eqs. (2.6.13) and (2.6.14), the S-matrix for the asymptotic idealized states φ_α can be written in the form[38]

$$S_{\alpha\beta} = < \Psi_\beta^- | \Psi_\alpha^+ > = \delta_{\alpha\beta} - 2\pi i\, \delta(E_\alpha - E_\beta) < \varphi_\beta | V | \Psi_\alpha^+ >. \qquad (2.6.15)$$

Since plane waves are spread all over the space and are of order 1 also in the region in which the potential is strong, the definition of scattering seems to lose its meaning. However, in a realistic scattering experiment one never deals with a stationary situation, since the target, the particle beam, etc., have to be prepared at a certain time, and the description of a scattering process is expected to be largely independent of what happened before the preparation of the experimental set-up. For example, for a particle scattering by a potential V (or by a target), since the potential realistically exists only for a finite time T, which can also be taken as the time in which the experiment takes place, it is reasonable to replace V by $e^{-\varepsilon|t|} V$ with $\varepsilon << T^{-1}$, and one expects that the results are independent of ε under these conditions. Clearly, T^{-1} and ε are related to the energy resolution of the detector involved in the experiment, and the result should become independent of ε, for ε much smaller than the typical energies entering in the scattering process.[39] It is instructive to compute the S matrix for the Coulomb scattering with the wrong choice of H_0 as the asymptotic dynamics, and see that in the perturbative expansion there are logarithmic divergences for $\varepsilon \to 0$.

6.4 Asymptotic condition

An alternative definition of the S-matrix can be done in the Heisenberg picture[40] in terms of asymptotic limits of operators, rather than of states. Such a strategy does not require an *a priori* identification of the asymptotic dynamics, nor the existence of the interaction picture (which is forbidden in relativistic quantum field theory). Moreover, the so obtained definition of the S matrix provides the basic step for a non-perturbative approach to scattering.

We start by considering the case in which there are no bound states, H_{as} has only continuous spectrum, and the Møller operators are unitary. From the asymptotic limit

---

[38] In fact, by using eq. (2.6.14) for Ψ_β^- and $H\Psi_\alpha^+ = E_\alpha \Psi_\alpha^+$, one has
$S_{\alpha,\beta} = (\varphi_b, \Psi_\alpha^+) + (\varphi_b, V(E_\beta + i\varepsilon - H)^{-1} \Psi_\alpha^+) = \delta_{\alpha\beta} - 2i\varepsilon/[(E_\beta - E_\alpha)^2 + \varepsilon^2](\varphi_\beta, V\Psi_\alpha^+)$.

[39] See M. Gell-Mann and M. L. Goldberger, Phys. Rev. **79**, 398 (1953).

[40] The Schrödinger picture makes implicit reference to states prepared at sharp times, i.e., by specifying the expectation values of a complete set of commuting observables (corresponding to simultaneous compatible experiments at a sharp time). The time evolution $U(t)$ is then defined by comparing states at different times $\Psi_t = U(t)\Psi_0$. This involves an idealization, since physical measurements always involve finite intervals of time, and moreover the so-defined states do not have simple transformation properties under the Lorentz group (by the non-covariance of simultaneity). Also, the covariance of the dynamics is not trivial in the Schrödinger picture. On the other hand, in the Heisenberg picture the Poincaré transformations are well defined as (implementable) automorphisms of the observables or of the field algebra.

of states in the Schrödinger picture, e.g., for $t \to -\infty$, for any (bounded) operator A, by using eq. (2.6.1), $\Psi^-(0) = \Omega^+ \varphi_-$, one has

$$(\Psi^-(0), e^{iHt} A e^{-iHt} \Psi^-(0)) = (\Psi^-(t), A\Psi^-(t)) \to$$

$$(e^{-iH_{as}t} \varphi_-, A e^{-iH_{as}t} \varphi_-) = (\Psi^-(0), \Omega^+ e^{iH_{as}t} A e^{-iH_{as}t} \Omega^{+*} \Psi^-(0)).$$

This gives the *weak* asymptotic behavior of operators in the Heisenberg picture

$$A_H(t) = e^{iHt} A e^{-iHt} \sim A_{in}(t) \equiv \Omega^+ e^{iH_{as}t} A e^{-iH_{as}t} \Omega^{+*} =$$

$$= e^{iHt}\Omega^+ A \Omega^{+*} e^{-iHt} = e^{iHt} A_{in}(0) e^{-iHt}, \tag{2.6.16}$$

where eq. (2.6.5) has been used in the last but one step.[41]

By the above relations Ω^+ intertwines between the Heisenberg operator $A_H(0) = A$ and $A_{in}(0)$, so that if $H_{as}(A_H)$ is a function of (the canonical variables) A_H at time zero, one has

$$H_{as}(A_{in}) \equiv \Omega^+ H_{as}(A_H) \Omega^{+*} \tag{2.6.17}$$

and

$$H(A_H) = \Omega^+ H_{as}(A_H) \Omega^{+*} = H_{as}(A_{in}), \tag{2.6.18}$$

i.e., the total Hamiltonian coincides with the asymptotic Hamiltonian as a function of the asymptotic variables A_{in}. Hence, the time evolution of the A_{in} variables is governed by the total Hamiltonian H, and corresponds to a free evolution given by the asymptotic Hamiltonian $H_{as}(A_{in})$, as a function of the asymptotic variables

$$A_{in}(t) = e^{iHt} A_{in}(0) e^{-iHt} = e^{iH_{as}(A_{in})t} A_{in}(0) e^{-iH_{as}(A_{in})t}.$$

A similar construction can be done for the asymptotic limit $t \to \infty$, leading to the introduction of the asymptotic variables A_{out},

$$A_{out}(0) \equiv \Omega^- A_H(0) \Omega^{-*} = \Omega^- \Omega^{+*} A_{in}(0) \Omega^+ \Omega^{-*}. \tag{2.6.19}$$

Furthermore, one has

$$(\Omega^- \varphi_\alpha, A_{out} \Omega^- \varphi_\beta) = (\Omega^+ \varphi_\alpha, A_{in} \Omega^+ \varphi_\beta).$$

Then, the states $\Psi_\alpha^- = \Omega^- \varphi_\alpha$ have the same spectral support with respect to the A_{out} as the states Ψ_α^+ have with respect to the A_{in}. Thus, in the Heisenberg picture the time-independent states $\Psi_\alpha^\pm = \Omega^\pm \varphi_\alpha$ can be interpreted as states characterized by their spectral support with respect to observables at times $t \to \mp\infty$, respectively; for this reason they are called *asymptotic in/out states*.

The S-matrix in the Heisenberg picture S_H is defined by

$$S_{\alpha\beta} = (\Psi_\alpha^-, \Psi_\beta^+) = (\Psi_\alpha^-, S_H \Psi_\beta^-) = (\varphi_\alpha, \Omega^{-*} S_H \Omega^- \varphi_\beta) =$$

$$= (\varphi_\alpha, \Omega^{-*} \Omega^+ \varphi_\beta),$$

---

[41] For canonical variables or fields described by unbounded operators, domain problems arise, and the weak limit should be understood for matrix elements on suitable domains (for the weak limit of quantum fields, see Chapter 6).

i.e.,

$$S_H = \Omega^+ \, \Omega^{-\,*} = \Omega^- \, S \, \Omega^{-\,*}, \qquad (2.6.20)$$

where $S = \Omega^{-\,*} \, \Omega^+$ is the S-matrix in the Schrödinger picture, discussed in Section 6.1 above. From eqs. (2.6.19), (2.6.20) one obtains

$$A_{out} = S_H^* \, A_{in} S_H.$$

Under the above assumptions the formulation of the S-matrix in the Heisenberg picture is equivalent to that in the Schrödinger picture; however, the first one can be disentangled from the existence of the interaction picture or of the Moeller operators and be given a more general validity, by relying only on the following structures
i) the Heisenberg operators approach asymptotic *in/out* operators

$$A_H(t) \sim A_{in/out}(t), \quad \text{for } t \to \mp\infty, \qquad (2.6.21)$$

ii) the time evolution of the $as = in/out$ operators is a free evolution,

$$A_{as}(t) = e^{iHt} \, A_{as}(0) \, e^{-iHt} = e^{iH_{as}(A_{as})t} A_{as}(0) e^{-iH_{as}(A_{as})t}, \qquad (2.6.22)$$

iii) the S-matrix is the unitary operator which intertwines between A_{in} and A_{out} (for brevity we omit the subscript H in S),

$$A_{out} = S^* \, A_{in} S. \qquad (2.6.23)$$

One of the big achievements of the non-perturbative approach to quantum field theory is the (rigorous) proof of the existence of the asymptotic limit of field operators and the existence of the S-matrix, which is a unitary operator if asymptotic completeness holds. Such an approach to the S-matrix is free from the problems of existence of the interaction picture and of the perturbative expansion.[42]

7   Wick theorem and Feynman diagrams

Given a product of creation and annihilation operators, the operation of *normal (or Wick) ordering*, denoted by N, corresponds to pulling all the creation operators to the left and all the destruction operators to the right and to multiplying by $(-1)^P$, where P is the parity of the permutation of fermion fields needed to obtain such a reordering, hereafter called the reordering parity:

$$N(A_1 \ldots A_n) = (-1)^P \, N(A_{i_1} \ldots A_{i_n}). \qquad (2.7.1)$$

[42] The definition of the S-matrix, by relying only on the asymptotic limit of the Heisenberg operators, has been proposed by H. Lehmann, K. Symanzik, and W. Zimmermann (Nuovo Cim. **1**, 425 (1955)). The general formulation of the scattering in quantum field theory goes under the name of the Haag–Ruelle theory, see R. Jost, *The General Theory of Quantized Fields*, Am. Math. Soc. 1965, Chap. VI; J. Glimm and A. Jaffe, *Quantum Physics. A Functional Integral Point of View*, Springer 1987, Chap. 13; N. N. Bogoliubov, A. A. Logunov, A. I. Oksak, and I. T. Todorov, *General Principles of Quantum Field Theory*, Kluwer 1990, Chap. 12. A brief account is given in Chapter 6, below.

Such a procedure extends by linearity to products of *free* fields through their decomposition into creation and annihilation operators. Clearly, the (Fock) vacuum expectation $<\ >_0$ of a normal product vanishes. The difference between a product of free fields and the corresponding normal product is called *contraction*; it is a c-number, since so are all commutators or anticommutators between creation and annihilation operators, and is denoted by $\underbrace{\varphi(x_1)\,\varphi(x_2)}$. Thus, for a scalar field, one has

$$\underbrace{\varphi(x_1)\,\varphi(x_2)} = \varphi(x_1)\,\varphi(x_2) - N(\varphi(x_1)\,\varphi(x_2)) = <\varphi(x_1)\,\varphi(x_2)>_0 =$$

$$= <\varphi^-(x_1)\,\varphi^+(x_2)>_0 = i\,\Delta^+(x_1, x_2).$$

Clearly, $\underbrace{\varphi(x)\,\varphi^-(y)} = 0$. We denote a normal product with a contraction inside by

$$N(A_1 \ldots \underbrace{A_j \ldots A_k} \ldots A_n) = (-1)^P \underbrace{A_j\,A_k}\, N(A_1 \ldots \widehat{A_j} \ldots \widehat{A_k} \ldots A_n), \qquad (2.7.2)$$

where the hat over A_l means that it has to be omitted.

Lemma 7.1 *(Wick recursive formula)*

$$N(A_1 \ldots A_n) = N(A_1 \ldots A_{n-1})\,A_n - \sum_{j=1}^{n-1} N(A_1 \ldots \underbrace{A_j \ldots A_n}). \qquad (2.7.3)$$

Proof. By linearity it is enough to discuss the cases in which the A_i are creation or destruction operators.

If A_n is a destruction operator, the proof is trivial because all the contractions vanish and A_n is already at the right place.

If A_n is a creation operator, all the creation operators A_j, $j \neq n$, can be brought to the left in each term of eq. (2.7.3) and then outside the normal product without affecting the equation, since they have vanishing contraction with A_n. Therefore one is left with the case in which all the A_j, $j \neq n$, are destruction operators. The proof is by induction. Clearly, eq. (2.7.3) holds for $n = 2$; we shall prove that if it holds for n it also hold for $n + 1$. In fact, by multiplying eq. (2.7.3) on the left by a destruction operator A_0 we get

$$A_0\,N(A_1 \ldots A_n) = A_0\,N(A_1 \ldots A_{n-1})\,A_n - \sum_{j=1}^{n-1} A_0\,N(A_1 \ldots \underbrace{A_j \ldots A_n}) =$$

$$= N(A_0 A_1 \ldots A_{n-1})A_n - \sum_{j=1}^{n-1} N(A_0 A_1 \ldots \underbrace{A_j \ldots A_n}), \qquad (2.7.4)$$

since the normal products on the right hand side contain only destruction operators (A_n being always contracted in the sum) and therefore A_0 can be brought inside. On the other hand, one has

$$A_0 N(A_1 \ldots A_n) = (-1)^P A_0 N(A_n A_1 \ldots A_{n-1}) = (-1)^P A_0 A_n N(A_1 \ldots A_{n-1})$$

$$= (-1)^P N(A_0 A_n) N(A_1 \ldots A_{n-1}) + (-1)^P \underbrace{A_0 A_n} N(A_1 \ldots A_{n-1}) =$$

$$= N(A_0 A_1 \ldots A_n) + N(\underbrace{A_0 A_1 \ldots A_n}), \tag{2.7.5}$$

where the last equality is obtained by noticing that if P' denotes the reordering parity for the interchange $A_0 A_n \to A_n A_0$, by eqs. (2.7.2) and (2.7.1) one has

$$N(A_0 A_n) N(A_1 \ldots A_{n-1}) = (-1)^{P'} N(A_n A_0) N(A_1 \ldots A_{n-1}) =$$

$$= (-1)^{P'} N(A_n A_0 \ldots A_{n-1}) = (-1)^P N(A_0 \ldots A_n). \tag{2.7.6}$$

Eqs. (2.7.4) and (2.7.5), imply eq. (2.7.3) for $n+1$.

Clearly, the Lemma remains valid if the normal product on the right-hand side contains contractions. Furthermore, the lemma says that the normal product can also be defined recursively, since it expresses the normal product of n fields in terms of normal products of $n-1$ and $n-2$ fields.

Theorem 7.2 *(Wick) The product of n fields can be written as a sum of normal products of such fields with all the possible contractions:*

$$A_1 \ldots A_n = N(A_1 \ldots A_n) + N(\underbrace{A_1 A_2}\, A_3 \ldots A_n) + \ldots + N(\underbrace{A_1 A_2 \ldots A_n}) + \ldots$$

$$+ N(A_1 \underbrace{A_2 A_3} \ldots A_n) + \ldots + N(\underbrace{A_1 A_2}\, \underbrace{A_3 A_4} \ldots A_n) + \ldots \tag{2.7.7}$$

Proof. The proof follows by induction. It is trivial for $n = 2$, and one obtains the equation for $n+1$ fields by multiplying the equation for n fields on the right by A_{n+1} and by using the extension of the Lemma to normal product with contractions. Thus, one gets

$$N(A_1 \ldots A_n) A_{n+1} = N(A_1 \ldots A_{n+1}) + \sum_j N(A_1 \ldots \underbrace{A_j \ldots A_{n+1}}),$$

where the normal product may include contractions which are not spelled out, and the sum runs over all indices j of fields which are not contracted in the original normal product.

The theorem can be extended to products of normal products. In fact, one has

$$N(A_1 \ldots A_k) N(A_{k+1} \ldots A_n) = (-1)^P A_{i_1} \ldots A_{i_k} (-1)^{P'} A_{i_{k+1}} \ldots A_{i_n},$$

where each group of factors is normal ordered. One can then apply the Wick theorem to the product on the right-hand side and obtain an expansion in terms of normal products and contractions, where all the *contractions inside each group vanish*, because any pair inside is normal ordered and its vacuum expectation vanishes. Thus, in the expansion only contractions between operators belonging to different groups appear.

It is worth noting that the normal product of free fields is less singular than the ordinary product. In fact, for a free scalar field (of mass m) the product $\varphi(x)\,\varphi(y)$ is singular in the limit $y \to x$ as displayed by its vacuum expectation (or two-point function), which diverges in that limit ($k_0 = \sqrt{\mathbf{k}^2 + m^2}$),

$$< \varphi(x)\,\varphi(y) >_0 = (2\pi)^{-3} \int d^3k\,(2k_0)^{-1} e^{-ik_o\,(x_0 - y_0) + i\mathbf{k}(\mathbf{x}-\mathbf{y})}. \tag{2.7.8}$$

On the other hand, the normal product has well-defined vacuum expectations with other products of fields in the limit $y \to x$. In fact, by the Wick theorem the product $\varphi(x_1)\ldots\varphi(x_j)\,N(\varphi(x)\,\varphi(y))\,\varphi(x_{j+1})\ldots\varphi(x_n)$ can be expanded in terms of normal products with all the possible contractions, and the only non-vanishing vacuum expectations are those of terms with all the fields contracted. Now, the contractions between $\varphi(x)$ or $\varphi(y)$ with $\varphi(x_l)$, $l = 1,\ldots n$ are well defined also in the limit $y \to x$, and the singular contraction between $\varphi(x)$ and $\varphi(y)$ has been subtracted out by the normal product.

In conclusion, for free fields the normal product allows to define the powers of fields at a point, called *Wick powers*, as operator-valued distributions; e.g., for a scalar field,

$$: \varphi^2 : (x) \equiv \lim_{y \to x} N(\varphi(x)\,\varphi(y)) = \lim_{y \to x} [\varphi(x)\,\varphi(y) - < \varphi(x)\,\varphi(y) >_0].$$

$$: \varphi^n : (x) = \lim_{y \to x} [: \varphi^{n-1} : (x)\,\varphi(y) - (n-1) < \varphi(x)\,\varphi(y) >_0 : \varphi^{n-2} : (x)].$$

The problem of defining powers of interacting fields is a much harder problem, which has been solved perturbatively, through perturbative renormalization.[43]

Wick theorem can be applied to the T-products (leading to *Feynman propagators*) which appear in the perturbative expansion of the S-matrix. For this purpose one first extends the T-product to products of fermionic fields by putting

$$T(A_1 \ldots A_n) = (-1)^P A_{i_1} \ldots A_{i_n}, \quad t_{i_1} > t_{i_2} > \ldots > t_{i_n}, \tag{2.7.9}$$

where P denotes the fermion reordering parity. Then, one defines the *chronological contraction* of two operators:

$$\underbrace{A_1\,A_2}_{T} \equiv T(A_1\,A_2) - N(A_1\,A_2) = < T(A_1\,A_2) >_0. \tag{2.7.10}$$

For a scalar field one has ($y \equiv x_1 - x_2$)

$$< T(\varphi(x_1)\,\varphi(x_2) >_0 = i\theta(y_0)\,\Delta^+(y) + i\theta(-y_0)\,\Delta^+(-y) =$$

$$= \frac{i}{(2\pi)^4} \int \frac{d^4k}{k^2 - m^2 + i\varepsilon}\,e^{-iky}. \tag{2.7.11}$$

---

[43] For the mathematical control of Wick powers of free fields, see A.S. Wightman and L. Gårding, Arkiv f. Fysik, **28**, 129 (1964). For the perturbative control of powers of fields, see W. Zimmermann, Brandeis Lectures, in *Lectures on Elementary Particles and Quantum Field Theory*, S. Deser *et al.* (eds.), MIT Press, 1971; J. H. Lowenstein, BPHZ Renormalization, in *Renormalization Theory*, G. Velo and A. S. Wightman (eds.), Plenum 1976.

The application of the Wick theorem to the right-hand side of eq. (2.7.9) and the properties (2.7.1) and (2.7.2) give an expansion in terms of N-products and chronological contractions, since the operators on the left-hand side are chronologically ordered. Thus, by using eqs. (2.7.9) and (2.7.7) and eq. (2.7.2) one obtains the Wick formula for the T-products $(t_{i_1} > t_{i_2} > \ldots > t_{i_n})$

$$T(A_1 \ldots A_n) = (-1)^P A_{i_1} \ldots A_{i_n} =$$

$$= (-1)^P \left[N(A_{i_1} \ldots A_{i_n}) + N(\underbrace{A_{i_1} A_{i_2}} \ldots A_{i_n}) + \ldots \right]$$

$$= (-1)^P \left[N(A_{i_1} \ldots A_{i_n}) + N(\underbrace{A_{i_1} A_{i_2}}_{T} \ldots A_{i_n}) + \ldots \right] =$$

$$= N(A_1 \ldots A_n) + N(\underbrace{A_1 A_2}_{T} A_3 \ldots A_n) + \ldots, \tag{2.7.12}$$

where the dots on the right-hand side denote the normal products with all possible chronologically ordered contractions.

7.1 Compton and electron–electron scattering; electron–positron annihilation

To see the effectiveness of the Wick theorem for the computation of the S-matrix elements, it is instructive to compute the lowest-order terms of the pion–pion scattering in a theory in which the pion is described by a (pseudo)scalar field φ with a (self-)interaction $\lambda : \varphi^4 : (x)$. The next order (λ^2) involves divergent integrals over the momentum running in the two contractions which join at the two end points, forming a "loop" and one needs renormalization.

Similarly, one may compute the *Compton scattering* $\gamma + e \to \gamma + e$, the *electron–electron scattering*, and the *electron–positron annihilation*, at the lowest order (e^2), the interaction being the electromagnetic interaction $e : \overline{\psi}(x)\, \gamma_\mu\, \psi(x) : A^\mu(x)$. The results for the corresponding differential cross-sections, which the reader is advised to derive,[44] are:

Compton scattering (in the laboratory frame in which the incoming electron is at rest),

$$\frac{d\sigma_{Compton}}{d\Omega} = \frac{\alpha^2}{4m^2} \left(\frac{k_f}{k_i}\right)^2 \left[\frac{k_f}{k_i} + \frac{k_i}{k_f} - 2 + 4\cos^2 \Theta\right],$$

where m is the electron mass, $\alpha \equiv e^2/(4\pi\hbar c)$, k_i, k_f are the momenta of the incident and emitted photon, and Θ is the angle between the directions of polarizations of the incident and emitted photon. For the quantization of the free electromagnetic potential in the Feynman-Gupta–Bleuler gauge, see Section 8.2 in Chapter 7.

---

[44] For help, see, e.g., F. Mandl, *Introduction to Quantum Field Theory*, Interscience 1959, Chap. 14; S. Schweber, *An Introduction to Relativistic Quantum Field Theory*, Harper and Row 1961, Sects. 14c, 14e; C. Itzykson and J.-C. Zuber, *Quantum Field Theory*, McGraw-Hill 1980, Sects. 5-2-1, 6-1-3, 5-2-2.

Electron–electron scattering, in the center of mass frame:

$$\frac{d\sigma_{e^-e^-}}{d\Omega} = \frac{\alpha^2}{4E^2} \frac{(p_1 \cdot p_2)^2}{|\mathbf{p}|^4} \left[\frac{4}{\sin^4 \theta} - \frac{3}{\sin^2 \theta} + \frac{|\mathbf{p}|^4}{(p_1 \cdot p_2)^2} \left(1 + \frac{4}{\sin^2 \theta}\right) \right],$$

where E is the energy of each electron, $|\mathbf{p}| = \sqrt{E^2 - m^2}$ the momentum, p_1, p_2 the four-momenta of the two electrons and θ the scattering angle.

Electron–positron annihilation, in the rest frame of the electron

$$\frac{d\sigma_{e^- + e^+ \to 2\gamma}}{d\Omega} = \frac{\alpha^2}{8|\mathbf{p}_2|} \frac{m + E_2}{(m + E_2 - |\mathbf{p}_2| \cos\theta)^2} \left[\frac{\omega_2}{\omega_1} + \frac{\omega_1}{\omega_2} + 2 - 4\cos^2 \Theta \right],$$

where ω_1, ω_2 denote the energies of the photons, $(E_2, \mathbf{p}_2)$ the positron four-momentum, and Ω the solid angle defined by the momentum of one of the photons and the momentum of the positron.

Again, in all the above processes, the next orders e^4 involves divergent integrals over momenta running over loops, and one needs renormalization.

3

Non-perturbative foundations of quantum field theory

1   Quantum mechanics and relativity

The investigation of the non-perturbative foundations of relativistic quantum field theory began soon after the success of the perturbative expansion of quantum electrodynamics, when the convergence of the series was seriously questioned, the theory was predicted to be afflicted by ghosts[1] and the high-energy behavior raised the question of the consistency of the theory. Furthermore, the mathematical problems of the interaction picture, and even the impossibility of using canonical quantization for the quantization of fields, as discussed in the previous chapter, indicated that a non-perturbative foundation of quantum field theory was an unavoidable issue.

The two major steps in this direction were the Lehmann, Symanzik, and Zimmermann approach (LSZ "axioms")[2] based on the asymptotic condition and the S-matrix elements, and the Wightman formulation of quantum field theory in terms of the so-called Wightman "axioms", which emphasize the spectral condition, relativistic invariance, and locality.[3] The Wightman approach, discussed below, proved to be more powerful, the first being derivable from it.[4] The term "axioms", even if correctly stressing the attention to the mathematical rigor and consistency, does not do justice to the fact that in both cases they represent the mathematical formulation of deep physical requirements.

The quantum-mechanical interpretation of the theory and its stability require the following **quantum-mechanical properties:**

QM1 (Hilbert space structure) The states are described by vectors of a separable Hilbert space $\mathcal{H}$.

---

[1] L. L. Landau, On the quantum theory of fields, in *Niels Bohr and the development of Physics*, W. Pauli *et al.* (eds.), McGraw Hill 1955.

[2] H. Lehmann, K. Symanzik, and W. Zimmermann, Nuovo Cim. **1**, 1425 (1955); for excellent accounts of the LSZ theory, see R. Hagedorn, *Introduction to Field Theory and Dispersion Relations*, Pergamon Press 1963, and S. Schweber, *Introduction to Relativistic Quantum Field Theory*, Harper and Row 1961, Sect. 18 b.

[3] R. F. Streater and A. S. Wightman, *P C T, Spin and Statistics and All That*, Benjamin 1980; R. Jost, *The General Theory of Quantized Fields*, Am. Math. Soc. 1965.

[4] K. Hepp, On the connection between the Wightman and the LSZ quantum field theory, in Brandeis Summer Inst. Theor. Phys. 1965, Vol. I, *Axiomatic Field Theory*, M. Chretien and S. Deser (eds.), Gordon and Breach 1966.

QM2 (Energy–momentum spectral condition) The spacetime translations are symmetries of the theory, and are therefore described in $\mathcal{H}$ by strongly continuous unitary operators $U(a), a \in \mathbf{R}^4$.

The spectrum of the generators P_μ is contained in the closed *forward cone* $\overline{V_+} = \{p_\mu : p^2 \geq 0, \, p_0 \geq 0\}$. There is a *vacuum state* vector Ψ_0, with the property of being the unique translationally invariant state in $\mathcal{H}$ (hereafter referred to as *uniqueness of the vacuum* in $\mathcal{H}$).

QM3 (Field operators) The theory is formulated in terms of fields $\varphi_k(x)$, $k = 1, \ldots N$, which are operator-valued tempered distributions in $\mathcal{H}$, with Ψ_0 a *cyclic* vector for the fields (i.e., by applying polynomials of the (smeared) fields to the vacuum one obtains a dense set $\mathcal{D}_0$ in $\mathcal{H}$).

Remarks. The separability of $\mathcal{H}$ actually follows from the cyclicity of the vacuum and temperedness, since $\mathcal{S}(\mathbf{R}^4)$ has a countable basis.[5]

The spectral condition is the relativistic version of the condition that the Hamiltonian is bounded from below, which guarantees the stability of the theory.

In terms of the spectral representation of $U(a) = \int e^{ipa} \, dE(p)$, the spectral condition reads,[6] $\forall \Phi, \, \Psi \in \mathcal{H}$,

$$\int da \, e^{-ipa} \, (\Phi, \, U(a) \, \Psi) = 0, \quad \text{if} \quad p \notin \overline{V}_+. \tag{3.1.1}$$

The cyclicity of the vacuum states that the fields provide a complete set of "dynamical variables" in terms of which one can describe all the states of $\mathcal{H}$.

The relativistic invariance of the theory is formalized by the following **relativistic properties**:

R1 (Relativistic covariance) The Lorentz transformations Λ are described by (strongly continuous) unitary operators $U(\Lambda(A))$, $A \in SL(2, C) = $ the universal covering group[7] of the restricted Lorentz group $L_+^\uparrow$ ($\det \Lambda = 1$, sign $\Lambda_0^0 > 0$), and the fields transform covariantly under the Poincaré transformations $U(a, \Lambda) = U(a) \, U(\Lambda)$:

$$U(a, \Lambda(A)) \, \varphi_i(x) \, U(a, \Lambda(A))^{-1} = S_{ij}(A^{-1}) \, \varphi_j(\Lambda x + a), \tag{3.1.2}$$

with $S(A)$ a finite dimensional representation of $SL(2, C)$.

R2 (Microcausality or locality) The fields either commute or anticommute at spacelike separated points

$$[\varphi_i(x), \, \varphi_j(y)]_\mp = 0, \quad \text{for} \quad (x - y)^2 < 0. \tag{3.1.3}$$

---

[5] Both the test-function spaces $\mathcal{S}$ and $\mathcal{D}$ are separable as topological spaces; for example, in d dimensions, consider the polynomials of d variables, with complex coefficients having rational real and imaginary parts, and for each n a C^∞ function of d variables $\chi^{(n)}$, with $\chi^{(n)} = 1$ inside the ball of radius n and vanishing outside the ball of radius $n + 1$; then, the polynomials in d variables, times the $\chi^{(n)}$, form a dense denumerable set.

[6] In Dirac notations, $\int dE(p) \, e^{ipa} \, (\ldots) = \sum_\alpha \int dp \, |p, \alpha > < p, \alpha| \, e^{ipa} \, (\ldots)$, where $|p, \alpha >$, $p \in \overline{V}_+$, is a complete set of (improper) eigenstates of the four-momentum.

[7] This accounts for the spinor representations; for simplicity, in the following the "label" A in $\Lambda(A)$ will not always be spelled out.

Remarks.   The invariance of the vacuum under $U(\Lambda)$ follows from the group law $U(\Lambda)^{-1} U(a) U(\Lambda) = U(\Lambda^{-1}a)$, which implies

$$U(a)U(\Lambda)\,\Psi_0 = U(\Lambda)\,U(\Lambda^{-1}a)\,\Psi_0 = U(\Lambda)\,\Psi_0,$$

and from the uniqueness of the translationally invariant state.

Eq. (3.1.2) guarantees the manifest covariance of the formulation, but is not strictly implied by relativistic invariance, since one could use non-covariant fields; it is an essentially technical condition, since it is more convenient to use dynamical variables with simple transformation properties under the symmetries of the theory.

The locality condition, eq. (3.1.3), deserves a special discussion. It holds for free fields (see Chapter 1, Section 5); in the interacting case, its validity for observable fields (such as the electromagnetic field $F_{\mu\nu}(x)$), and for the polynomial algebra $\mathcal{A}$ generated by them is required by Einstein causality; according to it, measurements of observables localized in spacelike separated regions must be always compatible (in the quantum-mechanical sense); namely, such observables must commute.

More precisely, if $\mathcal{A}(\mathcal{O})$ denotes the (polynomial) algebra generated by the observable fields smeared with test functions with support in the bounded region $\mathcal{O}$, typically a double cone, and $\mathcal{O}'$ denotes the set of points which are spacelike to every point of $\mathcal{O}$ (briefly, the *spacelike complement* of $\mathcal{O}$), Einstein causality requires that

$$[\,\mathcal{A}(\mathcal{O}),\,\mathcal{A}(\mathcal{O}')\,] = 0.$$

In general, the field algebra $\mathcal{F}$, generated by polynomials of the (smeared) fields, properly contains $\mathcal{A}$, and the validity of eq. (3.1.3) for unobservable fields, such as the fermion fields (or the charged fields), is an extrapolation with respect to the physical requirements, and may be justified as an essentially technical requirement. In fact, the locality condition for the field algebra $\mathcal{F}$ guarantees that the observable operators constructed in terms of unobservable local fields, such as the current density, the momentum density, and so on, automatically commute at spacelike separated points, i.e., satisfy Einstein causality. It is also fair to say that most of our present wisdom on QFT, including the full control of the low-dimensional cases, comes from models formulated in terms of local (in general unobservable) fields.

Moreover, the cyclicity of the vacuum with respect to a field algebra $\mathcal{F}$ which satisfies locality—briefly, a *local field algebra*—implies that the local states $\mathcal{D}_0 = \mathcal{F}\,\Psi_0$ are dense. This means that the states of the theory have a localization property. In fact, quite generally an *operator B is localized in the double cone $\mathcal{O}$* if

$$[\,B,\,\mathcal{A}(\mathcal{O}')\,] = 0;$$

and if $B = \varphi(f)$, with φ a local field (and the field algebra is irreducible), the localization region of $\varphi(f)$ is given by the support of the test function f.

Moreover, as emphasized by Doplicher, Haag, and Roberts,[8] from an operational point of view, a *state ω is localized in $\mathcal{O}$* (in the DHR sense) if

$$\omega(A) = (\Psi_0, A\,\Psi_0), \quad \forall A \in \mathcal{A}(\mathcal{O}'). \tag{3.1.4}$$

---

[8] See R. Haag, *Local Quantum Physics: Fields, Particles, Algebras*, Springer 1996, Chaps. III, IV.

In order to see the relation with the locality of the fields, we consider a state ω represented by a vector Ψ of the form

$$\Psi = U\Psi_0, \quad U^*U = UU^* = 1,$$

with U localized in $\mathcal{O}$—typically, a "function" of the local fields smeared with test functions with support in $\mathcal{O}$. Then, the state Ψ is localized in $\mathcal{O}$ in the DHR sense: in fact, by locality, for every observable $A \in \mathcal{A}(\mathcal{O}')$, we have

$$\omega(A) = (\Psi, A\,\Psi) = (\Psi_0, U^*AU\,\Psi_0) = (\Psi_0, A\,\Psi_0),$$

i.e., the representation of $\mathcal{A}(\mathcal{O}')$ given by ω is equivalent to the vacuum representation (technically $\rho_U(A) \equiv U^*AU$ is a *local morphism* of $\mathcal{A}$).

In conclusion, the locality property of the field algebra $\mathcal{F}$, with respect to which the vacuum is cyclic, means that one may have a description of the states of $\mathcal{H}$ in terms of localized states.

An alternative and more economical approach to relativistic quantum mechanics is to use only (bounded) local observables (*algebraic quantum field theory*). However, since not all the physical states belong to the vacuum sector of the observable algebra, $\mathcal{H}_0 \equiv \overline{\mathcal{A}\,\Psi_0}$, unobservable fields have to be constructed as intertwiners between inequivalent representations of the observable algebra (see Haag's book).

One of the main virtues of the Wightman approach is its strong link with the conventional wisdom of quantum field theory, including perturbation theory, which crucially relies on the use of non-observable fields, such as the vector potential and the electron field in QED. It also proved successful for constructive QFT. For these reasons, the investigation of the mathematical structures of the vacuum expectation values of the field algebra, which are at the basis of Wightman approach, has a direct impact on the conventional and the constructive approach.

2   Properties of the vacuum correlation functions

From QM1–QM3, R1, and R2 one deduces the following properties of the vacuum expectation values (*Wightman functions*) of a scalar field. As a consequence of QM3 (and Schwartz nuclear theorem):

W1 $\mathcal{W}(x_1, \ldots x_n) \equiv (\Psi_0, \varphi(x_1) \ldots \varphi(x_n)\,\Psi_0)$ are tempered distributions.

For brevity, we shall use a multivector notation $\mathcal{W}(x) = \mathcal{W}(x_1, \ldots x_n)$, $x = (x_1, \ldots x_n)$ and denote the vacuum expectations by $< \varphi(x_1) \ldots \varphi(x_n) >$.

W2 (Covariance) QM2 and R1 give

$$\mathcal{W}(x_1, \ldots x_n) = W(\xi_1, \ldots \xi_{n-1}) \equiv W(\xi) = W(\Lambda\xi), \quad \xi_j \equiv x_{j+1} - x_j. \tag{3.2.1}$$

W3 (Spectral condition) The support of the Fourier transform $\tilde{W}$ of W is contained in the product of forward cones, i.e.,

$$\tilde{W}(q_1, \ldots q_n) = 0, \quad \text{if, for some } j, \ q_j \notin \overline{V}_+. \tag{3.2.2}$$

In fact, denoting by $\tilde{W}^{(j)}$ the Fourier transform of W with respect to the jth variable, one has (using eq. (3.1.1))

$$(2\pi)^2\, \tilde{W}^{(j)}(\ldots q_j, \ldots)\, e^{iq_j\xi_j} = \int da\, e^{-iq_j a}\, W(\xi_1, \ldots \xi_j + a, \ldots \xi_n) =$$

$$\int da\, e^{-iq_j a} < \varphi(x_1)\ldots\varphi(x_j)\, U(a)\, \varphi(x_{j+1})\ldots\varphi(x_{n+1}) > = 0, \quad \text{if } q_j \notin \overline{V}_+.$$

W4 (Locality) R2, with the choice of the commutator (as required by Theorem 2.2 in Chapter 4), gives

$$\mathcal{W}(x_1, \ldots x_j, x_{j+1}, \ldots x_n) = \mathcal{W}(x_1, \ldots x_{j+1}, x_j, \ldots x_n), \quad \text{if } (x_j - x_{j+1})^2 < 0. \quad (3.2.3)$$

The Hilbert space structure of QM1 gives:
W5 (Positivity) For any terminating sequence $\underline{f} = (f_0, f_1, \ldots f_N)$, $f_j \in \mathcal{S}(\mathbf{R}^{4j})$, one has

$$\sum_{j,k} \int dx\, dy\, \bar{f}_j(x_j, \ldots x_1)\, f_k(y_1, \ldots y_k)\, \mathcal{W}(x_1, \ldots x_j; y_1, \ldots y_k) \geq 0. \qquad (3.2.4)$$

This is equivalent to the positivity of the norm of any state of the form

$$\Psi_{\underline{f}} = f_0\Psi_0 + \varphi(f_1)\,\Psi_0 + \varphi(f_2^{(1)})\,\varphi(f_2^{(2)})\Psi_0 + \ldots, \qquad (3.2.5)$$

where $\underline{f} = (f_0, f_1, \ldots f_N)$, $f_j = \prod_{k=1}^{j} f_j^{(k)}(x_k)$, $f_j^{(k)} \in \mathcal{S}(\mathbf{R}^4)$.

The uniqueness of the translationally invariant state is equivalent to:
W6 (Cluster property) For any spacelike vector a and for $\lambda \to \infty$

$$\mathcal{W}(x_1, \ldots x_j, x_{j+1} + \lambda a, \ldots x_n + \lambda a) \to \mathcal{W}(x_1, \ldots x_j)\, \mathcal{W}(x_{j+1}, \ldots x_n), \qquad (3.2.6)$$

where the convergence has to be understood in the sense of distributions, i.e., after smearing with test functions in $\mathcal{S}$.

Eq. (3.2.6) says that the correlation function of two monomials of (smeared) fields, briefly called clusters, factorizes (i.e., a decorrelation takes place) in the limit of infinite spacelike distance between the two clusters. This property plays a crucial role for the existence of asymptotic (free) fields and therefore for the construction of the S-matrix, as clarified by the Haag–Ruelle theory, discussed in Chapter 6 below.[9] It corresponds to a sufficient fall-off of the potential in potential scattering (see Chapter 2, Section 6, and Chapter 6 below), and it is one of the basic axioms of the so-called S-matrix theory.[10] Property (3.2.6) is related to the independence of events associated to two clusters

[9] See R. Jost, *The General theory of Quantized Fields*, Am. Math. Soc. 1965, Chap. VI; A. S. Wightman, Recent achievements of axiomatic field theory, in *Theoretical Physics*, Trieste 1962, IAEA 1963.
[10] See, e.g., R. J. Eden, P. V. Landshoff, and D. I. Olive, *The Analytic S-matrix*, Cambridge University Press 1966.

at infinite spacelike distance; the rate by which such a decorrelation is reached has actually been related to the decay rate of the interaction strength or "force" between the two clusters.[11]

The cluster property implies that if two clusters B_1, B_2 are localized in the bounded regions $\mathcal{O}_1$, $\mathcal{O}_2$, respectively, the state vectors $B_1 \Psi_0$, $B_2 \Psi_0$ become orthogonal, apart from their vacuum component, in the limit in which the spacelike separation between their localization regions becomes infinite.

Thus, for $\lambda \to \infty$, a spacelike, $U(\lambda a) B_2 \Psi_0$, will appear as the vacuum state, as seen by $B_1 \Psi_0$:

$$(B_1 \Psi_0, U(\lambda a) B_2 \Psi_0 - <B_2> \Psi_0) \to 0.$$

Under the assumption that the energy spectrum has a gap μ above zero and the vacuum is unique (e.g., if the theory describes massive particles), a proof of the cluster property, which uses only Lorentz covariance, can be found in Jost's book (*The General Theory of Quantized Fields*, Am. Math. Soc. 1965, pp. 68–71), together with the proof that the limit is reached exponentially fast, the energy gap providing the exponential decay constant. A proof based on the spectral condition, uniqueness of the vacuum, and locality (without assuming an energy gap, nor using Lorentz covariance) was given by Araki, Hepp, and Ruelle[12] with a sharp characterization of the decay, which is exponentially fast, namely, $\exp[-\mu r]/r^{3/2}$, if there is an energy or mass gap $\mu > 0$, and is like the Coulomb force r^{-2}, if there is no mass gap.

The following simple argument, even if not complete, may help in grasping the validity of the cluster property.

Since $U(\Lambda)^{-1} U(\lambda a) U(\Lambda) = U(\lambda \Lambda a)$, if a is spacelike one can find a Λ such that $a' \equiv \Lambda a = (a_0' = 0, \mathbf{a}')$; therefore the study of the matrix elements of $U(\lambda a)$ between generic vectors is reduced to the study of $< \Phi| U(\lambda \mathbf{a})| \Psi >$. By inserting a complete set of proper (i.e., corresponding to pure point spectrum) and improper (i.e., corresponding to continuous spectrum) eigenstates of P_μ,[13] one can write the scalar product (using the Dirac notation and $d\Omega_{p^2} \equiv d^3 p / \sqrt{p^2 + \mathbf{p}^2}$), as

$$\sum_n e^{-i\lambda \mathbf{a} \cdot \mathbf{p}_n} < \Phi|p_n >< p_n|\Psi > + \int dp^2 d\Omega_{p^2} < \Phi|p >< p|\Psi > e^{-i\lambda \mathbf{a} \cdot \mathbf{p}}.$$

The second term vanishes in the limit $\lambda \to \infty$ by the Riemann–Lebesgue lemma for the $d\Omega_{p^2}(\mathbf{p})$ integration and by the dominated convergence theorem for the integration in dp^2 (more generally, because the absolutely continuous part of the spectral measure is by definition continuous with respect to the Lebesgue measure and therefore

---

[11] H. Araki, Ann. Phys. **11**, 260 (1960).

[12] H. Araki, K. Hepp, and D. Ruelle, Helv. Phys. Acta **35**, 164 (1962).

[13] This corresponds to the decomposition of the energy–momentum spectral measure as a sum of its discrete and absolutely continuous parts, the (continuous) singular part being excluded by Lorentz covariance; formally:

$$\int dE(p) \, (\ldots) = (\sum_n |p_n >< p_n| + \int dp^2 \, d\Omega_{p^2} |p >< p|) \, (\ldots).$$

the Riemann–Lebesgue lemma applies to the integration with respect the spectral measure).

Furthermore, the discrete spectrum can consist only of the (possibly degenerate) point $p = 0$; otherwise by applying Lorentz transformations to an eigenvector Ψ_{p_n}, $p_n \neq 0$, one would obtain a continuous set of eigenvectors $\Psi_{\Lambda p_n}$, contrary to separability. In conclusion, denoting by $\mathcal{P}_0$ the projection on the subspace V_0 of $p_n = 0$ eigenvectors, one has

$$\lim_{\lambda \to \infty} (\Phi, U(\lambda \mathbf{a})\, \Psi) = (\Phi, \mathcal{P}_0\, \Psi). \tag{3.2.7}$$

This implies the equivalence between the cluster property and the uniqueness of the translationally invariant state.

The above simple properties of the Wightman functions, especially their Poincaré covariance, make them the privileged objects for the study of quantum field theory. In fact, as realized by Feynman, the vacuum expectation values of (time-ordered) fields allow for a much more tractable perturbative expansion than the Rayleigh–Schrödinger expansion in terms of non-covariant matrix elements.

Actually, from the experience of perturbation theory, it appears that the knowledge of the (time-ordered) vacuum correlation functions is all that is needed for the computation of the S-matrix elements (see Chapter 2, Section 1). Indeed, quite generally, by the cyclicity of the vacuum, the generic matrix elements of field operators $(\Phi, \varphi(x)\, \Psi)$ can be approximated by vacuum expectation values as much as one likes.

The realization and the proof that the vacuum expectation values or Wightman functions, satisfying W1-W6, completely define a relativistic quantum field theory, is the fundamental result at the basis of Wightman formulation of quantum field theory.[14]

3   Quantum mechanics from correlation functions

In the previous section we derived properties of the Wightman functions from the properties which characterize relativistic quantum mechanics. A relevant question of principle is when a set of correlation functions (no matter how they have been obtained) identify a relativistic quantum mechanics, namely, a Hilbert space and a quantum-mechanical structure satisfying QM1–QM3, R1-R2. It may very well be that such correlation functions have been obtained through intermediate steps in which the connection with Hilbert space expectations of field operators on a vacuum state is lost.

For example, the perturbative computations of Green's functions, e.g., the field propagators, start with the free ones, which have a simple quantum-mechanical interpretation in terms of vacuum expectations of free fields, but it is not obvious that such an interpretation survives at each order of the perturbative calculations. One may ask whether the Feynman diagrams of the perturbative expansion merely provide a relativistic S-matrix, i.e., matrix elements between asymptotic states, or a

<hr>

[14] A. S. Wightman, Phys. Rev. **101**, 860 (1956).

full relativistic quantum mechanics, namely, matrix elements of quantum fields and states at finite (not asymptotic) times.

In the non-perturbative constructive approach to quantum field theory models, the correlation functions are obtained by removing the volume cutoff and the ultraviolet or lattice cutoff in an (unphysical) butchered theory. It is therefore crucial to have criteria which guarantee that the resulting functions can be interpreted as vacuum expectation values of fields, in a relativistic quantum mechanics.

The relevance of this type of question became clear in the mid-1950s, when experience with soluble models, such as the Lee model, seemed to indicate that the quantum-mechanical interpretation is not always guaranteed, and that serious problems may arise due to the occurrence of "ghosts", i.e., states of "negative norm".[15]

We now sketch the proof of the (Wightman) reconstruction of relativistic quantum mechanics from the Wightman functions. In the quantum-mechanical formulation of field theory, a distinguished role is played by the states obtained by applying polynomials in the smeared fields to the vacuum, briefly called the *local states*, identified by the terminating sequences $\underline{f} = (f_0, f_1, \ldots f_N)$, $f_j \in \mathcal{S}(\mathbf{R}^{4j})$; see, e.g., eq. (3.2.5). For the reconstruction of the Hilbert space, it is therefore natural to start with the vector space $\mathcal{D}_0$ of the above terminating sequences.

A *representation of the Poincaré group* is naturally defined on $\mathcal{D}_0$ by

$$T_{\{a,\Lambda\}}\, \underline{f} = (f_0, f_1^{\{a,\Lambda\}}, \ldots, f_N^{\{a,\Lambda\}}), \tag{3.3.1}$$

$$f_j^{\{a,\Lambda\}}(x_1, \ldots, x_j) \equiv f_j(\Lambda^{-1}(x_1 - a), \ldots, \Lambda^{-1}(x_j - a)). \tag{3.3.2}$$

One may also define (smeared) *field operators* on $\mathcal{D}_0$; $\forall h \in \mathcal{S}(\mathbf{R}^4)$ we define

$$h \cdot \underline{f} = (0, hf_0, hf_1, \ldots, hf_N)$$

and

$$\varphi(h)\underline{f} = h \cdot \underline{f}. \tag{3.3.3}$$

The defined field-operators *transform covariantly under the Poincaré group*,

$$T_{\{a,\Lambda\}}\, \varphi(h)\, T_{\{a,\Lambda\}}^{-1} = \varphi(h^{\{a,\Lambda\}}). \tag{3.3.4}$$

The vector $\underline{f}_0 = (f_0, 0, 0, \ldots)$ is *cyclic* with respect to the polynomial algebra of the (smeared) fields, and is invariant under the Poincaré transformations. The Wightman functions define an *inner product* on $\mathcal{D}_0$:

$$< \underline{f}, \underline{g} >= \sum_{j,k} \int dx\, dy\, \bar{f}_j(x_j, \ldots x_1)\, g_k(y_1, \ldots y_k)\, \mathcal{W}(x_1, \ldots x_j; y_1, \ldots y_k) \tag{3.3.5}$$

---

[15] G. Källen and W. Pauli, Dank. Mat. Fys. Medd. **30**, No.7 (1955); G. Källen, Consistency problems in quantum electrodynamics, CERN Report 57–43 (1957); G. Källen, Renormalization theory, in Brandeis Summer Institute in Theoretical Physics 1961, *Lectures in Theoretical Physics*, Benjamin 1962, p. 169.

and one has

$$W(x_1, \ldots x_n) = <\underline{f}_0, \, \varphi(x_1) \ldots \varphi(x_n) \underline{f}_0>.$$

For brevity, it is convenient to introduce the following notations:

$$W(\underline{f}) \equiv f_0 + \sum_{n=1}^{N} W_n(f_n), \tag{3.3.6}$$

where W_n denotes the-n point Wightman function; $\underline{f} \times \underline{g}$ denotes the vector with components

$$(\underline{f} \times \underline{g})_j \equiv \sum_{i=0}^{j} f_i(x_1, \ldots x_i)\, g_{j-i}(x_{i+1}, \ldots x_j), \tag{3.3.7}$$

and $\underline{f}^*$ denotes the vector with components

$$(\underline{f}^*)_j \equiv \overline{f}_j(x_j, \ldots x_1). \tag{3.3.8}$$

With these notations and putting $W_0(f_0) \equiv f_0$, one may write

$$<\underline{f}, \underline{g}> = W(\underline{f}^* \times \underline{g}) = \sum_{n,m=0}^{\infty} W_{n+m}(f_n^* \times f_m). \tag{3.3.9}$$

The Poincaré covariance of the Wightman functions implies that the Poincaré transformations preserve the inner product

$$< T_{\{a,\Lambda\}}\underline{f}, \, T_{\{a,\Lambda\}}\underline{g}> = <\underline{f}, \underline{g}>.$$

The *quantum-mechanical interpretation* is provided by the positivity property W5, since then the inner product on $\mathcal{D}_0$ is non-negative and $\mathcal{D}_0$ becomes a pre-Hilbert space. By a standard procedure (taking quotients and completions) one obtains a Hilbert space $\mathcal{H}$, which is separable since so is $\mathcal{D}_0$, as a consequence of the separability of $\mathcal{S}$, and $\mathcal{D}_0$ is dense in $\mathcal{H}$.

The Poincaré transformations are then represented by densely defined operators, which preserve the scalar product and are therefore unitary. The *spectral condition* holds as a consequence of W3, and the locality property W4 implies that on $\mathcal{D}_0$ the field operators *commute at spacelike separations*.

The *uniqueness of the translationally invariant state* follows from the cluster property W6.

As a consequence of the (Wightman) reconstruction theorem, in order to exhibit a relativistic quantum field theory model, it is enough to give a set of Wightman functions satisfying W1–W6. This turned out to be a very hard problem, apart from the non-interacting case, also because it is difficult to satisfy the positivity condition, which has a non-linear structure, in contrast with the other properties, which have a linear structure.

4   General properties

From the general physical requirements, transcribed in the mathematical properties W1–W6, relevant structural properties have been derived, some of them having direct experimental consequences.

First, W1–W6 provide a non-perturbative substitute for canonical *quantization*. As we shall see below (Chapter 4, Section 1), W1–W6 imply canonical quantization for a field obeying a free field equation, but exclude it for a general class of interacting fields (see Chapter 2, Section 4).

Another important consequence of W1–W6 is the existence of asymptotic fields which provide a *non-perturbative* definition of the *S-matrix*, which is unitary if asymptotic completeness holds (*Haag–Ruelle scattering theory*). Also, the LSZ asymptotic condition and the LSZ reduction formulas can be derived from W1–W6, as proved by Hepp (see his Brandeis lectures); furthermore, one can prove *dispersion relations* for scattering amplitudes, yielding experimentally measurable relations.[16]

By exploiting the analyticity properties of the Wightman functions, general results have been derived. The connection between spin and statistics, which is at the basis of Pauli principle (namely, in the alternative of anti/commutation relations at spacelike points, fields carrying (half-)integer spin must (anti/)commute), is a (crucial) consequence of Lorentz covariance (*spin–statistics theorem*).[17] The (up to now experimentally established) *PCT* symmetry also follows from W1–W6, being in particular related to local commutativity (*PCT theorem*).

Finally, as a consequence of the spectral condition, Lorentz covariance, and locality, the Wightman functions have an analytic continuation to the so-called Euclidean points, and from this one derives the existence and the general properties of *Euclidean quantum field theory*. This is at the basis of the functional integral approach to QFT and of the non-perturbative approaches developed in the last decades (such as the lattice approach to gauge theories, the constructive strategy, etc.).

The above results exploit the analyticity properties of the Wightman functions, whose derivation will be briefly sketched below.

4.1   Spectral condition and forward tube analyticity

The relativistic spectral condition implies that the Wightman functions are the boundary values of analytic functions.

For square integrable functions this is an easy consequence of the properties of the Laplace transform. In fact, if $\tilde{f}(k) \in L^2(\mathbf{R}^4)$ and supp $\tilde{f} \subseteq \overline{V}_+$, then $k\eta \equiv k_\mu \eta^\mu > 0$, $\forall \eta \in V_+ \equiv \{k; k^2 > 0, k_0 > 0\}$, and $e^{-k\eta}\tilde{f}(k) \in L^2(\mathbf{R}^4)$; hence, the Laplace transform

$$(L\,\tilde{f})(x + i\eta) \equiv \int d^4k \, e^{ik(x+i\eta)} \, \tilde{f}(k) \tag{3.4.1}$$

<hr>

[16] A. Martin, The rigorous analyticity–unitarity program and its success, in *Quantum Field Theory*, P. Breitenlohner and D. Maison (eds.), Springer 1999, p. 127.

[17] A. S. Wightman, Am. J. Phys. **67**, 742 (1999); Elect. J. Diff. Eqs., Conf. **04**, 207 (2000).

exists and defines a C^1 function of the complex variable $z = x + i\eta$, analytic in the *forward tube*

$$\mathcal{T} \equiv \mathbf{R}^4 + i\, V_+, \tag{3.4.2}$$

since the Cauchy–Riemann equations are satisfied there. Moreover, $f(x)$ can be obtained as the (L^2) limit of $(L\,\tilde{f})(x + i\eta)$ as $\eta \to 0$, in V_+.

Such relations extend to tempered distributions[18] and therefore to the Wightman functions, as a consequence of the spectral condition, eq. (3.2.2). In conclusion, there are analytic functions

$$W(\zeta_1, \ldots, \zeta_n) \equiv W(\zeta), \quad \zeta_j = \xi_j + i\eta_j, \quad \eta_j \in V_+, \tag{3.4.3}$$

analytic in the $4n$-dimensional *forward tube*

$$\mathcal{T}_n \equiv \mathbf{R}^{4n} + i\,\Gamma, \quad \Gamma \equiv (V_+)^n \tag{3.4.4}$$

and $W(\xi)$ can be recovered as the boundary (distributional) limit of $W(\zeta)$, when $\operatorname{Im}\zeta = \eta \to 0$, in Γ.

Correspondingly, one obtains analyticity properties for the Wightman functions $\mathcal{W}(x_1, \ldots, x_n)$; namely, the existence of analytic functions $\mathcal{W}(z) \equiv \mathcal{W}(z_1, \ldots, z_n) = W(z_2 - z_1, \ldots, z_n - z_{n-1})$, analytic in the region

$$\sigma_n \equiv \{z \in \mathbf{R}^{4n};\ \operatorname{Re} z_j \in \mathbf{R}^4,\ \operatorname{Im}(z_{j+1} - z_j) \in V_+\}. \tag{3.4.5}$$

4.2 Lorentz covariance and extended analyticity

By Lorentz covariance, the analyticity domain of the Wightman functions can be extended from the forward tube to the *extended tube*

$$\mathcal{T}_n^{ext} \equiv \{w,\ w = \Lambda\zeta;\ \zeta \in \mathcal{T}_n,\ \Lambda \in L_+(C)\} = L_+(C)\mathcal{T}_n, \tag{3.4.6}$$

where $L_+(C)$ denotes the group of proper *complex Lorentz transformations*, i.e., the group of complex linear transformations Λ which preserve the product of (two) four-vectors and satisfy $\det\Lambda = 1$.[19]

The idea is to continue the Wightman functions to the points of $\mathcal{T}_n^{ext}$ by putting

$$W(\Lambda\,\zeta) = W(\zeta), \quad \forall \zeta \in \mathcal{T}_n,\ \Lambda \in L_+(C). \tag{3.4.7}$$

One has to check the consistency of this definition.

---

[18] The Laplace transform exists and defines an analytic function, because $\forall \eta \in V_+^n$, $e^{-q\eta}\,\tilde{W}(q) \in \mathcal{S}(\mathbf{R}^{4n})'$; for a detailed proof see R. F. Streater and A. S. Wightman, *PCT, Spin and Statistics and All That*, Benjamin 1964, Chap. II, Theors. 2.5–2.10. For a handy account see F. Strocchi, *Selected Topics on the General Properties of Quantum Field Theory*, World Scientific 1993, Sect. 2.1.

[19] The restricted Lorentz group $L_+^\uparrow$ is related to $SL(2, C)$, (the group of 2×2 matrices A with $\det A = 1$), through the correspondence: $\forall A \in SL(2, C)$, $A\underline{x}A^* \to \Lambda(A)x$, $\Lambda(A) \in L_+^\uparrow$, $\underline{x} \equiv x^\mu\tau_\mu$, $\tau_0 = \mathbf{1}$, $\tau_i = $ the Pauli matrices; one has $xy = \frac{1}{2}[\det(\underline{x} + \underline{y}) - \det\underline{x} - \det\underline{y}]$. $L_+(C)$ is related to $SL(2, C) \times SL(2, C)$ through the correspondence: $\forall A, B \in SL(2, C)$, $A\underline{x}B^T \to \Lambda(A, B)x$, $\Lambda(A, B) \in L_+(C)$.

As a first step, Lorentz covariance implies that

$$W(\zeta) = W(\Lambda\zeta), \quad \forall \zeta \in \mathcal{T}_n, \ \Lambda \in L_+^\uparrow, \tag{3.4.8}$$

and, since $\Lambda\zeta$ is analytic in the Lorentz parameters, such an equation has an analytic continuation to complex Λ_c, in some complex neighborhood of some real Λ, in particular of the identity, as long as ζ and $\Lambda_c\zeta$ belong to $\mathcal{T}_n$.

As a second step, one proves that eq. (3.4.8) extends to all complex Λ_c, provided ζ, $\Lambda_c\zeta \in \mathcal{T}_n$; in fact, in this case Bargmann's lemma[20] assures that there exists a continuous curve of proper complex Lorentz transformations $\{\Lambda_c(t), 0 \leq t \leq 1\}$, such that $\Lambda_c(0) = 1$, $\Lambda_c(1) = \Lambda_c$ and $\Lambda_c(t)\zeta \in \mathcal{T}_n$, for all $0 \leq t \leq 1$. Then, the unique analytic continuation is obtained by the standard method of successive continuations (all involving complex Λ_c in suitable neighborhoods of the identity) applied to a set of overlapping neighborhoods, covering the given curve.

Finally, if $w = \Lambda_c\zeta \notin \mathcal{T}_n$, one has to check that the analytic continuation given by eq. (3.4.7) is single-valued, i.e., that, if $w = \Lambda_1\zeta_1 = \Lambda_2\zeta_2$, ζ_1, $\zeta_2 \in \mathcal{T}_n$, then the two continuations $W_1(w) = W(\zeta_1)$ and $W_2(w) = W(\zeta_2)$ define the same function. In fact, $\zeta_2 = \Lambda_2^{-1}\Lambda_1\zeta_1 \equiv \Lambda_c\zeta_1$, $\Lambda_c \in L_+(C)$, so that the second step applies and $W(\zeta_1) = W(\zeta_2)$.

A very important property of the extended tube $\mathcal{T}_n^{ext}$ is that it contains real points, whereas $\mathcal{T}_n$ does not, since $\mathrm{Im}\,\zeta_j \in V_+ \not\ni 0$. The real points of $\mathcal{T}_n^{ext}$, called *Jost's points*, are characterized by Jost's theorem, according to which *a real point* $\zeta = \zeta_1, \ldots, \zeta_n$ *belongs to* $\mathcal{T}_n^{ext}$ *iff all the four-vectors of the form*

$$w = \sum_j \lambda_j \zeta_j, \quad \lambda_j \geq 0, \ \sum \lambda_j > 0, \tag{3.4.9}$$

are spacelike. This implies that all the ζ_j must be spacelike, and that the set of Jost points is not empty (e.g., $\zeta_j = \xi$, $j = 1, \ldots n$, $\xi^2 < 0$).

The proof is very simple for $n = 1$. If a point z belongs to $\mathcal{T}_1^{ext}$, there exists a $\zeta = \xi + i\eta \in \mathcal{T}_1$ and a $\Lambda \in L_+(C)$, such that $z = \Lambda\zeta$ and the reality (of z and therefore) of $z^2 = (\Lambda\zeta)^2 = \xi^2 - \eta^2 + 2i\,\xi\eta$ requires $\xi\eta = 0$. Since $\eta \in V_+$, this implies that ξ is spacelike, and therefore so is also z, i.e., eq. (3.4.9) holds. Conversely, if z is real and spacelike, then by a (real) Lorentz transformation it can be reduced to the form $(z_0, z_1, 0, 0)$, with $z_1 > |z_0|$, and then by the complex Lorentz transformation

$$z_0 \pm z_1 \to e^{\pm i\alpha}(z_0 \pm z_1) \equiv z_0' \pm z_1', \quad \sin\alpha > 0$$

is sent in $\mathcal{T}_1$, since $\mathrm{Im}\,z' = \sin\alpha\,(z_1, z_0, 0, 0)$, $z_1 > 0$, $z_1^2 > z_0^2$.

For $n > 1$, if $z = (z_1, \ldots, z_n) \in \mathcal{T}_n^{ext}$, then $z = \Lambda\zeta$, $\zeta = (\zeta_1, \ldots \zeta_n) \in \mathcal{T}_n$, for some complex Λ. For any set of λ_i as in eq. (3.4.9), let

$$w \equiv \sum \lambda_i z_i = \Lambda \sum \lambda_i \zeta_i \equiv \Lambda v.$$

Since each $\zeta_i \in \mathcal{T}_1$, and $\mathcal{T}_1$ is convex, $v \in \mathcal{T}_1$ and, by the above argument, the reality of w implies that v is spacelike and so is w, i.e., eq. (3.4.9) holds. For the converse, given a

---

[20] For the proof of the lemma, see Streater and Wightman, *PCT, Spin and Statistics, and All That*, pp. 67–70.

real $z = (z_1, \ldots, z_n)$, eq. (3.4.9) implies that in a suitable frame $z_i' = (z_i'^{(0)}, z_i'^{(1)}, 0, 0)$, with $z_i'^{(1)} > |z_i'^{(0)}|$, $\forall i$ (for the details of the proof, see Streater and Wightman, *PCT, Spin and Statistics, and All That*, p. 72); then the complex Lorentz transformation used in the $n = 1$ case sends all the z_i' to $\mathcal{T}_1$, and therefore $z' \in \mathcal{T}_n^{ext}$.

4.3 Locality and permuted extended analyticity

A larger analyticity domain containing the Euclidean points (see Chapter 5), is obtained as a consequence of locality.

By the above results, the Wightman functions $\mathcal{W}(x_1, \ldots, x_n) = W(\xi_1, \ldots, \xi_{n-1})$ have an analytic continuation $\mathcal{W}(z_1, \ldots, z_n)$ to the domain[21]

$$\sigma_n^{ext} = \{z;\ \zeta = (\zeta_1 = z_2 - z_1, \ldots, \zeta_{n-1}) \in \mathcal{T}_{n-1}^{ext}\}.$$

The permutation $\pi = P(j, j+1)$ which interchanges x_j and x_{j+1} induces the following transformation on the difference variables: $\zeta \to \pi(\zeta)$, with

$$\pi(\zeta)_k = \zeta_k, \quad k \neq j-1, j, j+1,$$

$$\pi(\zeta)_{j-1} = \zeta_j + \zeta_{j-1}, \quad \pi(\zeta)_j = -\zeta_j, \quad \pi(\zeta)_{j+1} = \zeta_{j+1} + \zeta_j \qquad (3.4.10)$$

and it is easy to see that the corresponding extended tubes $\mathcal{T}_{n-1}^{ext}$ and

$$\mathcal{T}_{n-1}^{ext,\pi} \equiv P(j, j+1)\,\mathcal{T}_{n-1}^{ext} = \{\pi(\zeta), \zeta \in \mathcal{T}_{n-1}^{ext}\}$$

have at least a neighborhood of a Jost point in common. For example, the point $\bar{\zeta}$ with

$$\bar{\zeta}_k = (0, b, 0, 0), \quad k \neq j-1, j, j+1, \quad \bar{\zeta}_{j-1} = (a, b, 0, 0),$$

$$\bar{\zeta}_j = (0, 0, \varepsilon, 0), \quad \bar{\zeta}_{j+1} = (-a, b, 0, 0), \quad 0 < |a| < b,$$

is a Jost point of both $\mathcal{T}_{n-1}^{ext}$ and of $\mathcal{T}_{n-1}^{ext,\pi}$, i.e., both $\bar{\zeta}$ and $\pi(\bar{\zeta})$ are Jost points of $\mathcal{T}_{n-1}^{ext}$. This property holds also for the points in a sufficiently small neighborhood of $\bar{\zeta}$.

Thus, the permuted Wightman function

$$\mathcal{W}_\pi(x_1, \ldots, x_n) \equiv \mathcal{W}(x_1, \ldots, x_{j-1}, x_{j+1}, x_j, \ldots, x_n) = W_\pi(\xi_1, \ldots, \xi_{n-1}),$$

which, as a consequence of the spectral condition and Lorentz covariance, has an analytic continuation $W_\pi(\zeta_1, \ldots, \zeta_{n-1})$ in the permuted tube $P(j, j+1)\mathcal{T}_{n-1}^{ext}$, by locality coincides with $W(\zeta_1, \ldots, \zeta_{n-1})$ in the neighborhood of $\bar{\zeta}$, which is a point of analyticity of both $\mathcal{T}_{n-1}^{ext}$ and $\mathcal{T}_{n-1}^{ext,\pi}$. Hence, $W_\pi(\zeta_1, \ldots, \zeta_{n-1})$ analytically continues $W(\zeta_1, \ldots, \zeta_{n-1})$ to a function which is analytic in $\mathcal{T}_{n-1}^{ext} \cup \mathcal{T}_{n-1}^{ext,\pi}$ and invariant under

---

[21] A real point $z \in \sigma_n^{ext}$ is characterized by the property that the corresponding ζ is a Jost point; clearly, as a consequence of eq. (3.4.9), all the differences $z_k - z_l = \zeta_l + \zeta_{l+1} + \ldots + \zeta_{k-1}$, $l < k$, are spacelike, briefly z is *totally spacelike*, but not every totally spacelike point z is a real point of σ_n^{ext}.

π (since so is in a neighborhood of $\bar{\zeta}$), so that the corresponding $\mathcal{W}(z_1, \ldots, z_n)$ is symmetric under the permutation π.

By repeating the argument for any permutation π, one obtains the analyticity of the Wightman functions $W(\zeta_1, \ldots, \zeta_{n-1})$ in the *permuted extended tube*

$$\mathcal{T}_{n-1}^{p,ext} = \cup_\pi \mathcal{T}_{n-1}^{ext,\pi}$$

and the analyticity and symmetry of the corresponding

$$\mathcal{W}(z_1, \ldots, z_n) = \mathcal{W}(\pi(z_1, \ldots, z_n)), \qquad z \in \sigma_n^{p.ext} \equiv \cup_\pi \sigma_n^{ext,\pi}. \tag{3.4.11}$$

In particular, for $\pi(z_1, \ldots, z_n) = (z_n, \ldots, z_1)$, one has

$$W(\zeta) = W(- \overleftarrow{\zeta}), \qquad \overleftarrow{\zeta} \equiv (\zeta_{n-1}, \ldots \zeta_1).$$

A very important property of the complex Lorentz group (which will allow the proof of the spin–statistics and PCT theorems) is that it connects the identity with the total inversion $\Lambda x = -x$, since

$$\Lambda(t) = \begin{pmatrix} \cosh(i\,t) & 0 & 0 & \sinh(i\,t) \\ 0 & \cos t & -\sin t & 0 \\ 0 & \sin t & \cos t & 0 \\ \sinh(i\,t) & 0 & 0 & \cosh(i\,t) \end{pmatrix} \in L_+(C), \quad \forall t,$$

and $\Lambda(\pi) = -1$. Quite generally, one has that $-1 = \Lambda(A = -1, B = 1)$ belongs to $SL(2, C) \times SL(2, C)$.

Thus, the extended tube contains all the points $\zeta = \xi + i\eta$, with $\xi \in \mathbf{R}^{4n}, \eta \in \Gamma = (V_+)^n$ as well as $-\zeta$, i.e., also all the points $\zeta = \xi + i\eta$, with $\eta \in -(V_+)^n$, and

$$W(\zeta) = W(-\zeta), \quad \forall \zeta \in \mathcal{T}_n^{ext}. \tag{3.4.12}$$

To illustrate the above analyticity domains, we consider the simplest case of the two-point function $\mathcal{W}(x_1, x_2) = W(\xi)$. The forward tube consists of the points $\zeta = \xi + i\eta$, with $\xi \in \mathbf{R}^4, \eta \in V_+$; the extended tube contains also the points $\zeta = \xi + i\eta, \eta \in -V_+$. By eq. (3.4.10), $\pi(\zeta) = -\zeta$, which is already in $\mathcal{T}_1^{ext}$. Therefore $\mathcal{T}_1^{p,ext} = \mathcal{T}_1^{ext}$ contains all the complex ζ except for the points ζ with $\zeta^2 \geq 0$, which correspond to the forward and backward cones (the Jost points are the real spacelike points $\zeta = \xi$, $\xi^2 < 0$). In the complex plane of the variable $w = \zeta^2$ one has analyticity except for the cut $w \geq 0$; the discontinuity across the cut gives the commutator function. It is instructive to check these analyticity properties for the two-point function of a free massive and massless scalar field.

4.4   Local structure of QFT

The analyticity properties discussed above imply that the knowledge of the Wightman functions of the fields localized in a given open set $\mathcal{O} \subset \mathbf{R}^4$, however small, completely determines the theory. In fact, if $\mathcal{F}(\mathcal{O})$ denotes the polynomial algebra of the fields smeared with test functions with support in $\mathcal{O}$, one has:

Theorem 4.1 *(Reeh–Schlieder) For any $\mathcal{O} \subset \mathbf{R}^4$, the vacuum is cyclic with respect to the algebra $\mathcal{F}(\mathcal{O})$.*

Moreover, if $A \in \mathcal{F}(\mathcal{O})$ and $A\,\Psi_0 = 0$, then $A = 0$.

Proof. The Fourier transform $\tilde{\Psi}_n(p, q)$ of the vector-valued distribution

$$\Psi_n(x_1, \xi) = \Psi_n(x_1, \xi_1, \ldots \xi_{n-1}) \equiv \varphi(x_1) \ldots \varphi(x_n)\Psi_0 \qquad (3.4.13)$$

vanishes unless $p_1, q_j \in \overline{V}_+$, so that its Laplace transform $\Psi_n(z, \zeta)$ is analytic in the forward tube $\mathcal{T}_n$. Now, for any vector Ψ orthogonal to $\mathcal{F}(\mathcal{O})\,\Psi_0$, one has $(\Psi, \Psi_n(x_1, \xi)) = 0$, for $x_1, \ldots, x_n \in \mathcal{O}$. This implies the vanishing of the analytic function $(\Psi, \Psi_n(z, \zeta))$ in an open set of the boundary and, by the edge of the wedge theorem,[22] its vanishing everywhere, i.e., $\forall\, x_1, \ldots x_n \in \mathbf{R}^4$. Then, since Ψ_0 is cyclic for $\mathcal{F}(\mathbf{R}^4)$, $\Psi = 0$.

Moreover, if $A \in \mathcal{F}(\mathcal{O})$ and $A\,\Psi_0 = 0$, then for any open set $\mathcal{O}'$ spacelike with respect to $\mathcal{O}$, one has

$$0 = (\mathcal{F}(\mathbf{R}^4)\,\Psi_0, \mathcal{F}(\mathcal{O}')\,A\,\Psi_0) = (\mathcal{F}(\mathbf{R}^4)\,\Psi_0, A\mathcal{F}(\mathcal{O}')\,\Psi_0) = 0,$$

and, by the density of $\mathcal{F}(\mathcal{O}')\,\Psi_0$, $A = 0$.

4.5   Quantization from spectral condition

Historically, the quantization of field equations was formally achieved through canonical quantization, just as in the quantization of finite dimensional systems. Such a procedure makes use of canonical fields, the so-called unrenormalized fields, which are related to the renormalized fields by a divergent (wave function) renormalization; thus, as discussed in Chapter 2, Section 4, the equal-time (canonical) commutation relations for interacting renormalized fields are spoiled, and cannot be used as a quantization condition for renormalized fields.

A natural question is whether and where a quantization condition is contained in the general requirements QM1–3 and R1–2, equivalently in the properties W1–W6. To this purpose, one may show that, as a consequence of the spectral condition, the field algebra cannot be commutative. Thus, W1–W6 imply quantization.

Since, as we shall see in the next chapter, W1–W6 imply that fields obeying free field equations satisfy canonical commutation relations, W1–W6 qualify as the way of defining *quantum* fields by providing the correct *substitute of canonical quantization* in the interacting case.

Proposition 4.1 *The spectral condition implies that the field algebra cannot be commutative.*

Proof. For simplicity, we consider the case of a hermitian scalar field $\varphi(x)$. If $[\varphi(x), \varphi(y)] = 0$, the two-point function satisfies $\tilde{W}(q) = \tilde{W}(-q)$ and the spectral condition requires $q = 0$, i.e., $W(x - y) = const$. As argued in the proof of Theorem 4.1 in Chapter 3, this implies a trivial theory.

---

[22] The edge of the wedge theorem is an extension of the Schwarz's reflection principle discussed in Chapter 1, Section 2.4; see Streater and Wightman, *PCT, Spin and Statistics, and All That*, Theor. 2-16.

The connection between spectral condition and quantization should not look surprising, since the need of non-commuting canonical variables q_i, p_i, for the description of atomic systems, could be argued solely by the requirement that the energy of the hydrogen atom, described by the Hamiltonian $H = p^2/2m - e^2/|q|$, be bounded from below. If the kinetic energy commutes with the potential energy, the spectrum of their sum is the sum of their spectra, and is inevitably unbounded below.

4

General non-perturbative results and examples

1 Free evolution implies canonical quantization

Because of the problems discussed in Chapter 2, Section 4, the non-perturbative (mathematical) foundations of QFT cannot rely on canonical quantization for the renormalized fields and in general W1–W6 qualify as a substitute of it.

In this section we will show that in the case of free evolution W1–W6 imply canonical quantization. As before, for simplicity we consider the case of a scalar hermitian field and show that if

$$(\Box + m^2)\varphi(x) = 0, \qquad (4.1.1)$$

then W1–W6 imply that

$$[\varphi(\mathbf{x}, t), \partial_0 \varphi(\mathbf{y}, t)] = i\delta(\mathbf{x} - \mathbf{y});$$

namely, one has the standard free field quantization (Jost and Schroer theorem[1]).

In fact, eq. (4.1.1) implies $\operatorname{supp} \tilde{\varphi}(p) \subseteq \{p; p^2 = m^2\}$, and therefore one can decompose the field as the sum of $\tilde{\varphi}^{\pm}(p)$, with support in the upper/lower hyperboloid, respectively.[2] The spectral condition implies $\varphi^- \Psi_0 = 0$. Furthermore, the (improper) state $\tilde{\varphi}^-(p)\,\tilde{\varphi}^+(q)\,\Psi_0$ has total momentum $p_+ \equiv p + q$, with $p^2 = q^2 = m^2$, $p_0 < 0$, $q_0 > 0$. p_+ is either spacelike or zero, since in the frame in which $\mathbf{p} = 0$, one has $p_+^2 = 2m^2 - 2q_0|p_0| \leq 0$. By the spectral condition, only $p_+ = 0$ is allowed, and by the uniqueness of the vacuum,

$$\varphi^-(x)\,\varphi(y)^+\,\Psi_0 = <\varphi^-(x)\,\varphi^+(y)>\,\Psi_0 = <\varphi(x)\,\varphi(y)>\,\Psi_0.$$

Putting $i\Delta^+(x - y) \equiv <\varphi(x)\,\varphi(y)>$, $\Delta(x) \equiv \Delta^+(x) - \Delta^+(-x)$, one has

$$[\varphi(x),\,\varphi(y)]\,\Psi_0 = i\Delta(x - y)\Psi_0 + [\varphi^+(x),\,\varphi^+(y)]\,\Psi_0. \qquad (4.1.2)$$

[1] R. Jost, Properties of the Wightman functions, in *Lectures on Field Theory and the Many-Body Problem*, E. R. Caianiello (ed.), Academic Press 1961, pp. 143–144.

[2] Such a decomposition is easily obtained if $m \neq 0$. In the case $m = 0$ a point singularity at the origin in momentum space can be excluded by the uniqueness of the vacuum and the condition $<\varphi> = 0$, which will be assumed in the sequel.

Now, the last term vanishes. In fact, since $\operatorname{supp} \tilde{\varphi}^+(p) \subseteq \overline{V}^+$, for any vector Ψ, $\operatorname{supp}(\Psi, \tilde{\varphi}^+(p_1)\tilde{\varphi}^+(p_2)\Psi_0) \subseteq \overline{V}^+ \cup \overline{V}^+$, so that the function

$$F(x,y) \equiv (\Psi, [\varphi^+(x), \varphi^+(y)]\Psi_0)$$

is the boundary value of an analytic function $F(z_1, z_2)$, $\operatorname{Im} z_i \in V^+$, which vanishes on the boundary when $(x - y)^2 < 0$, since, by locality, so do the other terms of eq. (4.1.2). Then, by the edge of the wedge theorem, $F(x,y) = 0$ (for any Ψ), and therefore

$$[\varphi^+(x), \varphi^+(y)]\Psi_0 = 0.$$

Finally, since $[\varphi(x), \varphi(y)] - i\Delta(x-y)$ is a local operator and it annihilates the vacuum, by Theorem 4.1 of the previous chapter, it is zero. W1–W6 dictate the explicit form of $\Delta^+(x)$, apart from a normalization factor, and canonical quantization follows.

It is easy and instructive to check the general properties of the Wightman functions, discussed in the previous chapter, in the case of a free scalar hermitian field. Without loss of generality, we take $<\varphi> = 0$.

We recall that

$$i\Delta^+(x) = (2\pi)^{-3} \int d^4p\, \theta(p_0)\, \delta(p^2 - m^2)\, e^{-ipx} = \tfrac{1}{2}(2\pi)^{-3} \int d\Omega_m(k)\, e^{-ikx}$$

The integral can be computed explicitly,[3] and in particular one has the following asymptotic behavior for large spacelike points $(m|\mathbf{x}| >> 1)$

$$i\Delta^+(\mathbf{x}, 0) \sim (m/(32\pi^2|\mathbf{x}|^3))^{1/2}\, e^{-m|\mathbf{x}|}\, (1 + O(1/m|\mathbf{x}|)),$$

i.e., an exponential decay of Yukawa type, in agreement with the discussion of the cluster property.

For a free massless scalar field, the analytic form of the two-point function can be easily computed, and one has

$$W(x) = -(2\pi)^{-2}\, (P(1/x^2) - i\pi\varepsilon(x_0)\, \delta(x^2)), \tag{4.1.3}$$

where ε denotes the antisymmetric step function. Its Laplace transform is

$$W(\zeta) = -(2\pi)^{-2}\, \zeta^{-2}, \quad \zeta = x + i\eta, \quad \eta \in V_+,$$

in agreement with the general results discussed above. It follows from eq. (4.1.3) that in this case the cluster property holds with a fall-off like the derivative of the Coulomb potential.

For a free hermitian scalar field, the n-point function is easily computed in terms of the two-point function; in fact, since $(\varphi^+)^* = \varphi^-$ one has

$$<\varphi(x_1)\ldots\varphi(x_n)> = <\varphi^-(x_1)\ldots\varphi(x_n)> = <[\varphi^-(x_1), \varphi(x_2)\ldots\varphi(x_n)]>$$

$$= \sum_j W(x_1 - x_j)\, \mathcal{W}(x_2, \ldots, \hat{x}_j, \ldots, x_n).$$

---

[3] J. Schwinger, Phys. Rev. **75**, 651 (1949), Appendix; N. N. Bogoliubov and D. V. Shirkov, *Introduction to the Theory of Quantized Fields*, Interscience 1959, Sect. 15.

Then, by using a recursive argument, for $n = $ odd the r.h.s. vanishes and for $n = 2m$ becomes

$$\sum_{pairs} W(x_{i_1} - x_{j_1}) \dots W(x_{i_m} - x_{j_m}),$$

where the sum is over all the ways of writing $1, \dots 2m$ as $i_1, \dots i_m$, $j_1 \dots j_m$, with $i_1 < i_2 < \dots < j_1 < \dots < j_m$.

2 Spin–statistics theorem

In the quantum description of atomic structures the Pauli exclusion principle plays a crucial role for the explanation of the Mendelejev's periodic table of the chemical elements,[4] so that an important foundational question is to possibly understand the origin of such a principle.[5]

The standard textbook discussion of the Pauli principle[6] involves two steps: the symmetry or antisymmetry of the N-particle wave functions of identical particles and the choice between the two alternatives according to the integer or half-integer spin, respectively.

Clearly, given a quantum system of N particles, the particles are identical iff all observables are invariant under permutations of the particle labels, briefly under particle exchange; then as states on the observables the states of N identical particles are symmetric under particle exchanges. Hence, *if N-particle pure states are identified by rays of $L^2(\mathbf{R}^{3N}, dx)$*, under a permutation P the wave function $\psi(x_1, \dots x_N)$ can only change by a phase $\varphi(P)$, which must therefore provide a one-dimensional representation of the group $\mathcal{P}$ of permutations. There are only two such representations: the trivial one $\varphi(P) = 1$ (symmetric case) or $\varphi(P) = \varepsilon(P)$, $\varepsilon(P) = $ the parity of P (antisymmetric case). The Pauli principle relates the choice between such alternatives to the spin of the particles.

In the quantum field theory framework, particles are associated to fields and the spin–statistics relations are encoded in the relations of the fields at spacelike separations. Thus, the transcription of symmetry or antisymmetry being the only possibilities for N-particles wave functions becomes the assumption that at spacelike separated points the fields either commute or anticommute. In the free case, the spin-statistics relation may be linked to the positivity of the energy (see Chapter 1, Section 4), but, apart from the limitation of the free case, this argument somewhat hides the crucial role of Lorentz covariance. In fact, as remarked by Wightman,[7] one can easily construct Euclidean invariant (non-relativistic) theories of particles with spin with the wrong spin-statistics connection.

---

[4] See e.g., R. P. Feynman, R. B. Leighton, and M. Sands, *The Feynman Lectures on Physics Quantum Mechanics*, Addison Wesley 1970, Sects. 4.7, 19.6.

[5] For a comprehensive review of the history of the Pauli exclusion principle and of the derivations of the spin–statistics theorem, see I. Duck and E. C. G. Sudarshan, *Pauli and the Spin-Statistics Theorem*, World Scientific 1997.

[6] P. M. A. Dirac, *The Principles of Quantum Mechanics*, Oxford University Press 1958, Chap. IX.

[7] A. S. Wightman, Am. J. Phys. **67**, 742 (1999).

The proof for the general case and the clarification of the crucial ingredients is provided by the spin–statistics theorem. The first step is the following:

Lemma 2.1 *Let φ, ψ be two fields whose Wightman functions satisfy W1–W6, with W4 suitably modified for anticommuting fields, then, if φ commutes (or anticommutes) with ψ at spacelike separations, it also does so with ψ^*.*

Proof. Given $f, g \in \mathcal{D}(\mathbf{R}^4)$, let a be a spacelike vector, such that the supports of f and of $g_a(x) \equiv g(x - a)$ are spacelike-separated. Then, if φ commutes or anticommutes with ψ, but does the converse with ψ^*, one has

$$0 \leq ||\psi(g_a)\,\varphi(f)\,\Psi_0||^2 = <\varphi(f)^*\psi(g_a)^*\,\psi(g_a)\varphi(f)>_0 =$$
$$= -<\varphi(f)^*\varphi(f)\,\psi(g_a)^*\,\psi(g_a)>_0.$$

Now, by the cluster property, when a goes to infinity in a spacelike direction, the right-hand side converges to

$$-<\psi(g)^*\psi(g)>_0<\varphi(f)^*\varphi(f)>_0 = -||\psi(g)\Psi_0||^2\,||\varphi(f)\,\Psi_0||^2 \leq 0,$$

in contradiction with the first inequality, unless $\varphi(f)$ or $\psi(g)$ annihilate the vacuum. Thus, by the Reeh–Schlieder theorem (Chapter 3, Section 4.4) either φ or ψ vanishes. It is worthwhile to stress the crucial role played by positivity, which may be in question in gauge theories (see Chapter 7).

The case of a scalar (not necessarily hermitian) field illustrates the simple idea and logic of the Spin-Statistics theorem, in the clearest possible way.

Theorem 2.2 *(Spin–statistics, scalar case) If a scalar (Wightman) field φ anticommutes at spacelike separations, then $\varphi\,\Psi_0 = 0$ and actually $\varphi = 0$ if all the fields either commute or anticommute (briefly if locality holds).*

Proof. If $W_1(\zeta)$, $W_2(\zeta)$ denote the analytic continuations of the two-point functions $<\varphi(x)\,\varphi^*(y)>_0$, $<\varphi^*(x)\,\varphi(y)>_0$, respectively, then, if at spacelike separations φ anticommutes with itself and therefore (by the above lemma) also with φ^*, one has at Jost points $\zeta = y - x$

$$W_1(\zeta) = -W_2(-\zeta). \tag{4.2.1}$$

On the other hand, by the invariance under the complex Lorentz group, eq. (3.4.12), one has $W_2(-\zeta) = W_2(\zeta)$, so that eq. (4.2.1) implies

$$W_1(\zeta) = -W_2(\zeta) \tag{4.2.2}$$

all throughout $\mathcal{T}_1^{ext}$, and, by letting Im $\zeta \to 0$ in V_+,

$$W_1(\xi) = -W_2(\xi), \quad \xi = y - x. \tag{4.2.3}$$

In conclusion, putting $\hat{f}(x) \equiv f(-x)$, $x' = -y$, $y' = -x$, one has

$$||\varphi(\hat{f})\Psi_0||^2 = \int dx\, dy\, W_2(y - x)\, \bar{f}(-x)\, f(-y) =$$

$$= \int dx'\, dy'\, W_2(y' - x')\, \bar{f}(y')\, f(x')$$

and by eq. (4.2.3),

$$||\varphi(f)^* \Psi_0||^2 + ||\varphi(\hat{f})\, \Psi_0||^2 = 0.$$

This implies $\varphi\, \Psi_0 = 0 = \varphi^*\, \Psi_0$ and, if locality holds for all fields, by the Reeh–Schlieder theorem, $\varphi = 0$.

For the general spin case we recall that a Poincaré covariant spinor field $\psi_{\alpha\dot{\beta}}$, $\alpha = \alpha_1 \ldots \alpha_j$, $\dot{\beta} = \dot{\beta}_1 \ldots \dot{\beta}_k$, transforms according to a finite-dimensional irreducible representation of $SL(2,C)$, eq. (3.1.2) with $S(A)$ given by one of the $\mathcal{D}^{(j/2,k/2)}(A)$; $j + k=$ even/odd corresponds to integer/half-integer spin, respectively (see Chapter 1, Section 6). The covariance of the Wightman functions under the Lorentz group now reads

$$W_\alpha(\xi) = S(A^{-1})_{\alpha\beta}\, W_\beta(\Lambda\xi), \qquad (4.2.4)$$

where, for brevity, α denotes the set of dotted and undotted indices appearing in the field product. Since $\mathcal{D}^{(j/2,k/2)}(A = -1) = (-1)^{j+k}$ and $\Lambda(A = -1) = 1$, eq. (4.2.4) implies

$$W_\alpha(\xi) = (-1)^{J+K}\, W_\alpha(\xi), \qquad (4.2.5)$$

where J and K denote the total number of, respectively, undotted and dotted indices in the set α. Then, if $W_\alpha \neq 0$,

$$J + K = \text{even}, \quad (-1)^J = (-1)^K.$$

Theorem 2.3   *(Spin–statistics general case) Let $\psi_{\alpha\dot{\beta}}$ be a spinor field transforming as $\mathcal{D}^{j/2,\,k/2}$, and therefore carrying an integer/half-integer spin, corresponding to $j + k = $ even/odd, then the wrong connection between spin and statistics, i.e., for integer/half-integer spin the field anticommutes/commutes at spacelike separations, implies that $\psi\, \Psi_0 = 0$ and, if locality holds, $\psi = 0$.*

Proof. As before, we denote by $W_{1\alpha}$, $W_{2\alpha}$ the analytic continuations of the two-point functions $< \psi_{\gamma\dot{\beta}}\, \psi^*_{\gamma\dot{\beta}} >$ and $< \psi^*_{\gamma\dot{\beta}}\psi_{\gamma\dot{\beta}} >$, respectively, with α denoting the set of undotted and dotted indices. Thus, for ζ real and spacelike (Jost point), the wrong connection between spin and statistics gives the generalization of eq. (4.2.1):

$$W_{1\alpha}(\zeta) + (-1)^{(j+k)}\, W_{2\alpha}(-\zeta) = 0. \qquad (4.2.6)$$

The $\mathcal{D}^{(j/2,k/2)}$ can be analytically continued to $SL(2,C) \times SL(2,C)$, corresponding to the complex Lorentz group, with the pair $(A, \bar{A})$ replaced by (A, B) in eq. (1.6.6). Then, the covariance under the complex Lorentz group, for $\Lambda(A = -1, B = 1) = -1 \in L_+(C)$, gives

$$W_{2\,\alpha}(-\zeta) = W_{2\,\alpha}(\Lambda(-1,1)\zeta) = S(-1,1)_{\alpha\,\beta}W_{2\,\beta}(\zeta) = (-1)^J\,W_{2\,\alpha}(\zeta).$$

Hence, since $\psi^*_{\gamma\,\dot\beta} = \psi_{\dot\gamma\,\beta}$, so that $J = j + k$, one has

$$W_{2\,\alpha}(-\zeta) = (-1)^{j+k}\,W_{2\,\alpha}(\zeta). \tag{4.2.7}$$

Thus, by combining eqs. (4.2.6) and (4.2.7) one has $W_{1\,\alpha}(\xi) = -W_{2\,\alpha}(\xi)$, and the argument proceeds as before.

It is worth adding a few comments concerning the argument on the symmetry and antisymmetry alternative, in order to make clear that this is not the only possibility, and that other statistics beyond the fermionic and the bosonic ones are possible.

By definition of identical particles, the permutation group leaves the observables pointwise-invariant, and therefore it plays the role of a (non-abelian) compact gauge group $\mathcal{G}$.

Thus, the problem of characterizing the physical states of identical particles is part of the general problem of understanding the physical consequences of the existence of a compact gauge group. Since by definition a gauge group is not seen by the observables, its only possible effect is to contribute to the classification of the representations of the algebra of observables $\mathcal{A}$.

Since by its very definition the identification of a gauge group $\mathcal{G}$ requires the introduction of a larger algebra $\mathcal{F} \supset \mathcal{A}$, called *field algebra*, on which $\mathcal{G}$ acts as a non-trivial group of automorphisms α_g, $g \in \mathcal{G}$, a (possible) strategy for understanding the physical effects of $\mathcal{G}$ is to look for the representations of $\mathcal{A}$ contained in the irreducible (Hilbert space) representations of $\mathcal{F}$.

Now, pure states on $\mathcal{A}$ (which are obviously gauge-invariant, since $(\alpha_g^*\omega)(A) \equiv \omega(\alpha_g(A)) = \omega(A)$) need not be pure states on $\mathcal{F}$, and therefore they need not to be represented by rays in the representation space $\mathcal{H}$ of an irreducible representation of $\mathcal{F}$. Actually, each irreducible representation $\pi_{\mathcal{G}}$ of $\mathcal{G}$ contained in $\mathcal{H}$ defines a factorial (i.e., with trivial center) representation of $\mathcal{A}$, which decomposes into equivalent representations. In fact, if $\Psi \in \mathcal{H}$ is a vector belonging to a unitary irreducible representation $\pi_{\mathcal{G}}$, the *mixed* state Ω on $\mathcal{F}$ defined by

$$\Omega(F) \equiv N \sum_{g \in \mathcal{G}} (U_g\Psi,\, F\,U_g\Psi),$$

with N a normalization constant fixed by the condition that $\Omega(\mathbf{1}) = 1$, defines a unique pure state ω on $\mathcal{A}$.

Thus, a pure state ω on $\mathcal{A}$ is given by a (not necessarily one-dimensional!) *subspace* of $\mathcal{H}$ carrying an irreducible representation $\pi_{\mathcal{G}}$ of $\mathcal{G}$; therefore, the only labeling of ω induced by the gauge group $\mathcal{G}$ is through the group invariants which identify $\pi_{\mathcal{G}}$.

For the description of N identical particles, one can take as a field algebra $\mathcal{F}$ the Heisenberg (or the Weyl) algebra generated by N coordinates $q_1, \ldots q_N$ and momenta $p_1, \ldots p_N$. Clearly, the observable algebra is generated by the functions of the canonical variables, which are symmetric under permutations. A (unitary) irreducible representation of the permutation group $\mathcal{P}$ may not be one-dimensional and, nevertheless, it identifies a permutation-invariant pure state on $\mathcal{A}$. Thus, a state on $\mathcal{A}$ is not necessarily described by a wave function, nor by a ray in $L^2(\mathbf{R}^{3N})$.

The irreducible representations of $\mathcal{P}$ are identified by the so-called Young tableaux,[8] which are labeled by the invariants (d, d'), with d, d' denoting the number of rows and columns respectively. Thus, such invariant pairs classify the irreducible representations of $\mathcal{A}$; *this is the way the gauge group shows up at the physical level.*

A state ω on $\mathcal{A}$ labeled by (d, d') defines a *parastatistics* of bosonic order d and of fermionic order d'. A para-boson (para-fermion) of order d (d') is such that the corresponding N-particle states are described by all the Young tableaux with d rows (d' columns).

A natural question is whether the above-mentioned strategy of looking for the representations of $\mathcal{G}$, defined by the representations of the larger field algebra $\mathcal{F}$, allows us to reach all the possible (physically relevant) representations of $\mathcal{A}$.

It is a deep result of algebraic local quantum field theory that in three space dimensions all the representations of $\mathcal{A}$ defined by states satisfying the general condition of localizability, in the DHR sense (see Chapter 3, Section 1, eq. (3.1.4)), are labeled by the invariants d, d' and can be obtained by using a larger field algebra $\mathcal{F}$, according to the strategy discussed above. Furthermore, a parastatistics of order p can be described by ordinary Bose and Fermi statistics, by introducing a hidden degree of freedom or, equivalently, an unbroken (global) gauge group.[9]

This structure becomes very important in gauge quantum field theory, where the $SU(3)$ color group is believed to be an unbroken (compact) gauge group; a quark state on the observables will then be described by a mixed state on the field algebra, corresponding to an irreducible representation of $SU(3)$ and therefore labeled by the $SU(3)$ invariants (λ, μ). The quark confinement is the statement that the only physically realized representations of the observable algebra are those corresponding to the (trivial) one-dimensional representation of $SU(3)$.

3 PCT theorem

The symmetry PCT corresponds to the product of space inversion P, $(\mathbf{x} \to -\mathbf{x})$, charge conjugation C, (charge $\to -$ charge), and time reversal T, $(t \to -t)$. Thus, its validity implies that each particle has its own antiparticle, with the same mass and spin.

In order to make clear the logic of the proof of the PCT theorem, we shall consider the case of a scalar field $\varphi(\mathbf{x}, t)$; then, the space inversion P, the charge conjugation C, and the time reversal T are described by the following substitution rules:

$$P : \varphi(\mathbf{x}, t) \to \varphi(-\mathbf{x}, t), \quad C : \varphi(\mathbf{x}, t) \to \varphi^*(\mathbf{x}, t), \tag{4.3.1}$$

$$T : \varphi(\mathbf{x}, t) \to \varphi^*(\mathbf{x}, -t), \quad PCT : \varphi(\mathbf{x}, t) \to \varphi(-\mathbf{x}, -t). \tag{4.3.2}$$

---

[8] See, e.g., A. O. Barut and R. Rączka, *Theory of Group Representations and Applications*, World Scientific 1986, Chap. 7, Sect. 5.

[9] R. Haag, *Local Quantum Physics*, Springer 1996, Chap. IV; K. Drül, R. Haag and J. Roberts, Comm. Math. Phys. **18**, 204 (1970); S. Doplicher and J. E. Roberts, Comm. Math. Phys. **131**, 51 (1990).

Quite generally, if a substitution rule $A \to \hat{A}$ is implemented by a unitary operator U, one has

$$(\Psi, \hat{A}\,\Psi) = (U\Psi, A\,U\Psi) = (\Psi, U^* A\,U\,\Psi), \quad \text{i.e.,} \quad \hat{A} = U^* A U,$$

whereas, if it is implemented by an antiunitary operator θ, one has

$$(\Psi, \hat{A}\,\Psi) = (\theta\Psi, A\,\theta\,\Psi) = (\theta\Psi, \theta\theta^* A\theta\Psi) = (\Psi, (\theta^* A\,\theta)^*\,\Psi), \qquad (4.3.3)$$

since $\theta^*\theta = 1 = \theta\theta^*$, $(\theta\Psi, \theta\,\Phi) = (\Phi, \Psi)$. Thus, $\hat{A} = \theta^* A^* \theta$.

As a consequence of the spectral condition, PCT must be described by an antiunitary operator.[10] In fact, if it is implemented by a unitary operator V, one has for the time translation operator $U(t) = e^{-iHt}$,

$$\widehat{U(t)} = U(-t) = V^* U(t)\,V, \quad \text{i.e.,} \quad V^* H V = -H,$$

so that if Ψ satisfies the spectral condition, i.e., $(\Psi, H\,\Psi) \geq 0$, $V\,\Psi$ violates it.[11] In the anti-unitary case, one has instead

$$\widehat{U(t)} = U(-t) = \theta^* U(t)^* \theta, \quad \text{i.e.,} \quad \theta^* H\,\theta = H.$$

The group law relation $U(a) \to \hat{U}(a) = U(-a)$, the uniqueness of the translationally invariant state and antiunitarity, eq. (4.3.3), imply $U(a)\,\theta\,\Psi_0 = \theta\,U(-a)^*\,\Psi_0 = \theta\,\Psi_0$, i.e., $\theta\,\Psi_0 = e^{i\alpha}\,\Psi_0$; the phase factor can be absorbed in the definition of θ (yielding also $\theta^*\,\Psi_0 = \Psi_0$).

PCT is a symmetry of the theory of a charged scalar field if the substitution rule (4.3.2) is implemented by an anti-unitary operator θ, such that $\theta^*\varphi(x)\theta = \varphi^*(-x)$. Therefore, PCT symmetry is equivalent to the following invariance property of the Wightman functions:

$$\mathcal{W}(x_1, \ldots, x_n) = (\theta\Psi_0, \theta\theta^*\varphi(x_1)\theta\theta^* \ldots \theta\theta^*\varphi(x_n)\theta\theta^*\Psi_0) =$$

$$= \overline{(\Psi_0, \theta^*\varphi(x_1)\theta \ldots \theta^*\varphi(x_n)\theta\Psi_0)} = \overline{(\Psi_0, \varphi^*(-x_1) \ldots \varphi^*(-x_n)\Psi_0)} =$$

$$= (\Psi_0, \varphi(-x_n) \ldots \varphi(-x_1)\Psi_0) = \mathcal{W}(-x_n, -x_{n-1}, \ldots, -x_1), \qquad (4.3.4)$$

---

[10] The occurrence (or non-occurrence) of the adjoint in the right-hand side of eqs. (4.3.1) and (4.3.2) is compatible with the corresponding (possible) symmetry being described by a unitary (or anti-unitary) operator. Actually, P and C may be implemented by unitary operators (if they are not broken); e.g.,

$$\hat{\varphi}(f) = \varphi^*(f) = U(C)^* \varphi(f)\,U(C),$$

whereas in the anti-unitary case one would have $\varphi^*(f) = \theta^*\varphi(f)^*\theta$.

[11] The transformation law $U(t) \to \hat{U}(t) = U(-t)$ under time reversal T and under PCT follows from the group relations of the corresponding elements α_t, g_T of the extended Poincaré group $g_T\,\alpha_t\,g_T = \alpha_{-t}$; see A. S. Wightman, L'invariance dans la mécanique quantique relativiste, in *Dispersion Relations and Elementary Particles*, Les Houches 1960, C. De Witt and R. Omnes eds., Wiley 1960, pp. 189–94.

i.e., $(\xi = (\xi_1, \dots \xi_{n-1}))$

$$W(\xi) = W(\overleftarrow{\xi}), \qquad \overleftarrow{\xi} \equiv \xi_{n-1}, \dots \xi_1. \tag{4.3.5}$$

In the non-perturbative (Wightman) approach to QFT, the validity of the *PCT* symmetry, and in particular the *existence of antiparticles*, is shown to be a consequence of covariance, spectral condition, and locality; actually, as proved below, if covariance and spectral condition hold, *PCT* symmetry is equivalent to the property of weak local commutativity (WLC), namely, that at the Jost points $\xi = \xi_1, \dots \xi_n$

$$W(\xi) = W(-\overleftarrow{\xi}), \tag{4.3.6}$$

a weaker condition than locality. Such equality extends to the analytic functions, i.e., $W(\zeta) = W(-\overleftarrow{\zeta})$.

Theorem 3.1 *(PCT theorem for a neutral scalar field) If a neutral scalar field satisfies covariance, spectral condition, and locality, then the corresponding Wightman functions are PCT-symmetric.*

Proof. Since the total inversion belongs to the complex Lorentz group, one has, in the extended tube, $W(\zeta) = W(-\zeta)$, and by WLC, eq. (4.3.6),

$$W(\zeta) = W(-\zeta) = W(\overleftarrow{\zeta}). \tag{4.3.7}$$

Then, going to the boundary $\mathrm{Im}\ \zeta_i \to 0$ in V_+, one obtains eq. (4.3.5), i.e., *PCT* invariance.

4 Appendix: PCT theorem for spinor fields

In order to realize P and T as linear transformations on spinors, as in the Dirac case, it is convenient to introduce a two-component spinor

$$\phi \equiv \begin{pmatrix} \xi_{\alpha\dot\beta} \\ \eta_{\dot\alpha\beta} \end{pmatrix} = \begin{pmatrix} \psi_1 \\ \psi_2 \end{pmatrix},$$

where $\eta_{\dot\alpha\beta}$ transforms under $SL(2,C)$ as the conjugate spinor $\xi^*_{\alpha\dot\beta} = \xi_{\dot\alpha\beta}$.[12] Thus, for a general spinor field the substitution rules are

$$P : \begin{pmatrix} \xi_{\alpha\dot\beta} \\ \eta_{\dot\alpha\beta} \end{pmatrix} (\mathbf{x}) \to \varepsilon_P \begin{pmatrix} 0 & (-1)^j \sigma \otimes \dots \sigma \\ (-1)^k \sigma \otimes \dots \sigma & 0 \end{pmatrix} \begin{pmatrix} \xi_{\alpha\dot\beta} \\ \eta_{\dot\alpha\beta} \end{pmatrix} (-\mathbf{x}), \tag{4.4.1}$$

---

[12] This is easily seen for the coordinate spinor $x_{\alpha\dot\beta} = (X)_{\alpha\dot\beta}$, $X = \sum x^\mu \tau^\mu$, $x_{\dot\alpha\beta} = (\bar X)_{\dot\alpha\beta}$, since, e.g., the space inversion reads $X \to \sigma \bar X \sigma^{-1}$, $\sigma = i\tau_2$, i.e.,

$$\begin{pmatrix} x_{\alpha\dot\beta} \\ x_{\dot\alpha\beta} \end{pmatrix} \to \begin{pmatrix} 0 & \sigma \otimes \sigma \\ \sigma \otimes \sigma & 0 \end{pmatrix} \begin{pmatrix} x_{\alpha\dot\beta} \\ x_{\dot\alpha\beta} \end{pmatrix},$$

where $\sigma \otimes \sigma$ means that the first σ acts on the first index and the second on the second. For the general case, see R. F. Streater and A. S. Wightman, 1964, Sect. 1.3.

$$C : \begin{pmatrix} \xi_{\alpha\dot\beta} \\ \eta_{\dot\alpha\beta} \end{pmatrix} \to \begin{pmatrix} 0 & 1 \\ 1 & 0 \end{pmatrix} \begin{pmatrix} \xi^*_{\alpha\dot\beta} \\ \eta^*_{\dot\alpha\beta} \end{pmatrix},$$
(4.4.2)

$$T : \begin{pmatrix} \xi_{\alpha\dot\beta} \\ \eta_{\dot\alpha\beta} \end{pmatrix}(\mathbf{x},t) \to \begin{pmatrix} \sigma \otimes \ldots \sigma & 0 \\ 0 & \sigma \otimes \ldots \sigma \end{pmatrix} \begin{pmatrix} \xi^*_{\alpha\dot\beta} \\ \eta^*_{\dot\alpha\beta} \end{pmatrix}(\mathbf{x},-t),$$
(4.4.3)

where $\varepsilon_P = i$ for $j + k = $ odd, $\varepsilon_P = 1$ for $j + k = $ even, the tensor product $\otimes$ involves $j + k$ factors $\sigma \equiv i\sigma_2$ (σ_2 being the second Pauli matrix), acting on the corresponding dotted or undotted spinor index. In conclusion,

$$PCT : \begin{pmatrix} \xi_{\alpha\dot\beta} \\ \eta_{\dot\alpha\beta} \end{pmatrix}(\mathbf{x},t) \to \bar\varepsilon_P \begin{pmatrix} (-1)^j & 0 \\ 0 & (-1)^k \end{pmatrix} \begin{pmatrix} \xi_{\alpha\dot\beta} \\ \eta_{\dot\alpha\beta} \end{pmatrix}(-\mathbf{x},-t).$$
(4.4.4)

Thus, PCT does not mix the two components ξ, η, and if it is a symmetry of the theory, there exists an antiunitary operator θ such that, by eqs. (4.3.3) and (4.4.4), for each component

$$\theta^* \psi_\alpha(\mathbf{x},t)\,\theta = (-1)^j\, i^{F(\psi)}\, \psi_\alpha^*(-\mathbf{x},-t),$$
(4.4.5)

where, for simplicity, α denotes the set of undotted and dotted indices of the (one-component) spinor ψ, $j = j(\psi)$ is the number of undotted indices, and $F(\psi) = 0$ if $j + k$ is even and $F(\psi) = 1$ otherwise; F has the meaning of the fermion number modulo 2. As before, the PCT invariance of the vacuum gives the following symmetry properties of the Wightman functions,

$$< \psi_{\alpha_1}(x_1) \ldots \psi_{\alpha_n}(x_n) > = i^F\, (-1)^J\, < \psi_{\alpha_n}(-x_n) \ldots \psi_{\alpha_1}(-x_1) >,$$
(4.4.6)

since, by eq. (4.2.5), $(-1)^J = (-1)^K$, and $F \equiv$ the sum of the $F(\psi_{\alpha_j}) = J + K \bmod 2$ may be taken even ($J + K$ is even by eq. (4.2.5)). By adopting a multi-index notation, we shall write eq. (4.4.6) as

$$W_\alpha(\xi) = i^F\, (-1)^J\, W_{\overleftarrow{\alpha}}(\overleftarrow{\xi}).$$
(4.4.7)

Theorem 4.1 *(PCT theorem general case) If a spinor field satisfies covariance, spectral condition, and locality, then the corresponding Wightman functions are PCT symmetric, i.e., eq. (4.4.7) holds.*

Proof. In a real neighborhood of a Jost point ζ, locality gives

$$W_\alpha(\zeta) = i^F\, W_{\overleftarrow{\alpha}}(-\overleftarrow{\zeta}),$$
(4.4.8)

since the reordering of the F fermions yields the factor

$$(-1)^{(F-1)+(F-2)+\ldots+1} = (-1)^{F(F-1)/2} = i^{F(F-1)} = i^{-F} = i^F,$$

$F = J + K$ being even. Now, invariance under the complex Lorentz group for $\Lambda(A = -1, B = 1)$ gives

$$W_\alpha(\zeta) = (-1)^J\, W_\alpha(-\zeta)$$

and eq. (4.4.7) follows.

5   Haag theorem

The impossibility of defining the interaction picture for relativistic interacting fields follows easily from the general principles discussed above.

Theorem 5.1   *(Generalized Haag theorem) Let $\varphi(\mathbf{x}, t)$ be a relativistic scalar field with vacuum state Ψ_0, satisfying*
i) at any time t is unitarily equivalent to a free field $\varphi_I(\mathbf{x}, t)$ of mass m, with corresponding Fock vector $\Psi_F(t)$

$$\varphi(\mathbf{x}, t) = V^*(t)\,\varphi_I(\mathbf{x}, t)\,V(t) \tag{4.5.1}$$

ii) $\Psi_0 = V(t)^*\Psi_F(t)$, then φ is a free field of mass m.

Proof. For any spacelike point x there is a Lorentz transformation $\Lambda\colon x \to x' = \Lambda x$ such that $x_0' = 0$; therefore, for $x - y$ spacelike, by Lorentz invariance, the two-point function of φ satisfies

$$W(y - x) = W(\mathbf{y}' - \mathbf{x}', 0) = (\Psi_F(t), \varphi_I(\mathbf{x}', t)\,\varphi_I(\mathbf{y}', t)\,\Psi_F(t)) =$$

$$= i\Delta^+(\mathbf{x}' - \mathbf{y}', 0; m^2) = i\Delta^+(x - y; m^2),$$

since Δ^+ is Lorentz-invariant. Thus, the two-point function coincides with the free two-point function at spacelike (Jost) points and therefore everywhere, by the analyticity properties of the Wightman functions. Then, $\|(\square + m^2)\varphi(x)\,\Psi_0\| = 0$ and by locality

$$(\square + m^2)\varphi(x) = 0,$$

i.e., φ is a free field of mass m (see Section 1).

As an application, we discuss the implication on the existence of the interaction picture in the theory of a relativistic scalar field φ with a polynomial $\mathcal{P}(\varphi)$ interaction. This means that at any time t, $\varphi(\mathbf{x}, t)$, $\partial_0\,\varphi(\mathbf{x}, t)$ is a well-defined irreducible set of operators unitarily equivalent to a free field $\varphi_I(\mathbf{x}, t)$ and its conjugate momentum $\partial_0\varphi_I(\mathbf{x}, t)$,

$$\varphi(\mathbf{x}, t) = V^*(t)\,\varphi_I(\mathbf{x}, t)\,V(t), \tag{4.5.2}$$

$$\partial_0\,\varphi(\mathbf{x}, t) = V^*(t)\,\partial_0\,\varphi_I(\mathbf{x}, t)\,V(t), \quad V^*(t) = e^{iHt}e^{-iH_0t}. \tag{4.5.3}$$

In fact, the time derivative of eq. (4.5.2) gives

$$\partial_0\,\varphi(\mathbf{x}, t) = V^*(t)\,\partial_0\,\varphi_I(\mathbf{x}, t)\,V(t) + iV^*(t)\,[\,(H - H_0)_I(t),\,\varphi_I(\mathbf{x}, t)\,]\,V(t),$$

and the second term on the right-hand side vanishes, since $H_{int} = H - H_0$ is only a function of $\varphi(t)$, so that $(H_{int})_I(t)$ is only a function of $\varphi_I(t)$ and commutes with it. Now, denoting by $U_i(\mathbf{a}, R)$, $i = 1, 2$, $\mathbf{a} \in \mathbf{R}^3$, $R \in SO(3)$, the two representations of the Euclidean group under which φ and φ_I, respectively, transform covariantly, one has that $U_1^*(\mathbf{a}, R)\,V^*(t)\,U_2(\mathbf{a}, R)\,V(t)$ commutes with the irreducible set of operators $\varphi(t)$, $\partial_0\,\varphi(t)$ and by irreducibility is a multiple of the identity $\omega(\mathbf{a}, R, t)$. Actually,

$\omega(\mathbf{a}, R, t)$ is independent of the group element $(\mathbf{a}, R)$ (because any continuous one-dimensional representation of the euclidean group is trivial). Thus, $\omega(\mathbf{a}, R, t) = \omega(0, \mathbf{1}, t) = 1$ and $U_2 V(t) = V(t) U_1$; hence, by the invariance of the vacuum under the Euclidean group,

$$U_2 V(t) \Psi_0 = V(t) U_1 \Psi_0 = V(t) \Psi_0.$$

Since the Fock vacuum Ψ_{0F} is the only state invariant under $U_2(\mathbf{a}, R)$, condition ii) holds and φ is a free field of mass m.

The above argument shows that from a mathematical point of view, the interaction-picture representation does not exist if the basic principles hold. Of course, the interaction picture exists and may be used as long as the UV and volume cutoffs are finite; the corresponding perturbative expansion can suggest the counter-terms for the non-perturbative strategy.

It is worthwhile to stress that the triviality argument discussed above crucially uses the analyticity properties related to locality and Lorentz covariance. In fact, there are non-relativistic models in which the interaction picture is well defined and the ground state coincides with the Fock no-particle state (the so-called models with persistent vacuum, such as the Yukawa model discussed in Chapter 2, Section 3.3). The problem in the relativistic case is that a local Hamiltonian cannot annihilate the Fock state; a polynomial of local fields, even if Wick ordered, will always contain terms with no annihilation operators. This is related to the phenomenon of vacuum polarization arising from pair creation—an unavoidable feature associated with relativistic locality.

6   Ultraviolet singularities and non-canonical behavior

As discussed in Chapter 2, relativistic fields necessarily give rise to ultraviolet singularities, so that the product of fields at the same point is divergent. In the free field case the fields have well-defined restrictions at sharp times, as operator-valued distributions in $\mathbf{x}$, so that the canonical commutation relations are well defined. However, even in the free case the products of fields at the same point requires a regularization or subtraction. In Chapter 1, Sections 4 and 5, we discussed the definition of field bilinears, such as the energy–momentum and the charge. Another notable example, with relevant physical implications, is the definition of the current of a free Dirac field.

6.1   Schwinger terms in current commutators

The conserved (vector) current $j_\mu(\mathbf{x}, t)$ of a free Dirac field is formally defined by

$$j_\mu(\mathbf{x}, t) = \bar{\psi}(\mathbf{x}, t) \, \gamma_\mu \, \psi(\mathbf{x}, t), \tag{4.6.1}$$

but the multiplication of field operators at the same point is ill defined and a regularization or subtraction is necessary. In fact, the use of the canonical anticommutation relations (at equal times) for the fermion fields

$$\{ \psi_\alpha(\mathbf{x}, t), \, \psi_\beta^*(\mathbf{y}, t) \} = \delta_{\alpha\beta} \, \delta(\mathbf{x} - \mathbf{y}), \quad \{ \psi_\alpha(\mathbf{x}, t), \, \psi_\beta(\mathbf{y}, t) \} = 0, \tag{4.6.2}$$

in the above formal expression of the current, yields the following equations:

$$[\,j_0(\mathbf{x},0),\,\psi(\mathbf{y},0)\,] = -\delta(\mathbf{x}-\mathbf{y})\,\psi(\mathbf{y},0), \tag{4.6.3}$$

$$[\,j_0(\mathbf{x},0),\,j_i(\mathbf{y},0)\,] = 0. \tag{4.6.4}$$

The second equation is inconsistent with the general principles of QFT, and therefore the above formal expression of the current is not reliable.[13] In fact, putting

$$j_\mu(f) = \int d^3x\, j_\mu(\mathbf{x},0)\, f(\mathbf{x}), \quad f \in \mathcal{S}(\mathbf{R}^3),$$

by eq. (4.6.4) and current conservation, $\partial^i j_i(f) = -\partial_0 j_0(f)$, one has

$$0 = i\,[\,j_0(f),\,\partial^i j_i(f)\,] = -i\,[\,j_0(f),\,\partial_0 j_0(f)\,] = [\,j_0(f),\,[\,H,\,j_0(f)\,]\,].$$

By taking the vacuum expectation, one obtains

$$< j_0(f)\,H\,j_0(f) >= 0$$

and, since the spectrum of H is positive (by the spectral condition), the only possibility is $H\,j_0(f)\Psi_0 = 0$. By the uniqueness of the vacuum, it follows that $j_0(f)\,\Psi_0 = \lambda\Psi_0$, $\lambda \in \mathbf{C}$ and actually $\lambda = 0$, since, by Lorentz invariance, $(\Psi_0,\,j_\mu(f)\,\Psi_0) = 0$. Finally, by the Reeh–Schlieder theorem $j_\mu = 0$.

In order to find the correct regularization, we study the singularity of the products $j_\mu(x;\varepsilon) \equiv \bar{\psi}(x+\varepsilon)\gamma_\mu\,\psi(x)$ when $\varepsilon \to 0$. It is enough to consider the matrix elements of such an operator on the dense set of local states obtained by applying polynomials $\mathcal{P}$ of the Dirac fields to the vacuum, i.e., the Wightman functions of monomials of the form $\mathcal{P}_1 j_\mu(x;\varepsilon)\,\mathcal{P}_2$. The expansion of such a monomial in terms of Wick products gives rise to contractions between the Dirac fields of the current and those of the polynomials, which are non-singular in the limit $\varepsilon \to 0$. The only singular contraction is that between the Dirac fields of the current

$$< \bar{\psi}(x+\varepsilon)\gamma_\mu\,\psi(x) >= -i(4\pi^2)^{-1}\,\varepsilon_\mu/(\varepsilon^2)^2.$$

Thus,

$$j_\mu(x) = \lim_{\varepsilon\to 0}\,[\,j_\mu(x;\varepsilon)- < j_\mu(x;\varepsilon) >\,] =: \bar{\psi}\,\gamma_\mu\,\psi : (x) \tag{4.6.5}$$

is a well-defined operator-valued distribution (*point-splitting regularization*).[14] By the Poincaré covariance of $< \bar{\psi}(x+\varepsilon)\gamma_\mu\,\psi(x) >$, the current defined by eq. (4.6.5) transforms correctly under the Poincaré group; moreover, it is a local (Wightman) field.

---

[13] J. Schwinger, Phys. Rev. Lett. **3**, 296 (1959); T. Goto and I. Imamura, Prog. Theor. Phys. **14**, 196 (1955).

[14] For a general discussion of the Wick products of free fields, see A. S. Wightman and L. Gårding, Arkiv f. Fysik, **28**, p. 129 (1964), esp. pp. 173–84.

One can use the so defined current to compute the equal-time commutators (4.6.3) and (4.6.4). For the first, using that $\{\psi_a(x), \psi_\beta(y)\} = 0$, $\{\psi_\alpha(x), \bar{\psi}_\beta(y)\} = -iS_{\alpha\beta}(x-y) = (i\gamma_\mu\partial^\mu + m)_{\alpha\beta}\, i\,\Delta(x-y)$, one has

$$[\,j_0(x),\, \psi_\alpha(y)\,] = -i(S(y-x)\,\gamma_0)_{\alpha\beta}\,\psi_\beta(x). \qquad (4.6.6)$$

Such a commutator allows a restriction at equal times, giving eq. (4.6.3)

$$[\,j_0(x),\, \psi_\alpha(y)\,]|_{x_0=y_0} = -\delta(\mathbf{x}-\mathbf{y})\,\psi_\alpha(\mathbf{y}). \qquad (4.6.7)$$

For the commutator between currents, one obtains

$$[\,j_0(x),\, j_i(y)\,] = \bar{\psi}(x)(i\gamma_0 S(x-y)\gamma_i)\psi(y) - \bar{\psi}(y)(i\gamma_i\, S(y-x))\gamma_0\psi(x). \qquad (4.6.8)$$

Such an operator-valued distribution[15] does not allow an equal time restriction, indicating that the current $j_\mu(x)$ is a well-defined Wightman field, but it does not have a sharp time restriction. The equal-time divergence is a c-number, therefore given by its vacuum expectation value, which has the form of the ith derivative of $\delta(\mathbf{x}-\mathbf{y})$ with a divergent coefficient. Such a term, unexpected by naive canonical commutations, is called a *Schwinger term* and clearly resolves the paradox discussed above; the occurrence of such terms is also relevant for the discussion of the algebra of currents.[16]

The impossibility of restricting the current at sharp times can also be argued by using the spectral representation of the vacuum expectation of the current commutator:[17]

$$< [\,j_\mu(x),\, j_\nu(y)\,] >= (\Box g_{\mu\nu} - \partial_\mu\,\partial_\nu)\int d\rho(\mu^2)\,i\,\Delta(x-y;\mu^2). \qquad (4.6.9)$$

The spectral measure can be calculated by inserting a complete set of states (only electron–positron pairs contribute) or by evaluating the vacuum expectation of $j_\mu(x)\,j_\nu(y)$ with the help of Wick theorem. The result is (the well-known *vacuum polarization* amplitude at lowest order)

$$\rho(p^2) = \frac{1}{3(2\pi)^2}(1 + \frac{2m^2}{p^2})(1 - \frac{4m^2}{p^2})^{1/2} \to_{p^2\to\infty} \frac{1}{3(2\pi)^2}. \qquad (4.6.10)$$

---

[15] In fact, the powers $(\Delta^+(x))^n$ are well-defined distributions, but they do not have sharp time restrictions.

[16] There is a vast literature on the equal time commutators of currents, especially under the motivation of a quantum field theory treatment of current algebra and the derived sum rules. See the reviews: S. Adler and R. Dashen, *Current Algebra*, Benjamin 1968; S. Adler, Perturbation Theory and Anomalies, in *Lectures on Elementary Particles and Quantum Field Theory*, 1970 Brandeis Summer Institute in Theoretical Physics, S. Deser *et al.* (eds.), MIT press 1970; S. B. Treiman, R. Jackiw, B. Zumino, and E. Witten, *Current Algebra and Anomalies*, World Scientific 1985, especially the contribution by R. Jackiw, Field theoretical investigations in current algebra, p. 81. However, the strategies proposed in the literature for the definition of currents and their equal-time commutators do not always agree. The recipe of calculating the equal-time commutators before the point-splitting limit is in our opinion not correct, the only justification being that of simplifying the computations. Actually, the two limits in general do not commute. For a careful treatment see J. Langerholc, DESY Report No. 66/24; quoted in R. A. Brandt, Phys. Rev. **166**, 1795 (1968), Sect. 3.C.

[17] The derivation is the same as in the proof of Theorem 4.1 in Chapter 2, the form of the differential operators being dictated by Lorentz covariance and current conservation.

If the equal time restriction of eq. (4.6.9) existed for $\mu = 0$, $\nu = i$, one would get

$$\int d\mu^2 \rho(\mu^2)\, \partial_i\, \delta(\mathbf{x} - \mathbf{y}),$$

which is, however, linearly divergent. This Schwinger term, which marks the deviation from the canonical behavior is a finite term in low dimensions.

The definition of the current in the interacting case is a much harder problem, since one does not *a priori* know the required subtraction. It is expected that the vacuum subtraction suffices, but this makes the definition of the current rather implicit, since in the interacting case the knowledge of the current is needed for the determination of the vacuum. A possibility is to improve the point-splitting procedure. Putting

$$j_\mu(x, \varepsilon) = \overline{\psi}(x + \varepsilon)\gamma_\mu\psi(x - \varepsilon), \quad \varepsilon^2 < 0,$$

one may take the following limit $\varepsilon \to 0$, keeping $\varepsilon_\mu(-\varepsilon^2)^{-1/2}$ bounded,

$$j_\mu(x) = \lim_{\varepsilon \to 0} \tfrac{1}{2}\left[j_\mu(x, \varepsilon) + j_\mu(x, -\varepsilon)\right]. \tag{4.6.11}$$

It is easy to check that in the free case such a prescription gives the same result as the Wick ordering of eq. (4.6.5), and there are indications that it may work also in the interacting case.[18]

In quantum electrodynamics one has the further constraint of gauge invariance, which is not shared by the regularization of eq. (4.6.11). In fact, under the c-number gauge transformation of the Dirac field

$$\psi(x) \to e^{i\Lambda(x)}\,\psi(x),$$

the right-hand side of eq. (4.6.11) becomes (for ε small enough so that $\Lambda(x + \varepsilon) - \Lambda(x - \varepsilon) \sim 2\varepsilon_\mu\, \partial^\mu \Lambda(x) \equiv 2\varepsilon\partial\Lambda$)

$$e^{-i2\varepsilon\partial\Lambda(x)}\, j_\mu(x, \varepsilon) + e^{i2\varepsilon\partial\Lambda(x)}\, j_\mu(x, -\varepsilon) \sim j_\mu(x, \varepsilon) + j_\mu(x, -\varepsilon) +$$

$$-2i\varepsilon\partial\Lambda(x)[\,j_\mu(x, \varepsilon) - j_\mu(x, -\varepsilon)\,].$$

The last term does not vanish when $\varepsilon \to 0$, because the operator in square brackets is singular ($\sim \varepsilon_\mu/(\varepsilon^2)^2$ for $\varepsilon \to 0$). A gauge-invariant point-splitting has been proposed:

$$\lim_{\varepsilon \to 0} \tfrac{1}{2}\, Z(\varepsilon)\,[\,j_\mu(x, \varepsilon, A) + j_\mu(x, -\varepsilon, A)\,], \tag{4.6.12}$$

$$j_\mu(x, \varepsilon, A) = N(\overline{\psi}(x + \varepsilon)\gamma_\mu \exp\left(ie \int_{x-\varepsilon}^{x+\varepsilon} A_\nu(\xi)d\xi^\nu\right) \psi(x - \varepsilon)), \tag{4.6.13}$$

---

[18] Such an improved point-splitting regularization should be used in the case of free field, even with the definition of the current in terms of fermion field commutator (W. Heisenberg, Z. Physik **90**, 209 (1934); G. Källen, *Quantum Electrodynamics*, Springer 1972, p. 60), since

$$\tfrac{1}{2}[\overline{\psi}(x), \gamma_\mu\,\psi(y)] =: \overline{\psi}(x)\gamma_\mu\psi(y) : + \tfrac{1}{2}\mathrm{Tr}(\gamma_\mu\, S^{(1)}(y - x)),$$

where $S^{(1)}(x) = -i(S^+(x) - S^-(x))$.

where A_μ denotes the electromagnetic potential, e the (renormalized) electric charge, $Z(\varepsilon)$ is a (possibly needed) wave function renormalization constant, and N denotes the generalized normal product recursively defined for fields A_i at spacelike separated points by

$$A_1 \ldots A_n = N(A_1 \ldots A_n) + \sum < A_{i_1} \ldots A_{i_\alpha} >_0 N(A_{i_{\alpha+1}} \ldots A_{i_n}). \qquad (4.6.14)$$

In perturbation theory a simpler prescription is expected to work with $j_\mu(x, \varepsilon, A)$ replaced by (M a suitable finite number)[19]

$$j_\mu(x, \varepsilon) - < j_\mu(x, \varepsilon) >_0 \sum_{n=0}^{M} (i^n e^n / n!) \, N((\int_{x-\varepsilon}^{x+\varepsilon} A_\nu(\xi) d\xi^\nu)^n). \qquad (4.6.15)$$

6.2 Axial current anomaly and $\pi_0 \to 2\gamma$ decay

By the Noether theorem for classical fields, the invariance of the Lagrangian under a one-parameter continuous group of transformations of the fields implies the existence of a conserved current, $j_\mu(x)$, $\partial^\mu j_\mu(x) = 0$. For the class of solutions with symmetric behavior at infinity, one further obtains a conserved charge $Q(t) = \int dx \, j_0(\mathbf{x}, t) = Q(0)$, i.e., a global conservation law; the failure of such a constant of motion characterizes the phenomenon of spontaneous symmetry breaking, which is at the basis the recent developments in theoretical physics.[20] In passing from the classical to the quantum theory, as a consequence of UV singularities, even the current conservation law may drastically change by the phenomenon of anomalies, as briefly discussed below.

For simplicity we consider the case of chiral symmetry in fermion electrodynamics. The classical Lagrangian for a massless Dirac field interacting with the electromagnetic field is symmetric under the chiral transformation laws (γ_5 antihermitian, $\alpha \in \mathbf{R}$):

$$\psi(x) \to e^{\alpha \gamma_5} \, \psi(x), \quad \overline{\psi}(x) \to \overline{\psi}(x) e^{\alpha \gamma_5}, \quad A_\mu \to A_\mu.$$

From the Noether theorem one obtains the following conserved axial current

$$j_\mu^5(\mathbf{x}, t) = i \, \overline{\psi}(\mathbf{x}, t) \, \gamma_5 \, \gamma_\mu \, \psi(\mathbf{x}, t), \quad \partial^\mu j_\mu^5(x) = 0.$$

The situation radically changes in the quantum case, where, as we have learned above, a point-splitting procedure is necessary in order to obtain the quantum field theory definition of such a current. We shall adopt the gauge-invariant point-splitting of eqs. (4.6.12) and (4.6.13), with γ_μ replaced by $\gamma_5 \gamma_\mu$; the so obtained analog of $j_\mu(x, \varepsilon, A)$ will be denoted by $j_\mu^5(x, \varepsilon, A)$. To display the effect of such a UV regularization, we consider the simpler case of a Dirac fermion interacting with an external (c-number) electromagnetic field $A_\mu(x)$, so that the exponential in eq. (4.6.13) is well defined and the Dirac equation for $m \neq 0$ reads

$$i\gamma\partial \, \psi = m\psi - e\gamma_\mu \, A^\mu \, \psi.$$

[19] W. Zimmermann, Comm. Math. Phys. **8**, 66 (1968) and references therein.
[20] For a review and a general discussion, see F. Strocchi, *Symmetry Breaking*, 2nd ed., Springer 2008.

One can now compute the four-divergence of $j_\mu^5(x, \varepsilon, A)$, for small ε, using the above Dirac equation and that

$$\int_{x-\varepsilon}^{x+\varepsilon} A_\nu(\xi)\, d\xi^\nu \sim 2\varepsilon^\nu A_\nu(x), \quad A^\mu(x+\varepsilon) - A^\mu(x-\varepsilon) \sim 2\varepsilon_\nu \partial^\nu A^\mu(x).$$

The result is

$$\partial^\mu j_\mu^5(x, \varepsilon, A) = j^5(x, \varepsilon, A) +$$

$$-ie\, j_\mu^5(x, \varepsilon, A)\,[\, A^\mu(x+\varepsilon) - A^\mu(x-\varepsilon) - \partial^\mu \int_{x-\varepsilon}^{x+\varepsilon} A_\nu(\xi)\, d\xi^\nu\,] =$$

$$= j^5(x, \varepsilon, A) - ie j_\mu^5(x, \varepsilon, A)\, 2\,\varepsilon_\nu\,[\, F^{\nu\,\mu}(x) + O(\varepsilon)\,],$$

where $j^5(x, \varepsilon, A)$ is the point-splitting regularized version of the naive divergence $2m\,\overline{\psi}\,\gamma^5\,\psi$ (the first two terms in square brackets arise from the Dirac equations for the point-split Dirac operators).

Now, the last term with $F^{\nu\,\mu}$ does not vanish in the limit $\varepsilon \to 0$, because $j_\mu^5(x, \varepsilon, A)$ diverges as ε^{-1}. In fact, to lowest order in e, one has

$$i \lim_{\varepsilon \to 0} 2\varepsilon_\nu\, < j_\mu^5(x, \varepsilon, A) >_0 = -(e/16\pi^2)\,\varepsilon_{\mu\nu\rho\sigma} F^{\rho\,\sigma}.$$

From the four-divergence calculated above, one can deduce the general divergence equation in the full quantum case:[21]

$$\partial^\mu j_\mu^5(x) = j^5(x) + (e^2/16\pi^2)\,\varepsilon_{\mu\nu\rho\sigma} F^{\rho\,\sigma}\, F^{\mu\,\nu}, \tag{4.6.16}$$

so that, in the massless limit, the (gauge-invariant) axial current is not conserved (*axial anomaly*). Such a non-canonical behavior is not a mere mathematical subtlety, but it plays a crucial role for explaining relevant physical phenomena, such as the $\pi_0 \to 2\gamma$ decay or the breaking of chiral symmetry in quantum chromodynamics.[22]

We briefly discuss the effect on the $\pi_0 \to 2\gamma$ decay. The transition is given (in Dirac notation) by $< \pi_0, in|2\gamma, out > = < 0|\varphi_{in}|2\gamma, out >$, where φ denotes a field whose asymptotic limits describe the asymptotic π_0 states (*pion interpolating field*).[23] According to the discussion of Chapter 2, Section 6.4 (see also Chapter 6, eqs. (6.1)–(6.3)), the (LSZ) asymptotic limits of field operators are given by (m_π denotes the pion mass)

$$\varphi_{f, in/out} = \lim_{x_0 \to \mp\infty} \varphi_f(x_0), \quad (\Box + m_\pi^2)f = 0, \quad f(\mathbf{x}, t) \in \mathcal{S}(\mathbf{R}^3), \tag{4.6.17}$$

---

[21] For the explicit calculation in external field, see R. Jackiw, 1985, pp. 132–4). The argument which derives the anomalous divergence from the external field calculation, through the method of the generating functional, goes back to Schwinger, Phys. Rev. **82**, 664 (1951). A general non-perturbative derivation of the anomaly can be obtained by the functional integral technique as pointed out by K. Fujikawa, Phys. Rev. **D21**, 2848 (1980); see also S. Weinberg, *The Quantum Theory of Fields*, Vol. II, Cambridge University Press 1996, Chapter 22.

[22] See the excellent review by R. Jackiw in *Current Algebra and Anomalies*, 1985, which we strongly recommend.

[23] As discussed in Chapter 6, it is enough that the field φ has non-vanishing matrix elements between the vacuum and the asymptotic π_0 states.

$$\varphi_f(x_0) \equiv i \int d^3x \, [\bar{f}(x)\partial_0 \, \varphi(x) - \partial_0 \bar{f}(x) \, \varphi(x)] \equiv i \int d^3x \bar{f}(x) \stackrel{\leftrightarrow}{\partial_0} \varphi(x),$$

where the test function f selects the mass of the asymptotic free field and its mass-shell restriction describes the corresponding one-particle wave function. Now, the surface integral

$$\int d^3x \, \partial^k \, [\bar{f}(x) < 0| \stackrel{\leftrightarrow}{\partial_k} \varphi(x) \, |2\gamma, out >]$$

vanishes, because $< 0|\varphi(\mathbf{x}, t)|2\gamma, out >= e^{i\mathbf{K}\cdot\mathbf{x}} < 0|\varphi(0, t)|2\gamma, out >$, with $\mathbf{K}$ the total momentum of the two γ's, and $f(\mathbf{x}, t) \in \mathcal{S}(\mathbf{R}^3)$; furthermore, $< 0|\varphi_{f,\,out} \, |2\gamma, out >= 0$. Then, the $\pi_0 \to 2\gamma$ decay is given by

$$\lim_{x_0 \to -\infty} < 0| \, \varphi_f(x_0)|2\gamma, out >= - \int_{-\infty}^{\infty} dx_0 \, \partial_0 < 0| \, \varphi_f(x_0)|2\gamma, out >=$$

$$= -i \int dx_0 \, d^3x \, \partial^\mu [\bar{f}(x) < 0| \stackrel{\leftrightarrow}{\partial_\mu} \varphi(x)|2\gamma, out >] =$$

$$= i \int d^4k \, \tilde{\bar{f}}(k) \, (k^2 - m_\pi^2) < 0| \, \tilde{\varphi}(k) \, |2\gamma, out > .$$

The so-called *Sutherland–Veltman (SV) theorem* states that the matrix element $< 0| \, \tilde{\varphi}(k) \, |2\gamma, out >$ vanishes for $k^2 = 0$, if the axial current anomaly vanishes and, under general regularity assumptions, this implies also the vanishing (or at least a very small value) of the $\pi_0 \to 2\gamma$ decay amplitude.

To this purpose, we denote by p, q the two-photon momenta $(p^2 = 0 = q^2,$ $k = p + q)$ and by $\varepsilon_\mu(p), \varepsilon_\nu(q)$ the corresponding polarizations, so that the amplitude takes the form

$$\varepsilon_\rho(p)\varepsilon_\sigma(q)T^{\rho\sigma}, \; T^{\rho\sigma} = \varepsilon^{\rho\sigma\alpha\beta}p_\alpha q_\beta T(k^2).$$

By the quark model (or by the PCAC hypothesis) one has

$$< 0| \, j^5(0)|\pi_0 >= m_\pi^2 \, f_\pi, \; f_\pi \approx 96 \text{Mev},$$

so that one may use $j^5/(m_\pi^2 f_\pi)$ as an interpolating field, and if the anomaly vanishes, $j^5 = \partial^\mu j_\mu^5$. Hence, we have to compute $k^\mu < 0|j_\mu^5|2\gamma >$, which is of the form $(-k^2 + m_\pi^2)k^\mu T_\mu^{\rho\sigma}(p, q)\varepsilon_\rho(p)\varepsilon_\sigma(q)$. Lorentz covariance, odd parity, Bose symmetry, and the photon transversality condition imply the following form for $T_\mu^{\rho\sigma}(p, q)$:

$$T_\mu^{\rho\sigma}(p, q) = \varepsilon^{\rho\sigma\alpha\beta}p_\alpha q_\beta k_\mu T_1(k^2) + \varepsilon_\mu^{\rho\sigma\alpha}(p_\alpha - q_\alpha)k^2 T_2(k^2)+$$

$$+(\varepsilon_\mu^{\rho\alpha\beta}q^\sigma - \varepsilon_\mu^{\sigma\alpha\beta}q^\rho)p_\alpha q_\beta T_3(k^2) + 2(\varepsilon_\mu^{\rho\alpha\beta}p^\sigma - \varepsilon_\mu^{\sigma\alpha\beta}p^\rho)p_\alpha q_\beta T_2(k^2),$$

where the invariant functions $T_i(k^2), i = 1, 2, 3$ are free of kinematical singularities and of dynamical singularities at $k^2 = 0$, to lowest order in the elecromagnetic coupling. Therefore,

$$k^\mu T_\mu^{\rho\sigma}(p, q) = \varepsilon^{\rho\sigma\alpha\beta}p_\alpha q_\beta \, k^2(T_1(k^2) - 2T_2(k^2))$$

vanishes as $k^2 \to 0$.

From the PCAC hypothesis, and from perturbative arguments one expects that (the matrix elements of) the field $\tilde{j}(k) \equiv (-k^2 + m_\pi^2)\tilde{\varphi}(k)$, which describes the source of the interaction for the interpolating field, has a smooth behavior in k, so that its restriction to $k^2 = m_\pi^2$ may be reasonably approximated by its restriction to $k^2 = 0$. Then, the SV theorem implies the vanishing (or at least a very small value) of the $\pi_0 \to 2\gamma$ decay amplitude.

The occurrence of the axial anomaly gives an extra contribution to $k^\mu T_\mu^{\rho\sigma}$; namely, $(e^2/8\pi^2) < 0|\varepsilon_{\alpha\beta\delta\gamma}F^{\delta\gamma}F^{\alpha\beta}|2\gamma >$, which can be calculated to lowest order in e yielding $(e^2/2\pi^2)\varepsilon^{\rho\sigma\lambda\nu}p_\lambda q_\nu$. Hence, one has

$$T(0) = \frac{e^2}{2\pi^2 f_\pi}$$

in agreement with the experimental value.

6.3 The derivative coupling model

The definition of products of fields at the same point is a prerequisite for defining interactions and for discussing the solutions; in general, an overall renormalization constant (called a *wave function renormalization constant*) is needed besides the necessary subtractions. The following simple model (*gradient coupling model*) displays i) the non-canonical behavior of the (renormalized) fields as a consequence of a divergent wave function renormalization, ii) the invariance under gauge transformations

$$\psi(x) \to e^{i\Lambda(x)}\,\psi(x), \quad \varphi(x) \to \varphi(x) + \Lambda(x), \tag{4.6.18}$$

and the need of a gauge-invariant regularization of the current.

The model is formally defined by the following Lagrangian

$$L = L_0(\varphi) + L_0(\psi) - g\overline{\psi}\gamma_\mu\,\psi\,\partial^\mu\varphi, \tag{4.6.19}$$

$$L_0(\varphi) = \tfrac{1}{2}\left[\partial_\mu\varphi\,\partial^\mu\varphi - m^2\,\varphi^2\right], \quad L_0(\psi) = \overline{\psi}\,(i\gamma_\mu\partial^\mu - M)\,\psi.$$

The corresponding equations of motion are

$$(i\gamma_\mu\partial^\mu - M)\psi = g\gamma_\mu\psi\,\partial^\mu\varphi, \quad (\Box + m^2)\,\varphi = 0, \tag{4.6.20}$$

since the (formal) current $j_\mu = \overline{\psi}\gamma_\mu\psi$ is conserved. Classically, the model is easily solved by a free field φ of mass m and by a Dirac field

$$\psi(x) = e^{-ig\varphi(x)}\,\psi_0(x),$$

with ψ_0 a free Dirac field of mass M. It is not difficult to find the class of initial data for which such a solution is well defined.

As one may expect from the discussion of the previous subsections, the (unavoidable) UV singularities of quantum fields require much more care for the definition of the equations and of the solutions. The general (Wightman) framework discussed above provides the clear guide for the solution of the mathematical problems which arise. In order to see the short-distance singularities, one may start by quantizing the model by means of canonical quantization, i.e., one requires that the scalar field $\varphi(\mathbf{x}, t)$ and its conjugate momentum $\pi(\mathbf{x}, t) = \partial_0\varphi(\mathbf{x}, t) - gj_0(\mathbf{x}, t)$ obey the canonical commutation relations (CCR),

$$[\varphi(\mathbf{x},t),\,\pi(\mathbf{y},t)] = i\delta(\mathbf{x}-\mathbf{y}), \quad [\varphi,\varphi] = 0 = [\pi,\pi], \tag{4.6.21}$$

and the fermion fields commute with φ and π and obey the anticommutation relations (ACR),

$$\{\psi_\alpha(\mathbf{x},t),\,\psi_\beta^*(\mathbf{y},t)\} = \delta_{\alpha\beta}\,\delta(\mathbf{x}-\mathbf{y}), \quad \{\psi_\alpha(\mathbf{x},t),\,\psi_\beta(\mathbf{y},t)\} = 0. \tag{4.6.22}$$

We have to find the necessary subtraction needed for the definition of the interaction term in the Dirac equation (in this case, the vacuum subtraction does not work since the vacuum expectation vanishes). The singularity of the product $\varphi(x+\varepsilon)\psi(x)$, when $\varepsilon \to 0$, can be extracted from the unequal-time commutator $[\varphi(x),\,\psi(y)]$ which is a solution of the Klein–Gordon equation in the variable x, and is therefore completely determined by its initial data at the time $x_0 = y_0$. Such data are given by the equal time relations $[\varphi,\,\psi] = 0$, and, from $[\pi,\,\psi] = 0$,

$$[\partial_0\,\varphi(\mathbf{x},t),\,\psi(\mathbf{y},t)] = g\,[j_0(\mathbf{x},t),\,\psi(\mathbf{y},t)] = -g\delta(\mathbf{x}-\mathbf{y})\,\psi(\mathbf{y},t). \tag{4.6.23}$$

In conclusion,

$$[\varphi(x),\,\psi(y)] = g\Delta(x-y)\,\psi(y) \tag{4.6.24}$$

and, since φ is a free field,

$$<\varphi(x)\,\psi(y)\,\overline{\psi}(z)> \;=\; <[\varphi^-(x),\,\psi(y)\overline{\psi}(z)]> \;=$$
$$= g\,(\Delta^+(x-y) - \Delta^+(x-z)) <\psi(y)\,\overline{\psi}(z)>.$$

Thus, a regularization of the interaction term is given by

$$(\gamma\,\partial\varphi\,\psi)_r(x) \equiv \lim_{x\to y}\,(\gamma^\mu\partial_\mu\,\varphi(x)\,\psi(y) - g(\gamma_\mu\,\partial^\mu\Delta^+)(x-y)\,\psi(y)) =$$
$$= \lim_{\varepsilon\to 0}\,([\gamma\partial\varphi(x+\varepsilon)\,\psi(x-\varepsilon)] + [\varepsilon \to -\varepsilon]) =: \gamma\partial\varphi\,\psi : (x).$$

In order to derive the solution of the model, we introduce the Wick exponential of the free field φ defined by

$$: e^{-ig\,\varphi} : (x) = \sum(-ig)^n (n!)^{-1} : \varphi^n : (x), \tag{4.6.25}$$

(for the definition of the Wick polynomials see Chapter 2, Section 7). The right-hand side defines a non-tempered Wightman field in four dimensions (whereas in two dimensions it is tempered). The Wick exponential of the free field φ can also be written as

$$: e^{-ig\varphi} : (x) = e^{-ig\varphi^+(x)} e^{-ig\varphi^-(x)},$$

since the Wick powers $: \varphi^n : (x)$ can also be defined as the limit of Wick ordered polynomials of UV regularized fields $\varphi_\Lambda(x)$

$$: \varphi^n : (x) = \lim_{\Lambda\to\infty}\, : \varphi_\Lambda^n(x) :$$

(Λ denoting a momentum space cutoff),[24] and

$$: e^{g\varphi_\Lambda(x)} := e^{g\varphi_\Lambda^+(x)}\, e^{g\varphi_\Lambda^-(x)}.$$

It is easy to see that

$$\partial_\mu : e^{g\varphi} : (x) = g : \partial_\mu\varphi\,(: e^{g\varphi} :)\,: (x) = g : \partial_\mu\varphi\, e^{g\varphi} : (x).$$

Then, the solution of the (regularized) Dirac equation is given by

$$\psi_r(x) =: e^{-ig\varphi} : (x)\,\psi_0(x), \tag{4.6.26}$$

with ψ_0 a free Dirac field of mass M. The product on the right-hand side is well defined by the same argument used for the product $\varphi\,\psi$, since now $[\,\varphi(x),\,\psi_0(y)\,] = 0$ and consequently $[: \varphi^n : (x),\,\psi_0(y)\,] = 0$. Furthermore,

$$\partial_\mu\varphi(x) : e^{g\varphi} : (y) =: \partial_\mu\varphi(x)\, e^{g\varphi}(y) : +ig\Delta^+(x-y) : e^{g\varphi}(y),$$

so that, as above, the Wick ordering provides the regularization of the interaction terms $(\gamma\partial\varphi\psi_r)_r =: \gamma\partial\varphi\psi_r :.$

The so obtained solution is a well-defined operator-valued non-tempered distribution, with well-defined Wightman functions satisfying the general properties W2–W6. In fact, the two-point function of the Wick exponential is given by

$$<: e^{ig\varphi} : (x) : e^{-ig\varphi} : (y) >= e^{g^2 i\Delta^+(x-y)},$$

which is not a tempered distribution. Thus, the model provides an example of a quantum field theory which requires a more general class of distributions than the standard ones.[25]

Moreover, the distribution $e^{g^2 i\Delta^+(x-y)}$ does not allow a restriction at equal times. Thus, one cannot write the equal-time anticommutation relations for the field $\psi_r(x)$ (*non-canonical behavior*).

Such a loss of canonical behavior can also be understood in terms of the conventional theory of renormalization. In fact, the Wick ordering subtraction needed for the definition of the interaction term and of the solution $\psi_r(x)$ can be interpreted as a renormalization. For this purpose, with the introduction of an UV cutoff Λ the solution is given by

$$\psi_\Lambda(x) = e^{-ig\varphi_\Lambda(x)}\psi_0(x),$$

but its correlation function do not converge to well-defined distributions in the limit of $\Lambda \to \infty$. To achieve such a convergence, subtractions or counter-terms must be introduced. This is obtained by the Wick ordering (subtraction) which defines the renormalized field with UV cutoff

---

[24] For a discussion of the properties of Wick products and exponentials, see J. Glimm and A. M. Jaffe, *Quantum Physics*, Springer 1987, pp. 188–9, and Sect. 9.1

[25] See B. Klaiber, Nuovo Cim. **36**, 165 (1965); the modifications of Wightman (mathematical) formulation in order to includes this type of fields with non-tempered UV singularities have been discussed by A. M. Jaffe, Ann. Phys. **32**, 127 (1965); Phys. Rev. **158**, 1454 (1967) and references therein. The case of non-tempered infrared singularities which appear in local gauge quantum field theories has been discussed by U. Moschella and F. Strocchi, Lett. Math. Phys. **24**, 103 (1992).

$$\psi_{r,\Lambda}(x) =:\, e^{-ig\varphi_\Lambda}\, :(x)\,\psi_0(x) = e^{\frac{1}{2}g^2 i\Delta_\Lambda^+(0)}\,\psi_\Lambda(x) \equiv Z_\Lambda^{-1/2}\,\psi_\Lambda(x).$$

This equation can be interpreted as the relation between the unrenormalized field ψ_Λ and the renormalized field $\psi_{r,\Lambda}$, with a divergent wave function renormalization constant Z_Λ in the limit $\Lambda \to \infty$. Formally,

$$\psi_r(x) = Z^{-1/2}\,\psi(x)$$

with $\psi(x)$ denoting the unrenormalized field as the formal limit of ψ_Λ.

Thus, the unrenormalized field ψ formally satisfies the canonical anticommutation relation (ACR), but its correlation functions are divergent, whereas the renormalized field ψ_r has well-defined correlation functions, but does not satisfy the ACR; formally,

$$\{\,\psi_r(\mathbf{x},t),\,\psi_r^*(\mathbf{y},t)\,\} = Z^{-1}\,\delta(\mathbf{x}-\mathbf{y}).$$

The gauge-invariant point-splitting regularization of the current, eq. (4.6.13), gives

$$j_\mu(x) =:\, \overline{\psi}_0(x)\,\gamma_\mu\,\psi_0(x)\, :, \quad [\,j_0(x),\,\psi_r(y)\,] = -iS(y-x)\,\gamma_0\psi_r(x),$$

since $[\,\varphi,\,\psi_0\,] = 0$.

5
Euclidean quantum field theory

---

1   The Schwinger functions

The technical convenience of considering field correlation functions at purely imaginary times was realized from the early times of (perturbative) renormalization theory and became popular under the name of Wick rotation.[1] Later, Schwinger and Nakano[2] pointed out that the imaginary time correlation functions, later called *Schwinger functions*, have very interesting properties and can be studied as solutions of Euclidean invariant differential equations, with singularities appearing only at coincident points, rather than on the whole forward and backward cones.

The (non-perturbative) approach to quantum field theory in terms of Schwinger functions became a standard strategic tool after Symanzik's realization[3] that the Schwinger functions have a functional integral representation, which provides a mathematically acceptable alternative to the problematic Feynman "integral" over real-time field histories, first appeared in the Matthews–Salam formula.[4] In fact, by Cameron's theorem, the Feynman "integration" over real-time histories does not have the property (σ additivity) which allows approximations by taking limits inside the "integral".[5] This makes its practical use, beyond its perturbative definition, rather problematic.

On the other hand, there are non-perturbative indications that the Schwinger functions may be expressible as moments of a σ-additive measure on the function space of classical fields (a candidate for a higher-dimensional analog of the Wiener measure). Furthermore, as recognized by Symanzik, the functional integral representation of the Schwinger functions provides their interpretation as the statistical mechanics correlation functions of suitable *classical* systems.[6]

Symanzik's ideas and program became the basis of constructive quantum field theory, because the construction and/or control of the Schwinger functions proved to

---

[1] F. J. Dyson, Phys. Rev. **75**, 1736 (1949); G. C. Wick, Phys. Rev. **96**, 1124 (1954).

[2] J. Schwinger, Phys. Rev. **115**, 721 (1959); T. Nakano, Prog. Theor. Phys. **21**, 241 (1959).

[3] K. Symanzik, Euclidean quantum field theory, in *Local Quantum Theory*, R. Jost ed. Academic Press 1969, p. 153.

[4] P. T. Matthews and A. Salam, Nuovo Cim. **2**, 120 (1955).

[5] For a discussion of this point, see, e.g., F. Strocchi, *An Introduction to the Mathematical Structure of Quantum Mechanics*, 2nd ed., World Scientific 2008, Chapter 6.

[6] For a very illuminating review of the main ideas and mathematical status of the functional integral formulation of quantum field theory, see A. S. Wightman, in *Renormalization Theory* (Erice School in Mathematical Physics 1975) G. Velo and A. S. Wightman (eds.), Reidel 1976, pp. 1–24.

be easier than that of the Wightman functions; as a result, the Euclidean techniques, promoted by the pioneering works of Nelson and Guerra,[7] superseded the Hamiltonian approach of the early constructive approach.[8] Most of the non-perturbative information on quantum field theory, especially gauge theories, have been obtained by using the functional integral representation, which, as explained above, relies on the Euclidean formulation.

In the following sections we will briefly outline the properties of the correlation functions at imaginary times (Schwinger functions), and the complete equivalence between the real-time or Minkowski quantum field theory and its Euclidean formulation.

The first problem is a mathematical basis of the heuristic prescription, by which Euclidean field theory is obtained by simply replacing t by it. Indeed, as argued below, the analyticity domain of the Wightman functions, discussed in Chapter 3, Section 4.3, contains the so-called Euclidean points, so that the knowledge of the correlation functions at real times determines their Euclidean continuation, and vice versa.

A euclidean point in complex form, or briefly a *complex euclidean point* $z \in \sigma_n^{p.ext}$ is a point of the form ($\tau_k \in \mathbf{R}$, $\mathbf{x}_k \in \mathbf{R}^3$)

$$z = (z_1, \ldots, z_n), \quad z_k = (i\tau_k, \mathbf{x}_k). \tag{5.1.1}$$

A Euclidean point in real form, or briefly a *(real) Euclidean point*, is of the form

$$y = (y_1, \ldots, y_n), \quad y_k = (\tau_k, \mathbf{x}_k). \tag{5.1.2}$$

The correspondence $z \leftrightarrow y$ is one-to-one, and complex Lorentz transformations Λ on z, $z \to \Lambda z = z' = (i\tau', \mathbf{x}')$, define $SO(4)$ rotations R_Λ on y, $y \to R_\Lambda y = y' = (\tau', \mathbf{x}')$, and vice versa, since

$$(Ry)^2 = \tau'^2 + \mathbf{x}'^2 = \tau^2 + \mathbf{x}^2, \Rightarrow z'^2 \equiv -(\tau')^2 - \mathbf{x}'^2 = -\tau^2 - \mathbf{x}^2 = z^2.$$

A *non-coincident Euclidean point* $y = (y_1, \ldots, y_n)$ (or equivalently, $z = (z_1, \ldots, z_n)$) has the property that $y_i \neq y_j, \forall i \neq j$.

Proposition 1.1 *The whole set of non-coincident complex Euclidean points is contained in the permuted extended domain* $\sigma_n^{p.ext} = \cup_\pi \sigma_n^{ext,\pi}$, *eq. (3.4.11).*

Proof. One has to show that for any such complex Euclidean point z there are a permutation P and a complex Lorentz transformation Λ such that $\Lambda P z \in \sigma_n$. In fact, if $y = (y_1, \ldots, y_n)$ denotes the corresponding real Euclidean point and $e_{ij} \equiv (y_i - y_j)/|y_i - y_j|$, one can find at least one unit vector n such that $n \cdot e_{ij} \neq 0$, $\forall i, j$, i.e., $n \cdot y_i \neq n \cdot y_j$, $\forall i, j$. In fact, $v \cdot e_{ij} \neq 0$ means that v belongs to the complement CI_{ij} of the plane I_{ij} of vectors orthogonal to e_{ij}; thus n must belong to $\cap_{ij} CI_{ij} = C(\cup_{ij} I_{ij})$, which is clearly non-void. Let P be a permutation such that

---

[7] E. Nelson, Jour. Funct. Analys. **12**, 97 (1973); A Quartic Interaction in Two Dimensions, in *Mathematical theory of Elementary Particles*, R. Goodman and I. Segal (eds.) MIT Press 1966, p. 69; F. Guerra, Phys. Rev. Lett. **28**, 1213 (1972).

[8] The basic reference textbooks are B. Simon, *The $P(\varphi)_2$ Euclidean (Quantum) Field Theory*, Princeton University Press 1974; J. Glimm and A. Jaffe, *Quantum Physics. A Functional Integral Point of View*, Springer 1987.

$n \cdot y_{P(1)} < n \cdot y_{P(2)} < \ldots < n \cdot y_{P(n)}$ and R a rotation such that $Rn = (1,0,0,0)$, then $y' \equiv (RPy)$ is chronologically ordered; namely, $\tau'_1 < \tau'_2 < \ldots < \tau'_n$, and the corresponding complex Euclidean point $z' = \Lambda_R P z \in \sigma_n$.

In terms of the difference variables a complex Euclidean point z corresponds to

$$\zeta = (\zeta_1, \ldots, \zeta_{n-1}), \quad \zeta_k = (i\,\eta_k, \boldsymbol{\xi}_k), \quad \eta_k \in \mathbf{R}, \quad \boldsymbol{\xi}_k \in \mathbf{R}^3. \tag{5.1.3}$$

The *Schwinger functions* are then defined at the non-coincident Euclidean points (with $y_i = (\tau_i, \mathbf{x}_i) \leftrightarrow z_i = (i\tau_i, \mathbf{x}_i)$) by

$$\mathcal{S}(y) = \mathcal{S}(y_1, \ldots, y_n) \equiv \mathcal{W}(z_1, \ldots, z_n) = \mathcal{W}(z). \tag{5.1.4}$$

The analyticity properties of the Wightman functions provide the mathematical basis of the heuristic prescription of defining the Euclidean field theory by replacing t by $i\,t$. In particular, it guarantees that such a transformation does not connect disjoint regions of analyticity, e.g., separated by cuts of non-analyticity, which would make the relation with the real-time theory rather loose (if not void). Moreover, the heuristic prescription misses the role of spectral condition, covariance, locality and the need of excluding the coincident points. Such a deep structural relation is one of the rewarding results of the analysis of the general properties of the vacuum expectation values.

2   Euclidean invariance and symmetry

From the properties of the Wightman functions one immediately derives:

Proposition 2.1   *The Schwinger functions are symmetric,*

$$\mathcal{S}(y_1, \ldots, y_n) = \mathcal{S}(y_{\pi(1)}, \ldots y_{\pi(n)}), \tag{5.2.1}$$

for any permutation π, and invariant under the full Euclidean group.

Proof. Symmetry follows from definition (5.1.4) and the symmetry of the analytic continuation of the Wightman functions in $\sigma_n^{p.ext}$, eq. (3.4.11). Invariance under translations follows from the similar property of the Wightman functions, and invariance under $SO(4)$ transformations is a direct consequence of the invariance of the Wightman functions under the complex Lorentz group ($SO(4) \ni R \leftrightarrow \Lambda_R \in L_+(C)$):

$$\mathcal{S}(Ry + \bar{y}) = \mathcal{W}(\Lambda_R z + \bar{z}) = \mathcal{W}(z) = \mathcal{S}(y). \tag{5.2.2}$$

Eq. (5.2.2) implies in particular that the Schwinger functions are functions of the difference variables $\xi_j^E = y_{j+1} - y_j = (\eta_j, \boldsymbol{\xi}_j)$:

$$\mathcal{S}(y) = S(\xi^E), \quad \xi^E = (y_2 - y_1, \ldots, y_n - y_{n-1}).$$

For the symmetry property of the Schwinger functions a crucial role is played by microscopic causality or locality. It is due to this property that the Schwinger functions can be regarded as the correlation functions of commuting (i.e., classical) variables, and this is at the basis of the strict relation between Euclidean quantum field theory (EQFT) and *classical statistical mechanics.*[9] For fermionic fields the Schwinger functions are antisymmetric and can be interpreted as correlation functions of classical Grassmann variables.

For the following discussion, restricted to the case of a hermitian scalar field, it is useful to derive the consequence on the Schwinger functions of the hermiticity condition of the Wightman functions:

$$\overline{\mathcal{W}(x_1,\dots,x_n)} = \overline{(\Psi_0,\,\varphi(x_1)\dots\varphi(x_n)\Psi_0)} = (\Psi_0,\varphi(x_n)\dots\varphi(x_1)\Psi_0) =$$

$$= \mathcal{W}(x_n,\dots,x_1).$$

Briefly,

$$\overline{\mathcal{W}(x)} = \mathcal{W}(\overleftarrow{x}), \quad \overline{W(\xi)} = W(-\overleftarrow{\xi}), \quad \overleftarrow{x} \equiv (x_n,\dots x_1). \tag{5.2.3}$$

This implies

$$\overline{\tilde{W}(q)} = \tilde{W}(\overleftarrow{q}). \tag{5.2.4}$$

Since, by eqs. (3.4.1), (3.4.3), and (5.1.4) the Schwinger functions at chronologically ordered points are the Laplace transforms of the Wightman functions, i.e., (in the multivector notation),

$$S(y) = S(\xi^E) = \int dq\, \tilde{W}(q)e^{-\eta q^0}e^{-i\boldsymbol{\xi}\cdot\mathbf{q}}, \quad \xi_j^E = (\eta_j,\,\boldsymbol{\xi}_j), \tag{5.2.5}$$

one has

$$\overline{S(y)} = \int dq\, \tilde{W}(\overleftarrow{q})e^{-\eta q^0}e^{+i\boldsymbol{\xi}\cdot\mathbf{q}} = S(\overleftarrow{r\,y}), \quad r\,y = (-\tau,\mathbf{x}). \tag{5.2.6}$$

For the following discussion, it is convenient to extend the reflection operation r to the test functions, by defining the *Osterwalder–Schrader reflection operator* ϑ:

$$(\vartheta f)(\tau,\mathbf{x}) \equiv f(-\tau,\mathbf{x}) \tag{5.2.7}$$

As displayed above, the transcription of the linear properties of the Wightman functions to the Schwinger functions is relatively easy. The non-trivial point is the identification of the counterpart of the non-linear positivity relation, as well as the characterization of the property of being the Laplace transforms of the Wightman

---

[9] Clearly the symmetry of the Wightman functions on $\sigma_n^{p.ext}$ does not imply the symmetry of the boundary values, since, e.g., by letting Im $\zeta_k = $ Im $(z_{k+1} - z_k) \to 0$, inside V_+, $k = 1,\dots n-1$, in

$$\mathcal{W}(z_1,\dots,z_j,z_{j+1},\dots z_n) = \mathcal{W}(z_1,\dots,z_{j+1},z_j,\dots z_n)$$

one obtains on the left-hand side $\mathcal{W}(x_1,\dots,x_j,x_{j+1},\dots x_n)$, but on the right-hand side one does not obtain $\mathcal{W}(x_1,\dots,x_{j+1},x_j,\dots x_n)$, because the latter is obtained by going to the boundary in the domain $P(j,j+1)\,\sigma_n^{ext}$, i.e., by letting Im $\zeta_j = $ Im $(z_{j+1} - z_j) \to 0$, inside V_-.

functions. This problem was solved by Osterwalder and Schrader, and allowed them to establish the complete equivalence between relativistic and Euclidean quantum field theory.[10]

3   Reflection positivity

The Euclidean counterpart of the positivity condition of the Wightman functions (property W5) is the *reflection positivity*, also called the *Osterwalder–Schrader (OS) positivity*. This property allows the reconstruction of the relativistic quantum field theory from the (Euclidean) Schwinger functions and leads to a correspondence between the local states of the relativistic theory and the states obtained by application of (commuting) Euclidean field operators to the vacuum. Most of the achievements of constructive QFT and the non-perturbative approach to QFT crucially rely on the Osterwalder–Schrader formulation.

The origin of the reflection positivity is easily seen by considering the two-point function at chronologically ordered points. Let

$$f(y) = f(\tau, \mathbf{x}) \in \mathcal{S}(\mathbf{R}^4), \quad \operatorname{supp} f \subset \mathbf{R}_+ \times \mathbf{R}^3 \equiv \mathbf{R}_+^4, \quad \mathbf{R}_+ \equiv (0, \infty),$$

then, by eq. (5.2.5) (with the notations of eqs. (3.3.9) and (5.2.7)),

$$\mathcal{S}(\vartheta f^* \times f) = \int dy_1\, dy_2\, \bar{f}(y_1)\, f(y_2)\, \mathcal{S}(ry_1, y_2) =$$

$$= \int dy_1\, dy_2\, \bar{f}(y_1)\, f(y_2) \int dq\, \tilde{W}(q)\, e^{-q_0(\tau_1 + \tau_2)}\, e^{-i\mathbf{q}\cdot(\mathbf{x}_2 - \mathbf{x}_1)} =$$

$$= \int dq\, \tilde{W}(q)\, |f^L(q)|^2 \geq 0,$$

where

$$f^L(q) \equiv \int dy\, f(y)\, e^{-q_0\tau}\, e^{-i\mathbf{q}\cdot\mathbf{x}}\,\big|_{q_0 > 0} \tag{5.3.1}$$

is the Laplace–Fourier transform of f (with respect to τ and $\mathbf{x}$ respectively).

For the general case, as in Chapter 3, we introduce the terminating sequences $\underline{f} = (f_0, f_1, \ldots f_n)$, of *functions of chronologically ordered positive variables*, $f_k \in \mathcal{S}_<(\mathbf{R}^{4k})$,

$$\mathcal{S}_<(\mathbf{R}^{4k}) \equiv \{ f_k \in \mathcal{S}(\mathbf{R}^{4k}), \ \ f_k = 0 \ \ \text{unless} \ \ 0 < \tau_1 < \tau_2 < \ldots < \tau_k \} \tag{5.3.2}$$

and extend the definition of eq. (5.2.7) to the multivariable case

$$(\vartheta f_k)(y) = f_k(ry). \tag{5.3.3}$$

Then, *reflection positivity* takes the form

$$\mathcal{S}(\vartheta \underline{f}^* \times \underline{f}) \geq 0. \tag{5.3.4}$$

---

[10] K. Osterwalder and R. Schrader, Comm. Math. Phys. **42**, 281 (1975); K. Osterwalder, Euclidean Green's Functions and Wightman Distributions, in *Constructive Quantum Field Theory*, G. Velo and A. S. Wightman (eds.), *Lecture Notes in Physics*, Vol. 25, Springer 1973.

To prove eq. (5.3.4) one notices that the Laplace transforms of the vector valued distributions $\Psi_n(x_1, \xi)$, eq. (3.4.13), yield well-defined vectors $\Psi_n(z_1, \zeta)$, since no smearing is necessary and at the non-coincident Euclidean points $z_1 = (i\tau_1, \mathbf{x}_1)$, $\zeta_k = (i\eta_k, \boldsymbol{\xi}_k)$, $\eta_k = \tau_{k+1} - \tau_k$, $(y = (\tau_1, \mathbf{x}_1)$, $\xi^E = (\eta, \boldsymbol{\xi}))$, they define the Euclidean state vectors

$$\Psi_n^E(y, \xi^E) = e^{-H\tau_1}\,\varphi(\mathbf{x}_1, 0)\,e^{-H(\tau_2 - \tau_1)}\,\varphi(\mathbf{x}_2, 0)\ldots e^{-H(\tau_n - \tau_{n-1})}\,\varphi(\mathbf{x}_n, 0)\Psi_0. \quad (5.3.5)$$

They may be interpreted as the result of the application of the Euclidean operators $\varphi^E(y_1)\varphi^E(y_2)\ldots\varphi^E(y_n)$ to the vacuum,

$$\varphi^E(y) \equiv e^{-H\tau}\varphi(\mathbf{x}, 0)e^{H\tau} \equiv \varphi(z) = \varphi(\bar{z})^*, \quad z \leftrightarrow y.$$

Thus, one has $(\bar{z}_{n-1} - \bar{z}_n = -\bar{\zeta}_{n-1})$

$$(\varphi(z_1)\ldots\varphi(z_n)\,\Psi_0,\ \varphi(z_1')\ldots\varphi(z_m')\,\Psi_0) =$$

$$= (\,\Psi_0,\ \varphi(\bar{z}_n)\ldots\varphi(\bar{z}_1)\,\varphi(z_1')\ldots\varphi(z_m')\,\Psi_0\,) = \mathcal{W}(\bar{z}_n, \ldots \bar{z}_1, z_1', \ldots z_m') =$$

$$= W(-\overset{\leftarrow}{\bar{\zeta}}, -\bar{z}_1 + z_1', \zeta'). \quad (5.3.6)$$

At time-ordered Euclidean points, the right-hand side becomes

$$S(-\overset{\leftarrow}{r\xi}, -ry_1 + y_1', \xi') = \mathcal{S}(ry_n, \ldots ry_1, y_1', \ldots y_m'),$$

since $r(y_{n-1} - y_n) = -r\xi_{n-1}$. Then, for functions of chronologically ordered variables, $f_n \in \mathcal{S}_<(\mathbf{R}^{4n})$, eq. (5.3.2), since $\vartheta f_n^* \times f_m \in \mathcal{S}(\mathbf{R}^{4(n+m)})$, by eq. (5.3.6),

$$(\Psi_n^E(f_n),\ \Psi_m^E(f_m)) = \mathcal{S}(\vartheta f_n^* \times f_m) \quad (5.3.7)$$

is a well-defined scalar product; eq. (5.3.4) follows from the positivity of such a scalar product of the Euclidean vectors $\Psi_n^E(f_n)$.

The recognition of this property represented a real breakthrough in Euclidean quantum field theory; it encodes the Hilbert space structure given by the positivity condition, and marks a distinctive feature with respect to ordinary classical statistical mechanics. This property proved very useful also for the discussion of explicit models, especially for the lattice approach to gauge theories.[11]

4 Cluster property

The cluster property derived in Chapter 3, Section 2, eq. (3.2.7), gives

$$\lim_{\lambda \to \infty} \mathcal{S}_{n+m}(ry_n, \ldots ry_1, y_1' + \lambda\mathbf{a}, \ldots y_m' + \lambda\mathbf{a}) = \lim_{\lambda \to \infty} (\Psi_n^E,\ U(\lambda\mathbf{a})\,\Psi_m^E) =$$

$$= (\Psi_n^E,\ \Psi_0)\,(\Psi_0,\ \Psi_m^E) = \mathcal{S}(ry_n, \ldots ry_1)\,\mathcal{S}(y_1', \ldots, y_m'), \quad (5.4.1)$$

$\Psi_n^E \equiv \Psi_n^E(y, \xi^E)$, eq. (5.3.5). The validity of the *cluster property of the Schwinger functions* for any Euclidean vector a follows from $SO(4)$ symmetry. As in the relativistic

---

[11] See, e.g., H. Grosse, *Models in Statistical Mechanics and Quantum Field Theory*, Springer 1988.

case, the validity of the cluster property for the Schwinger functions is equivalent to the uniqueness of the vacuum.

5   Laplace transform condition

For the equivalence between Minkowski and Euclidean quantum field theory, a relevant issue is the complete characterization of the relation between the distributional properties of the Wightman and Schwinger functions.

By the analyticity properties of the former, the latter are C^∞ functions at non-coincident Euclidean points. This implies that for all test functions $f_n(y_1, \ldots, y_n) \in \mathcal{S}(\mathbf{R}^{4n})$ vanishing at coincident points together with all their derivatives $f_n^{(k)}$, briefly

$$f_n \in \mathcal{S}_{\neq}(\mathbf{R}^{4n}) = \{f_n \in \mathcal{S}(\mathbf{R}^{4n}), \quad f_n^{(k)} = 0, \ \forall k, \ \text{at coincident points}\},$$

the integrals $\mathcal{S}_n(f_n) = \int d^{4n}y\, \mathcal{S}(y_1, \ldots, y_n)\, f_n(y_1, \ldots, y_n)$ are absolutely convergent and define continuous linear functionals on $\mathcal{S}_{\neq}(\mathbf{R}^{4n})$ (a closed subspace of $\mathcal{S}(\mathbf{R}^{4n})$, equipped with the relative topology); briefly

$$\mathcal{S}(y_1, \ldots, y_n) \in \mathcal{S}_{\neq}(\mathbf{R}^{4n})'. \tag{5.5.1}$$

Such a property (together with those spelled out previously) allows to prove that $\mathcal{S}_n$ is a Laplace transform separately in each variable, but not that it is a joint Laplace transform of tempered distributions in all the variables, contrary to what was originally believed.[12]

The identification of a simple condition which allows to prove that the Schwinger functions are the Laplace transform of tempered distributions (*Laplace transform condition*) has been a debated question in the literature. For a regular function of one variable $\tilde{f}(p) \in \mathcal{S}(\mathbf{R}_+) = \{\tilde{f} \in \mathcal{S}(\mathbf{R}), \quad \text{supp}\, \tilde{f} \subset \overline{\mathbf{R}_+}\}$, the Laplace transform f^L, defined by

$$f^L(y) = \int dp\, e^{-py}\, \tilde{f}(p)\, |_{\overline{\mathbf{R}_+}}$$

belongs to $C^\infty(\mathbf{R}_+)$ and it is of fast decrease at infinity, so that $f^L \in \mathcal{S}(\mathbf{R}_+)$; actually f^L has a continuous extension together with all its derivatives to $\mathcal{S}(\overline{\mathbf{R}_+})$, since $\tilde{f} \in \mathcal{S}(\mathbf{R}^4)$.

For distributions

$$\tilde{W} \in \mathcal{S}(\overline{\mathbf{R}_+})' \equiv \{\tilde{W} \in \mathcal{S}(\mathbf{R})', \quad \text{supp}\, \tilde{W} \subset \overline{\mathbf{R}_+}\},$$

the Laplace transform may have polynomial singularities at $y = 0$, and one has:

---

[12] K. Osterwalder and R. Schrader, Comm. Math. Phys. **31**, 83 (1973); for the discussion of this delicate point see V. Glaser, Comm. Math. Phys. **37**, 257 (1974), B. Simon, *The $P(\varphi)_2$ Euclidean (Quantum) Field Theory*, Princeton University Press 1974; K. Osterwalder, in *Constructive Quantum Field Theory*, G. Velo and A. S. Wightman (eds.), *Lecture Notes in Physics*, Vol. 25, Springer 1973. Erice Lectures, 1973.

Theorem 5.1   *Let $\tilde{W} \in \mathcal{S}(\overline{\mathbf{R}_+})'$, then its Laplace transform has the following properties:*

$$S(y) = \int dq\, e^{-qy}\, \tilde{W}(q)\, |_{\mathbf{R}_+} \in \mathcal{S}(\mathbf{R}_+)' \tag{5.5.2}$$

and $\forall f \in \mathcal{S}(\mathbf{R}_+)$

$$S(f) = \tilde{W}(f^L) \tag{5.5.3}$$

$$|S(f)| \le \| f^L \|_{\mathcal{S}(\overline{\mathbf{R}_+})}, \tag{5.5.4}$$

where $\| \ \|_{\mathcal{S}(\overline{\mathbf{R}_+})}$ is some seminorm of $\mathcal{S}(\overline{\mathbf{R}_+})$, i.e., a seminorm of the form $\sup_{x \ge 0} |x^\alpha D^\beta f|$.

Conversely, if $S \in \mathcal{S}(\mathbf{R}_+)'$ and eq. (5.5.4) holds, then there exists a unique $\tilde{W} \in \mathcal{S}(\overline{\mathbf{R}_+})'$ such that S is the Laplace transform of $\tilde{W}$.

By the nuclear theorem, Theorem 5.1 has an extension to the multivariable case, with $\mathbf{R}_+$ replaced by $\mathbf{R}_+^{4(n-1)} \equiv (\mathbf{R}_+ \times \mathbf{R}^3)^{(n-1)}$. This provides a possible Laplace transform condition for the Schwinger functions in order that they define a relativistic quantum field theory. Unfortunately, eq. (5.5.4) is not easy to check, and possible alternatives have been proposed in the literature.[13]

An instructive model of Euclidean quantum field theory is the Euclidean formulation of quantum mechanics (*Euclidean quantum mechanics*), where all the above structures are reproduced in a simple way; actually, quantum mechanics can be viewed as a quantum field theory in $0 + 1$ spacetime dimensions.[14]

6   From Euclidean to relativistic QFT

The possibility of discussing quantum field theory models in terms of their Euclidean formulation relies on the complete equivalence with the relativistic formulation, and the conceptual problem arises of when a set of Schwinger functions (no matter how they have been obtained, even approximately, by functional integral, lattice regularization etc.) defines a set of Wightman functions with the correct relativistic properties. Such a problem is crucial also from a practical point of view, as a guide for a constructive non-perturbative approach.

The problem has been solved by Osterwalder and Schrader, who identified the Euclidean counterparts of the relativistic properties W1–W6, called the *Euclidean* or *O-S axioms*, so that their fulfilment by the Schwinger functions guarantees that they define a relativistic quantum field theory. For a hermitian scalar field, the Euclidean

---

[13] For a more detailed discussion of these (technical) points, see K. Osterwalder, Erice Lectures, and B. Simon book quotes in footnote 12; for a handy account, see F. Strocchi, *Selected Topics on the General Properties of Quantum Field Theory*, World Scientific 1993, esp. Sect. 3.5. An alternative, called the weak spectral condition, has been proposed by Yu. M. Zinoviev, Comm. Math. Phys. **174**, 1 (1995); and in *Constructive Physics*, V. Rivasseau (ed.), Springer 1995.

[14] For a sketchy account, see F. Strocchi, *An Introduction to the Mathematical Structure of Quantum Mechanics*, 2nd ed., World Scientific 2008, Sects. 6.5 and 6.6.

axioms are the following ones, in close correspondence with the related relativistic properties:

OS1 (Distributional properties) $S(\xi^E) \in \mathcal{S}(\mathbf{R}_+^{4(n-1)})'$ and hermiticity, eq. (5.2.6).
OS2 (Euclidean invariance) Eq. (5.2.2).
OS3 (Laplace transform condition) Eq. (5.5.4), with $\mathbf{R}_+$ replaced by $\mathbf{R}_+^{4(n-1)}$.
OS4 (Symmetry) Eq. (5.2.1).
OS5 (Reflection positivity) Eq. (5.3.4).
OS6 (Cluster property) Eq. (5.4.1).

The set of Schwinger functions satisfying OS1–OS6 define a Hilbert space $\mathcal{H}$ (actually the same Hilbert space defined by the Wightman functions, see below) and an abelian algebra of Euclidean fields; furthermore, by the Laplace transform condition, one obtains the corresponding Wightman functions, which satisfy W1–W6 as a consequence of OS1–OS6. In this way, one has a *Euclidean reconstruction theorem*.

The proof is the same as for the relativistic case. The analog of the vector space $\mathcal{D}_0$ is the vector space $\mathcal{D}_<$ generated by the terminating sequences

$$\underline{f} = (f_0, \ f_1(y_1), \dots f_n(y_1, \dots y_n), 0, \dots), \quad f_j \in \mathcal{S}_<(\mathbf{R}^{4j}). \tag{5.6.1}$$

$\mathcal{D}_<$ is the carrier of a representation $E(a, R)$ of the proper Euclidean group given by

$$E(a, R)\underline{f} = \underline{f}^{\{a,R\}}, \quad f_j^{\{a,R\}}(y) \equiv f_j(R^{-1}(y - a))$$

and the Euclidean field operators are defined on $\mathcal{D}_<$ by

$$\varphi^E(h)\underline{f} = h \times \underline{f} = (0, hf_0, hf_1, \dots hf_n, \dots), \quad h \in \mathcal{S}_<(\mathbf{R}^4).$$

The inner product

$$< \underline{f}, \underline{g} >^E \equiv \sum_{n,m} \mathcal{S}_{n+m}(\vartheta f_n^* \times f_m) \tag{5.6.2}$$

is non-negative thanks to OS reflection positivity, and therefore defines a pre-Hilbert structure on $\mathcal{D}_<$. The Hilbert space $\mathcal{H}$ is then obtained as in the relativistic case.

The Laplace transform condition guarantees the existence of tempered distributions $\tilde{W}(q_1, \dots, q_{n-1})$ with supp $\tilde{W} \subseteq \overline{(\mathbf{R}_+ \times \mathbf{R}^3)^{(n-1)}}$, such that $\mathcal{S}(\xi_1^E, \dots, \xi_{n-1}^E)$ are their Laplace transforms, i.e., eq. (5.2.5) holds. The Wightman functions are given by

$$\mathcal{W}(x_1, \dots, x_n) = \int dq_1 \dots dq_{n-1} \, e^{i \sum_k (x_{k+1} - x_k)q_k} \, \tilde{W}(q_1, \dots, q_{n-1}). \tag{5.6.3}$$

Furthermore, $\mathcal{S}(Ry) = \mathcal{S}(y)$, $\forall R \in SO(4)$, implies $W(q) = W(\Lambda_R q)$, with Λ_R the corresponding Lorentz transformation (see Section 1 above), since $(\eta, \boldsymbol{\xi}) \to R(\eta, \boldsymbol{\xi}) \leftrightarrow \zeta = (i\eta, \boldsymbol{\xi}) \to \Lambda_R \zeta$ and $e^{-\eta q_0 - i\boldsymbol{\xi} \cdot \mathbf{q}} = e^{i\zeta q}$.

Thus, Poincaré covariance of the $\mathcal{W}$'s follows from Euclidean invariance (OS2) and restricts the support of the $\tilde{W}$'s to $\overline{V_+^{(n-1)}}$, the maximal Lorentz-invariant subspace of $\overline{(\mathbf{R}_+ \times \mathbf{R}^3)^{(n-1)}}$ (spectral condition); locality follows from symmetry (OS4) and the cluster property from OS6.

Positivity follows from OS5. In fact, given $g_i(\mathbf{x}_i) \in \mathcal{S}(\mathbf{R}^4)$ and $h_i(t_{i+1} - t_i) \in \mathcal{S}(\mathbf{R})$, $i = 1, \ldots n$, such that the Fourier transform of h_i, restricted to positive energy is the Laplace transform of a test function $v_i(s) \in \mathcal{S}(\mathbf{R}_+)$,

$$\tilde{h}(q_0)\,|_{q_0 \geq 0} = \int ds\, e^{-q_0 s}\, v(s),$$

one considers the functions

$$f_n(x_1, .., x_n) = \prod_i g_i(\mathbf{x}_i) h(t_{i+1} - t_i), \quad f_n^E(y_1, .., y_n) = \prod_i g_i(\mathbf{x}_i) v_i(\eta_i), \qquad (5.6.4)$$

where $\eta_i = \tau_{i+1} - \tau_i$. Then, it is easy to see that

$$\mathcal{W}_{2n}(f_n^* \times f_n) = \mathcal{S}_{2n}(\vartheta f_n^{E\,*} \times f_n^E) \geq 0. \qquad (5.6.5)$$

The positivity property extends to linear combinations of such test functions, and by a density argument to any terminating sequence $\underline{f}$, $f_j \in \mathcal{S}(\mathbf{R}^{4j})$.

From the above discussion it follows that a state of the system can be described either in terms of the commutative Euclidean field algebra or in terms of the non-commutative relativistic field algebra. In fact, if $\Psi_n(x, \xi)$, $\Psi_n^E(y, \xi^E)$ denote the improper vector and the vector, respectively obtained by applying to the vacuum relativistic fields and time ordered euclidean fields (see eqs. (3.4.13) and (5.3.5)), with f_n, f_n^E related by eqs. (5.6.4), one has

$$\Psi_n^E(f_n^E) = \Psi_n(f_n). \qquad (5.6.6)$$

7   Examples

The technical convenience of working with the Schwinger functions, rather than with the Wightman functions, becomes clear if one compares the analyticity properties of the first with those of the latter. For this purpose it is instructive to consider the two-point function of a scalar field. For the relativistic two-point functions, singularities occur on the light-cone, $\zeta^2 = 0$, and on the cones $\zeta^2 > 0$, where there is a cut. On the other hand the (complex) Euclidean points correspond to $\zeta = (i\eta, \boldsymbol{\xi})$, so that $\zeta^2 = -\eta^2 - \boldsymbol{\xi}^2 \leq 0$, and all such point are analyticity points except the coincident point $\zeta = 0$. Thus, in the Euclidean formulation all the singularity regions of the relativistic case reduce to a single point, the origin.

To visualize the resulting picture it is convenient to draw a two-dimensional relativistic space (η, ξ), and recognize that the euclidean space is given by the plane orthogonal to the time axis, corresponding to the Wick rotation $\eta \to i\eta$. Such a plane intersects the singularity regions $\zeta^2 \geq 0$ only at the origin.

Another instructive computation is to determine the two-point Euclidean function for a free scalar field of mass m. By using eq. (5.2.5), for time-ordered points $\tau_2 - \tau_1 = \eta > 0$ one has

$$S(y_1, y_2) = S(\xi^E) = \tfrac{1}{2}(2\pi)^{-3} \int d^3q\, e^{-i\mathbf{q} \cdot \boldsymbol{\xi}}\, e^{-\eta\sqrt{\mathbf{q}^2 + m^2}} (\mathbf{q}^2 + m^2)^{-1/2}. \qquad (5.7.1)$$

The symmetry of the Schwinger functions gives $S(\xi^E) = S(-\xi^E)$ and the invariance under the three-dimensional rotations yields $S(\eta, \boldsymbol{\xi}) = S(\eta, -\boldsymbol{\xi})$, so that

$$S(\eta, \boldsymbol{\xi}) = S(-\eta, \boldsymbol{\xi})$$

and therefore the Schwinger functions for arbitrary Euclidean points is given by eq. (5.7.1), with η replaced by $|\eta|$.

Such an explicit expression can be cast into a manifestly Euclidean invariant form by using the identity

$$\pi \frac{e^{-|\eta|\,\omega}}{\omega} = \int dq_4 \, \frac{e^{-i\eta\,q_4}}{q_4^2 + \omega^2}, \quad \omega > 0,$$

which follows easily by closing the integration contour in the upper (respectively lower) complex q_4 plane corresponding to $\eta < 0$ (respectively $\eta > 0$). The result is

$$S(\xi^E) = (2\pi)^{-4} \int d^4 q \, \frac{e^{-iq\,\xi^E}}{q^2 + m^2}, \quad q^2 = q_4^2 + \mathbf{q}^2. \tag{5.7.2}$$

The two-point function can be expressed in terms of the modified Bessel function K_1:[15]

$$S(\xi) = (2\pi)^{-2} m|\xi|^{-1} K_1(m|\xi|) \sim_{|\xi|\to\infty} \tfrac{1}{2} m^2 e^{-m|\xi|}/(2\pi m|\xi|)^{3/2}, \quad m \neq 0,$$

$$S(\xi) = (2\pi)^{-2} |\xi|^{-2}, \quad m = 0.$$

It is worth remarking that the free Euclidean field does not satisfy the free Euclidean equation, since

$$(-\Delta + m^2)S(\xi) = \delta^4(\xi).$$

8 Functional integral representation

As for the ordinary quantum mechanics, one may look for a functional integral representation of the Schwinger functions. The corresponding functional measure should provide the higher-dimensional generalization of the Wiener measure. Already in the free case, some new important technical differences arise. We start by considering the free scalar field of mass m in s spacetime dimensions. The formal expression for the functional measure (the one widely used in the theoretical physics literature) is (for simplicity we omit the upper index E to denote the Euclidean field)

$$d\mu_0(\varphi) = e^{-\frac{1}{2}\int[(\nabla\varphi)^2 + m^2\varphi^2]d^s x}\, \mathcal{D}\varphi \equiv e^{-A^E}\, \mathcal{D}\varphi, \tag{5.8.1}$$

but the symbol $\mathcal{D}\varphi$ cannot define an integration measure over the field configurations, because there is no infinite-dimensional analog of the Lebesgue measure. A mathe-

---

15 For the derivation, see J. Arsac, *Fourier Transforms and the Theory of Distributions*, Prentice-Hall 1961, Sect. 5.7, and H. Bateman, *Tables of Integral Transforms*, McGraw-Hill 1954. For a discussion of the properties of the two-point Schwinger function, see J. Glimm and A. Jaffe, 1987 pp. 159ff.

matically acceptable way to define $d\mu_0$ is through the relation between its (inverse) Fourier transform and the two-point Schwinger function

$$\int e^{i\varphi(f)}\, d\mu_0(\varphi) = e^{-\frac{1}{2}<f,C f>} \equiv G(f), \quad f \in \mathcal{S}(\mathbf{R}^4), \tag{5.8.2}$$

$$<f,C f> \equiv <\varphi(f)\,\varphi(f)> = <f,(-\Delta+m^2)^{-1}f>. \tag{5.8.3}$$

The n-point functions are obtained as moments of the measure $d\mu_0$, which is Gaussian as a consequence of the Gaussian form of the right-hand side of eq. (5.8.2). In fact, for $n =$ even, one has

$$(-i\frac{d}{d\alpha})^n\, G(\alpha f)|_{\alpha=0} = \int \varphi(f)^n\, d\mu_0(\varphi) = (n-1)!!\; <f,C f>^{n/2}, \tag{5.8.4}$$

and for n odd, the moments vanish.

A relation with the above formal expression, eq. (5.8.1), is obtained by considering a finite number N of "points" in $\mathcal{S}(\mathbf{R}^s)'$: $q_j = \varphi(f_j)$, $j = 1, \ldots N$. Then, by introducing the matrix $C_{ij} \equiv <f_i, C f_j>$ and its inverse C^{-1}, one has

$$d\mu_0 = (\det C^{-1})^{1/2}\, (2\pi)^{-N/2} e^{-\frac{1}{2}\sum_{i,j} q_i C_{ij}^{-1} q_j} \prod_{k=1}^{N} dq_k. \tag{5.8.5}$$

The mathematical basis of the definition (5.8.2) is given by Minlos' theorem, which generalizes the Bochner theorem on the characterization of characteristic functions of random variables to functionals on $\mathcal{S}(\mathbf{R}^n)$.[16]

Theorem 8.1   *A functional $G(f)$ defined on $\mathcal{S}(\mathbf{R}^s)$ satisfying*
i) $G(0) = 1$
ii) continuity in the topology of $\mathcal{S}(\mathbf{R}^s)$
iii) positive definiteness

$$\sum_{i,j=1}^{N} \bar{z}_i\, z_j\, G(\bar{f}_i - f_j) \geq 0, \quad \forall z_i \in \mathbf{C},\; f_i \in \mathcal{S}(\mathbf{R}^s),\; i = 1, \ldots N, \tag{5.8.6}$$

is the (inverse) Fourier transform of a (unique) regular Borel probability measure $d\mu$ on $\mathcal{S}(\mathbf{R}^s)'$, and conversely, the (inverse) Fourier transform of a regular Borel probability measure $d\mu$ on $\mathcal{S}(\mathbf{R}^s)'$ defines a functional on $\mathcal{S}(\mathbf{R}^s)$ satisfying i)–iii).

The functional measure $d\mu_0$ represents a higher-dimensional generalization of the functional measure dU associated to the Ornstein–Uhlenbeck process, i.e., to the positive (quantum-mechanical) kernel $e^{-H_0'\tau}$, $H_0' \equiv -\Delta + x^2/2$. However, new mathematical problems arise.

As in the quantum-mechanical case, the set of (tempered) distributions for which the classical (Euclidean) action is finite has zero functional measure (since $\mathcal{D}\varphi$ does

---

[16] For a discussion of this strategy and the proof of the theorem, see B. Simon, *Functional Integration and Quantum Physics*, Academic Press 1979, esp. Sect. 2; J. Glimm and A. Jaffe, *Quantum Physics: A Functional Integral point of View*, Springer 1987, Sect. 3.4, Sect. 6.2 and Appendix A.6.

not define a measure). *The relevant distributions $\varphi^E(x) \in \mathcal{S}(\mathbf{R}^s)'$ are not continuous,* and locally they are not even signed measures, i.e., the set of such distributions has zero measure. The set of distributions with measure 1 is the set of distributions φ^E which satisfy the following local regularity: let Δ_{s-1} denote the Laplacian in any $s-1$ of the s variables of $\mathbf{R}^s$, then

$$[(-\Delta_{s-1} + 1)^{\frac{1}{2}(-s/2+1-\alpha)} \varphi^E](x), \quad 0 < \alpha \leq 1/2,$$

is locally Hölder continuous for all (Hölder) exponents $\alpha' < \alpha$.[17]

Other much more serious problems arise for the discussion of the perturbation of the free measure. The situation is much more difficult than in the quantum-mechanical case. Here, even simple polynomial interactions require (non-perturbative) renormalization. The point is that for spacetime dimensions $s > 1$, the singular behavior of the field configurations on which the measure is concentrated makes the definition of their polynomial functions problematic. The constructive strategy is to work with regularized field configurations

$$\varphi_K(x) = \int d^s y \, h_K(x - y) \, \varphi^E(y), \quad \mathcal{S}(\mathbf{R}^s) \ni h_K(x) \to_{K \to \infty} \delta(x).$$

Thus, φ_K is C^∞ and polynomially bounded, and K plays the role of an ultraviolet cutoff. In general, one needs also a volume cutoff to give meaning to the interaction. Equivalently, one may use a lattice-space regularization by introducing field configurations defined on the (discrete) points of a (square symmetric) finite lattice V.

With the introduction of the "ultraviolet" and "volume" cutoffs one has reduced the problem to a quantum-mechanical one, involving only a finite number of degrees of freedom. Then, the functional integral measure

$$d\mu_{K,V} = Z_{KV}^{-1} e^{-A_{int,V}(\varphi_K)} d\mu_0, \quad Z_{KV} = \int d\mu_0 \, e^{-A_{int,V}(\varphi_K)},$$

where $A_{int,V}$ is the contribution of the interaction to the action in the volume V, is mathematically well defined.

The problem is then to find possible counter-terms to add to the Euclidean action such that
i) the corresponding Schwinger functions have well-defined limits when $K, V \to \infty$, satisfying OS1–OS6,
ii) the coupling constant defined by such correlation functions does not vanish.

This program defines the so-called non-perturbative renormalization and has proved successful in low-dimensional ($s = d + 1 \leq 3$) quantum field theory models.[18] Unfortunately, the realistic $s = 4$ case is not under control in general (except for

---

[17] A function is Hölder continuous with exponent γ if $|f(x) - f(y)| \leq$ Const $|x - y|^\gamma$. For an overview of these problems, we strongly recommend the splendid Orientation by A. S. Wightman in *Renormalization Theory*, G. Velo and A. S. Wightman (eds.) (Erice School 1975), Reidel 1976, pp. 1–24.
[18] For an account of these hard results, see J. Glimm and A. Jaffe, 1987.

the triviality of the φ^4 model), and remains a big challenge to mathematicians and mathematical physicists.

On the other hand, one should remark that most of the non-perturbative and structural results on quantum field theory models, including gauge theories and their applications to elementary particle physics, have been obtained by using the Euclidean lattice formulation with finite ultraviolet cutoff.

6
Non-perturbative S-matrix

1 LSZ asymptotic condition in QFT

In the perturbative approach the structure of the (asymptotic) scattering states is assumed *a priori* on the basis of the particle content of the free theory (as enforced by the interaction picture). The need of a non-perturbative definition of the S-matrix in QFT, without relying on the interaction representation, emerges from the problems discussed in Chapter 2 and becomes crucial for the discussion of the *infrared problem* in QED and for the *problem of confinement* in QCD. In both cases, the deep theoretical question is which are the asymptotic states, i.e., the possible outcomes of a scattering process. A quantum field theory should (at least in principle) be able to predict the possible asymptotic particle states, without relying on the perturbative ansatz, which might be in conflict with the non-perturbative solution.

The basic step in the direction of a non-perturbative definition of the S-matrix is due to Lehmann, Symanzik, and Zimmermann (LSZ).[1] As mentioned in Chapter 2, Section 6.4, the idea is to replace the definition of the S-matrix in terms of the asymptotic limit of the states in the Schrödinger picture (which involves the unitary operators of the interaction picture), with the definition of the S operator as the intertwiner between the *in* and *out* asymptotic limits of the field operators in the Heisenberg picture.

The advantage is that the control of the field asymptotic limits does not require the splitting of the Hamiltonian into a "free" and an interaction part. The mathematical problem is then to give a precise meaning to the intuitive idea that for asymptotic times $(t \to \mp\infty)$ the interacting fields should approach asymptotic (*in/out*) (free) fields.

We consider for simplicity the case of a scalar (hermitian) field $\varphi(x)$. In general, an interacting field may have non-trivial matrix elements between the vacuum and different one-particle states (the two-point spectral measure $d\rho(m^2)$ of the field may contain more that one δ functions $\delta(m^2 - m_i^2)$). Therefore, one cannot expect that $\varphi(\mathbf{x}, t)$ has a free field of given mass as its asymptotic limit, and one must in some way isolate the mass for which one explores the existence of the asymptotic limit.

For this purpose, given a normalized positive-frequency *smooth solution* f of the free Klein–Gordon equation with mass m, i.e., of the form

$$f(\mathbf{x}, t_0) = (2\pi)^{-3/2} \int d^4 p \, \delta(p^2 - m^2) \, \theta(p_0) e^{-ipx} g(\mathbf{p}), \quad g(\mathbf{p}) \in \mathcal{D}(\mathbf{R}^3), \qquad (6.1.1)$$

<hr>

[1] H. Lehmann, K. Symanzik, and W. Zimmermann, Nuovo Cim. **1**, 425 (1955), hereafter referred to as LSZ I; *ibid.* **6**, 319 (1955) (LSZ II).

let

$$\varphi_f(x_0) \equiv i \int d^3x \, [\, \bar{f}(\mathbf{x}, x_0)\partial_0\varphi(\mathbf{x}, x_0) - \partial_0\bar{f}(\mathbf{x}, x_0)\,\varphi(\mathbf{x}, x_0)\,] \qquad (6.1.2)$$

$$\equiv i \int d^3x \, \bar{f}(\mathbf{x}, x_0) \overset{\leftrightarrow}{\partial_0} \varphi(\mathbf{x}, x_0).$$

The *LSZ asymptotic condition* states that there are free (asymptotic) fields of mass m, $\varphi_{as}(x)$, $as = in/out$ such that,

$$\text{weak} - \lim_{\tau \to \mp\infty} \varphi_f(t + \tau) = \varphi_f^{in/out}(t) = \varphi_f^{in/out}, \qquad (6.1.3)$$

the right-hand side being independent of time as a consequence of the free evolution with mass m obeyed by both f and φ^{as}.

A few technical remarks are in order, for a mathematically consistent definition, without running into divergences. First, since the field operators are unbounded, the operator weak limit is not defined and should therefore be understood as the weak limit of the vectors $\varphi_f(t)\, D$, with D a suitable dense domain containing the vacuum vector.

Since one will need the asymptotic limit of polynomials of the fields, one meets the technical difficulty that on one side a natural candidate is the domain $\mathcal{D}_0$ of the local states defined by the Wightman functions (see Chapter 3, Sections 1, 3), where the polynomials of the interacting fields are well defined; on the other side, the natural domain for polynomials of the asymptotic fields is $\mathcal{D}^{as}$, obtained by applying the asymptotic fields $\varphi^{in/out}$ to the vacuum,[2] and it is not obvious that the interacting fields are well defined there.

A second more delicate problem is the existence of the integrals $\varphi_f(t)$, since, by the inevitable ultraviolet singularities, the fields are operator-valued distributions in $\mathcal{D}(\mathbf{R}^4)'$ or in $\mathcal{S}(\mathbf{R}^4)'$, in general not allowing restrictions at sharp times, and therefore, as remarked by Greenberg and Wightman,[3] a regularization is necessary to avoid divergences.

The mathematical control of the asymptotic condition should not be regarded as a merely pedantic problem, since it does not only provide a non-perturbative definition (and possibly some hint on a non-perturbative computation) of the S-matrix, but it

---

[2] Actually, it is more convenient to consider the set of asymptotic states in which the wave packets of the (asymptotic) particles do not overlap in velocity space, since, as shown by K. Hepp, Brandeis lectures 1965, on such states one can also derive the LSZ reduction formulas from the asymptotic condition.

The limit of $\varphi_f(t)\,D$ cannot be in the strong sense, since otherwise

$$< \varphi_f(t + \tau)\,\varphi_f(t + \tau) > \to < \varphi_f^{in/out}(t)\,\varphi_f^{in/out}(t) >,$$

for $\tau \to \mp\infty$ and, by the time-independence of such vacuum expectations, the two-point function of φ would coincide with that of a free field.

[3] Quoted in LSZ II, p. 321.

also allows for the derivation of the so-called LSZ reduction formulas, which proved very useful for the physical applications.[4]

In the LSZ approach, the asymptotic condition is postulated, leaving open the problem of its relation with the properties of QFT; therefore, for the reasons discussed above, a very important issue is its *derivation from the general principles of QFT*, starting with a more rigorous reformulation of it.

2   Haag–Ruelle scattering theory (massive case)

A solution of the above important problems has been obtained by the Haag–Ruelle theory (see below) and by Hepp revisitation of the LSZ approach within the Wightman framework,[5] in the case of *mass gap*. This means that the spectrum of the generators of the spacetime translations, as determined by the Wightman functions and the reconstruction theorem, satisfies the following:

Mass gap condition. Above the isolated point $p = 0$ (corresponding to the vacuum state) there is a gap up to an *isolated hyperboloid* $p^2 = m^2$ (mass gap), with finite multiplicity (one-particle states), followed by a continuous spectrum starting at $p^2 = (2\,m)^2$ (two-particle states). Clearly it is enough that such a mass gap condition holds in the subspace characterized by the conserved quantum numbers of the asymptotic states under investigation.

As we shall see, the Haag–Ruelle theory does not only provide a basis for the LSZ strategy, but it also clarifies the crucial role of the cluster property for the existence of the asymptotic states and therefore of the S-matrix.

2.1   One-body problem

The first step in the Haag–Ruelle (HR) theory is the solution of the so-called one-body problem, i.e., the regularization of eq. (6.1.2).

The existence of free asymptotic fields of mass m should be related to the existence of an isolated hyperboloid $p^2 = m^2$ in the two-point spectral function of the field φ.

Let Ψ_f denote the one-particle states labeled by the wave functions $f \in \mathcal{S}(\mathbf{R}^3)$ and $\mathcal{H}_1$ the subspace of such states; by the cyclicity of the vacuum with respect to the local field algebra $\mathcal{F}$ there must be an element $B \in \mathcal{F}$, typically a sum of terms of the form

$$\int d^4x_1 \ldots d^4x_n \, g(x_1, \ldots x_n) \, \varphi(x_1) \ldots \varphi(x_n), \quad g \in \mathcal{S}(\mathbf{R}^{4n}),$$

[4] See, e.g., R. Hagedorn, *Introduction to Field Theory and Dispersion Relations*, Pergamon Press 1963, and S. Schweber, *Introduction to Relativistic Quantum Field Theory*, Harper and Row 1961, Sect. 18 b.

[5] R. Haag, Phys. Rev. **112**, 669 (1958); see also *Local Quantum Physics*, Springer 1996, Sect. II.4; D. Ruelle, Helv. Phys. Acta, **35**, 147 (1962). For general accounts, see also A. S. Wightman, in *Theoretical Physics*, Lectures presented at the Seminar on Theoretical Physics, Trieste 16 July–25 August 1962, IEAA Vienna 1963, Sect. 2.4; R. Jost, *The general Theory of Quantized Fields*, AMS 1962, Chap. VI; N. N. Bogoliubov, A. A. Logunov, and I. T. Todorov, *Introduction to Axiomatic Quantum Field Theory*, W. A. Benjamin 1975, Chap. 13; N. N. Bogolubov, A. A. Logunov, A. I. Oksak, and I. T. Todorov, *General Principles of Quantum Field Theory*, Kluwer Academic 1990, Chap. 12.

which interpolates between the vacuum and the one particle states, i.e., such that

$$(\Psi_0, B\,\Psi_f) \neq 0. \tag{6.2.1}$$

One of the relevant points of the Haag–Ruelle theory is that asymptotic states may be obtained through the asymptotic limits of elements of the field algebra and not exclusively by the asymptotic limits of the basic local covariant fields which generate $\mathcal{F}$. This realization has important practical consequences, as we have seen in the discussion of the $\pi_0 \to 2\gamma$ decay, in Chapter 4, Section 6.2.

For simplicity, in the following we consider the case when $B = A(g)$, $g \in \mathcal{S}(\mathbf{R}^4)$, with $A(x)$ a local covariant field, e.g., $\varphi(x)$, as was assumed in the original LSZ formulation, eqs. (6.1.2) and (6.1.3).

Then, one isolates the one particle hyperboloid by defining

$$\tilde{A}_1(p) \equiv \tilde{A}(p)\,\tilde{h}(p^2), \tag{6.2.2}$$

with $\tilde{h} \in C^\infty$, $0 \leq \tilde{h} \leq 1$, $\tilde{h}(p^2) = 1$ in a small neighborhood of $\{p^2 = m^2,\, p_0 > 0\}$, $\tilde{h}(p^2) = 0$ outside. A_1 is not a free field, since $\operatorname{supp}\tilde{A}(p)$ is not contained in the spectrum of P_μ and $(p^2 - m^2)\tilde{A}_1(p) \neq 0$; however, $(p^2 - m^2)\tilde{A}_1(p)\Psi_0 = 0$. It is convenient to normalize A in such a way that the two-point function of $A_1(x)$ is that of a free field of mass m; in particular, $< A_1 >_0 = 0$.

The field $A_1(x)$ provides the regularized substitute of the field $\varphi(x)$ for the LSZ asymptotic condition (eq. (6.1.3)), and one may consider the asymptotic limits of the operators

$$A_f(t) \equiv i \int d^3x\, \bar{f}(\mathbf{x}, t)\, \overset{\leftrightarrow}{\partial_0}\, A_1(\mathbf{x}, t). \tag{6.2.3}$$

In fact, since $\tilde{h} \in \mathcal{S}(\mathbf{R})$ in the variable p_0 ($\mathbf{p}$ fixed), one has that $\forall \tilde{f} \in \mathcal{S}(\mathbf{R}^3)$, $\tilde{f}(\mathbf{p})\tilde{h}(p^2)e^{ip_0 t} \in \mathcal{S}(\mathbf{R}^4)$, and the same is true for all its time derivatives; therefore $A_f(t)$ defined by eq. (6.2.3) is well defined on $\mathcal{D}_0$, C^∞ in t, and satisfies

$$A_f(t)\Psi_0 \in \mathcal{H}_1, \quad A_1^*(g)\Psi_0 = 0,$$

(the last equation follows from the spectral condition, which forbids the existence of states with $p_0 < 0$).

The technical improvement with respect to eq. (6.1.2) is the selection of (a neighborhood of) the mass shell at the level of $A(x)$, the three-dimensional smearing with the smooth solution f of the free Klein–Gordon equation being not enough for selecting the mass shell.

The smearing with f in eq. (6.2.3) effectively subtracts the free time evolution from $A_1(\mathbf{x}, x_0)$, which would lead to oscillating time factors; in fact $A_f(t)$ is independent of t if $A(x)$ is a free field. As we shall see, such an improvement, together with a refined cluster property, gives the existence of the asymptotic multiparticle states as a *strong* limit of vectors of the form $\Phi(t) = A_{f_1}(t)\ldots A_{f_n}(t)\Psi_0$.

2.2   Large time decay of smooth solutions

The existence of the asymptotic limit, and therefore of the S-matrix, relies on two basic ingredients: one is the large t decay of the smooth solutions of the free Klein–Gordon equation, and the other is (a refined version of) the cluster property for the spacelike asymptotic behavior of the correlation functions (see Section 2.3 below).

The important estimates for the large time decay are the following

$$|t|^{3/2}\max_{\mathbf{x}}|f(\mathbf{x},t)| < C, \tag{6.2.4}$$

$$\int d^3x\,|f(\mathbf{x},t)| \le C\,(1+|t|^{3/2}), \tag{6.2.5}$$

$$|t|^{1/2}\max_{\mathbf{x}}|\mathbf{x}|\,|f(\mathbf{x},t)| < C, \tag{6.2.6}$$

(with C a suitable constant), which follow from the following bound:

$$|\lambda|^{3/2}\,|f(\lambda u)| \le C,\quad u=(\mathbf{x},t),$$

if for some $\lambda \in (0,\infty)$, λu intersects the support of $\delta(p^2-m^2)g(\mathbf{p})$ and otherwise $f(\lambda u)$ has a fast decrease. Such a bound can be (heuristically) justified by the stationary phase dominance of the integral[6]

$$f(\mathbf{x},t) = (2\pi)^{-3/2}\int d^3p\,(2\omega(\mathbf{p}))^{-1}g(\mathbf{p})\,e^{i\mathbf{p}\mathbf{x}-i\omega(\mathbf{p})t},\quad g(\mathbf{p})\in\mathcal{D}(\mathbf{R}^3),$$

for large t, $\mathbf{v}\equiv\mathbf{x}/t$ fixed. In fact, the solution of the stationarity condition

$$(\partial/\partial p_i)(\mathbf{p}\mathbf{x}-\omega(\mathbf{p})t) = x_i - p_i\,t(\mathbf{p}^2+m^2)^{-1/2} = 0$$

is $\mathbf{p}=m\mathbf{v}/\sqrt{1-v^2}$, yielding

$$f(\mathbf{x},t) \sim Ct^{-3/2}e^{-imt/\gamma}\gamma^{3/2}\,g(m\gamma\mathbf{v}) + O(t^{-5/2}),\quad \gamma\equiv(1-\mathbf{v}^2)^{-1/2}.$$

For a more detailed and complete proof we refer to the literature.[7]

2.3   Refined cluster property

The refined cluster property involves the truncated correlation functions, recursively defined by

$$< A(x_1) >=< A(x_1) >_T \equiv < 1 >_T,$$

$$< A(x_1)\,A(x_2) >\equiv< 1,2 >=< 1,2 >_T + < 1 >_T< 2 >_T$$

$$< 1,2,3 >=< 1,2,3 >_T + < 1,2 >_T< 3 >_T + < 1 >_T< 2,3 >_T +$$

$$+ < 1,3 >_T< 2 >_T + < 1 >_T< 2 >_T< 3 >_T,\quad etc.$$

<hr>

[6] R. Haag, *Local Quantum Physics*, Springer 1996, p. 89.
[7] D. Ruelle 1962; R. Jost 1962, Chap. VI, Sects. 3,4; A. S. Wightman, Recent achievements of axiomatic field theory, in *Theoretical Physics*, IEAA Vienna 1963, Sect. 2.5.

In general,

$$< 1, \ldots n > = \sum \prod < j_1, \ldots j_k >_T, \qquad (6.2.7)$$

where the sum is over all partitions of $1, \ldots n$ into non-empty subsets $j_1, \ldots j_k$, and the product is over the truncated correlation functions of the subsets (all x_i's occurring in the subsets in the order in which they occur in $1, \ldots n$). The expansion of a generic correlation function into truncated ones explicitly shows all the possible vacuum contributions in the intermediate states.

The refined cluster property states that in a theory with a mass gap, the truncted (Wightman) functions $\mathcal{W}_T(x_1, \ldots x_n) \equiv < A(x_1) \ldots A(x_n) >_T$ fall off faster than any inverse power in the spacelike directions, i.e., $\forall f \in \mathcal{S}(\mathbf{R}^{4n})$,

$$\int d^4 x_1 \ldots d^4 x_n \, f(x_1, \ldots x_n) \, \mathcal{W}_T(x_1 + \mathbf{a}_1, \ldots x_n + \mathbf{a}_n) \in \mathcal{S}(\mathbf{R}^{3(n-1)}). \qquad (6.2.8)$$

Clearly, by translational invariance the above integral is only a function of the difference variables $\boldsymbol{\alpha}_j \equiv \mathbf{a}_{j+1} - \mathbf{a}_j$.

Eq. (6.2.8) is a refinement of the standard form of the cluster property, which states the decay when the n-points are split into two spacelike separated clusters; here, the fall-off faster than any inverse power is stated for any division of the n-points into several spacelike separated clusters. The physical meaning is rather simple if the (smeared) fields describe (local) observables, since the factorization of the expectation values corresponds to the absence of correlations for infinite spacelike separations, and the rate of fall-off is a sign of mass gap (exponential decay).

This kind of consideration led Haag[8] to consider, from the start, *quasi-local operators* $B(x)$, essentially characterized by the property that in the case of mass gap their truncated correlation functions satisfy, $\forall f \in \mathcal{S}(\mathbf{R}^{4n})$,

$$\int dx_1 \ldots dx_n \, f(x_1, \ldots x_n) \, < B(x_1 + \mathbf{a}_1) \ldots B(x_n + \mathbf{a}_n) >_T \in \mathcal{S}(\mathbf{R}^{3(n-1)}). \qquad (6.2.9)$$

By a general result of the theory of distributions,[9] this implies that

$$< B(\mathbf{x}_1, t) \ldots B(\mathbf{x}_n, t) >_T \equiv F(\boldsymbol{\xi}_1, \ldots \boldsymbol{\xi}_{n-1}) \in O'_C(\mathbf{R}^{3(n-1)}), \qquad (6.2.10)$$

$(\boldsymbol{\xi}_j = \mathbf{x}_j - \mathbf{x}_i)$, i.e. they are bounded by $\prod_{k=1}^{n-1} c_k (1 + \boldsymbol{\xi}_k^2)^{-2}$. Haag's ideas were later developed by Ruelle, who proved that in the case of mass gap, every local (Wightman) field $A(x)$ defines a quasi-local operator, i.e., eq. (6.2.9) holds.[10]

Since

$$\tilde{f}(p_1, \ldots p_n) \prod_{i=1}^{n} \tilde{h}_i(p_i^2) \in \mathcal{S}(\mathbf{R}^{4n}), \quad \forall f \in \mathcal{S}(\mathbf{R}^{4n}),$$

<hr>

[8] R. Haag, in *Colloque Int. sur les Problèmes Mathématiques de la Théorie Quantique des Champs*, Lille 1957, CNRS, Paris, pp. 151–62; Phys. Rev. **112**, 669 (1958).

[9] L. Schwartz, *Théorie des distributions*, Vol.II, Hermann 1959, Théor. IX.

[10] D. Ruelle 1962; R. Jost 1962, Chap. VI, second auxiliary theorem.

eq. (6.2.9) is satisfied also by $A_1(x)$ (defined in eq. (6.2.2)), and this implies the crucial estimate

$$| < A_{f_1}(t) \ldots A_{f_n}(t) >_T | \leq C \, |t|^{-3(n-2)/2}. \tag{6.2.11}$$

In fact, eq. (6.2.10) implies that

$$\left| \int d^3x_1 \ldots d^3x_n \, f_1(\mathbf{x}_1,t) \ldots f_n(\mathbf{x}_n,t) < A_1(\mathbf{x}_1,t) \ldots A_1(\mathbf{x}_n,t) >_T \right| \leq$$

$$\leq \int d^3x_1 |f_1(\mathbf{x}_1,t)| \int d^3\xi_1 \ldots d^3\xi_{n-1} \prod_{i=2}^{n} \max|f_i(\mathbf{x}_i,t)| \prod_{k=1}^{n-1} c_k (1 + \boldsymbol{\xi}_k^2)^{-2}$$

$$\leq C'(1 + |t|^{3/2}) \, |t|^{-3(n-1)/2} \, \left(\int d^3\xi \, (1 + \boldsymbol{\xi}^2)^{-2} \right)^{n-1} \sim C_1 \, |t|^{-3(n-2)/2}.$$

The same bound holds if some of the f_j's are replaced by its time derivative, since a time derivative of a smooth function is also a smooth solution. As discussed in the next subsection, the *cluster property is* the really crucial and physically *essential property for the existence of* the asymptotic states and of the S-matrix.[11]

2.4   The asymptotic limit

The complete control of the asymptotic limit, and consequently the non-perturbative definition of the S-matrix, is provided by the following

Theorem 2.1   *(Haag–Ruelle) The following asymptotic limits exist*

$$strong - \lim_{t \to \mp\infty} (\Phi(t) \equiv A_{f_1}(t) \ldots A_{f_n}(t) \, \Psi_0) = \Phi^{in/out}(f_1, \ldots f_n), \tag{6.2.12}$$

independently of the Lorentz frame chosen in eq. (6.2.3), and define asymptotic in/out states.
   The operators A_f^{as}, as $= in/out$, defined by

$$A_f^{as} \Phi^{as}(f_1, \ldots f_n) = \Phi^{as}(f, f_1, \ldots f_n), \tag{6.2.13}$$

define two free (scalar hermitian) fields $A^{as}(\mathbf{x}, x_0)$, with

$$A_f^{as} = i \int d^3x \bar{f}(\mathbf{x},t) \overset{\leftrightarrow}{\partial_0} A^{as}(\mathbf{x},t), \tag{6.2.14}$$

which transform covariantly under the Poincaré group

$$U(a, \Lambda) A^{as}(x) U(a, \Lambda)^{-1} = A^{as}(\Lambda x + a) \tag{6.2.15}$$

and satisfy

$$A^{out}(-x) = \theta^* A^{in}(x)\theta, \tag{6.2.16}$$

---

[11] A more general strategy, which covers the case of a sharp hyperboloid $p^2 = m^2 \neq 0$ immersed in a spectrum of massless particles, has been proposed by W. Dybalski (Lett. Math. Phys. **72**, 27 (2005)). For an updated overview, see D. Buchholz and S. J. Summers, Scattering in relativistic quantum field theory. Fundamental concepts and tools, in *Encyclopedia of Mathematical Physics*, J.-P. Françoise *et al.* (eds.), Elsevier 2006.

where θ is the PCT operator of the interacting theory.

Proof. It is enough to show that (for large $|t|$)

$$|t|^{3/2}\,||d\Phi(t)/dt||\ \leq C,\tag{6.2.17}$$

since then,

$$||\Phi(t_1)-\Phi(t_2)||=||\int_{t_1}^{t_2}d\tau\,d\Phi(\tau)/d\tau||\leq\int_{t_1}^{t_2}d\tau||d\Phi(\tau)/d\tau||\leq$$

$$\leq C\int_{t_1}^{t_2}d\tau\,\tau^{-3/2}=C'\,|t_1^{-1/2}-t_2^{-1/2}|\longrightarrow 0,\quad\text{as }|t_1|,\,|t_2|\to\infty.$$

For this purpose one remarks that $||d\Phi(t)/dt||^2$ can be expanded in products of truncated correlation functions, each containing an even number of fields $A_f(t)$, two of which are derived with respect to t. The terms which consist only of products of two-point truncated functions vanish, since $A_f(t)\Psi_0$ is independent of time. The other terms involve at least either two three-point or one higher-point truncated function; therefore, by the estimate (6.2.11) they decrease at least as $|t|^{-3}$, and therefore the bound (6.2.17) holds.

Also, the second part of the theorem follows from the estimate (6.2.17). For this purpose we consider the scalar product of two vectors,

$$\Phi(t)=A_{f_1}(t)\ldots A_{f_n}(t)\,\Psi_0,\quad \Psi(t)=A_{g_1}(t)\ldots A_{g_m}(t)\,\Psi_0$$

and expand it in terms of truncated functions, as above. By eq. (6.2.11), in the limit $t\to\mp\infty$, only those terms survive which consist of products of two-point functions. On the other hand, since the two-point function of $A_1(x)$ is that of a free field, this expansion coincides with that corresponding to states of a free scalar field theory. One can then introduce the field A_f^{as} by eq. (6.2.13). The covariance of A^{as} follows from the property that the limit $t\to\mp\infty$ of $\Phi(t)$ is independent of the Lorentz frame, i.e., for $|t|\to\mp\infty$, $||\Phi_\Lambda(t)-\Phi(t)||\to 0$, where $\Phi_\Lambda(t)$ is defined as $\Phi(t)$ with the replacements $f_j(x)\to f_j(\Lambda^{-1}x)$, $A_1(x)\to A_1(\Lambda^{-1}x)$.[12] Eq. (6.2.16) follows easily from $\theta^* A(x)\theta=A(-x)$.

The solution of the one-body problem trivially yields the existence of one-particle asymptotic states $A_f(t)\,\Psi_0=\Phi^{as}(f)$, which are stable states as a consequence of the isolated hyperboloid at the bottom of the spectrum. The non-trivial result of Theorem 2.1 is the asymptotic limit of many particle states and the related definition of the asymptotic fields, i.e., of the S-matrix. It is instructive to compare the proof with that given for potential scattering (Chapter 2, Section 6.1); in both cases the large time behavior is governed by the large-distance fall-off of the "interaction", and the exponential decay of the cluster property, due to the mass gap, plays the same role as the short range of the potential. In conclusion, the Haag–Ruelle theory clarifies the *strong connection between the cluster property and the existence of the S-matrix.*

---

[12] It is enough to show that the infinitesimal variations of Φ under the Lorentz boosts vanish in the asymptotic limit, again due to eq. (6.2.11). The details of the simple argument can be found in R. Jost 1962, Chap. VI, Lemma 12.

2.5   The S-matrix and asymptotic completeness

Since A^{out} is a free scalar field, there exists a PCT operator θ^{out} such that

$$\theta^{out} * A^{out}(x)\, \theta^{out} = A^{out}(-x). \tag{6.2.18}$$

By combining this equation with eq. (6.2.16) we obtain the existence of an operator $S^* \equiv \theta^{out}\,\theta^*$ such that

$$A^{out}(x) = S^* A^{in}(x)\, S, \tag{6.2.19}$$

i.e., S is the S-matrix.

The PCT operator θ^{out} is uniquely defined if the vacuum is cyclic for the *out*-asymptotic fields (and therefore, by eq. (6.2.16), is cyclic also for the *in*-fields). This property is known as *asymptotic completeness*.

W7. Asymptotic completeness. Denoting by $\mathcal{H}^{in}$ the subspace spanned by the *in*-asymptotic states, the condition states that

$$\mathcal{H}^{in} = \mathcal{H}. \tag{6.2.20}$$

Such a condition is not implied by the Wightman axioms,[13] and it has to be added to them for a complete description of the states in terms of asymptotic states.

If the asymptotic completeness holds, the S-matrix is not only an isometric operator between $\mathcal{H}^{in}$ and $\mathcal{H}^{out}$, but a *unitary* operator in $\mathcal{H}$ (*unitarity of the S-matrix*).

The existence of an isometric S-matrix can also be argued by using the mass gap condition: by eq. (6.2.13) the vacuum is annihilated by $A_f^{as\,*}$, and therefore defines a Fock representation of the asymptotic fields. Thus, the representations of the asymptotic fields are unitarily equivalent, and the S-matrix is the intertwining operator between the two representations.

3   Buchholz scattering theory (massless particles)

The scattering theory discussed in the previous section strongly relied on the exponential fall-off of the correlation functions for large spacelike separations (mass gap condition). Actually, in the past such a property has been regarded as a necessary condition for the possible definition of the S-matrix and the existence of long-range correlations as an essential difficulty for the LSZ formulation. Indeed, in such a case one meets the so-called infrared problem, whose non-perturbative solution has been partly discussed in the QED case,[14] but it is still out of (mathematical) control in quantum chromodynamics (confinement problem).

---

[13] This is the case of a generalized free field; see, e.g., N. N. Bogoliubov, A. A. Logunov, and I. T. Todorov, *Introduction to Axiomatic Quantum Field Theory*, W. A. Benjamin 1975, Sects. 12.5 and 13.5.

[14] J. Fröhlich, G. Morchio, and F. Strocchi, Ann. Phys. **119**, 241 (1979); G. Morchio and F. Strocchi, Nucl. phys. **B211**, 471 (1983), **B232**, 547 (1984); G. Morchio and F. Strocchi, Infrared problem, Higgs phenomenon and long-range interactions, Lectures at the Erice School, 1985, in *Fundamental Problems of Gauge Field Theory*, G. Velo and A. S. Wightman (eds.), Plenum Press 1986, p. 301.

The extension of the Haag–Ruelle theory to massless particles has been obtained by Buchholz, who proved the existence of asymptotic fields corresponding to massless particles, by exploiting Huyghens' principle and locality. Here we present a brief and sketchy account of Buchholz scattering theory for massless particles, referring to his important papers for a detailed and complete analysis.[15]

3.1   Huyghens' principle and locality

In $3+1$ dimensional spacetime, the solutions of the free wave equation propagate without dispersion. This means that not only the *propagation is hyperbolic*, i.e., if the initial data $f(\mathbf{x}, 0)$, $\dot{f}(\mathbf{x}, 0)$ have support in a bounded region K, then

$$f(\mathbf{x}, t) = 0, \quad \text{in} \quad \mathcal{O}', \tag{6.3.1}$$

where $\mathcal{O}'$ denotes the spacelike complement of the double cone $\mathcal{O}$ subtended by K, but one further has

$$f(\mathbf{x}, t) = 0, \quad \text{in} \quad \mathcal{O}_\pm, \tag{6.3.2}$$

where $\mathcal{O}_\pm$ denotes the forward/backward cone with apex respectively at the top/ bottom of $\mathcal{O}$ (hereafter called the *future/past tangent* of $\mathcal{O}$). Property (6.3.2) is peculiar of the free wave equation (in $3 + 1$ dimensions), and it means that the influence of the initial data propagate with fixed velocity, i.e., without dispersion (*Huyghens' principle*).

This property will allow a localization of the asymptotic fields, and will make possible the exploitation of locality as a (partial) substitute of the strong form of the cluster property.

For any bounded region $\mathcal{O}$, typically a double cone, we shall consider the local field algebra $\mathcal{F}(\mathcal{O})$ generated by the local Wightman fields localized in $\mathcal{O}$, i.e., smeared with test functions with support in $\mathcal{O}$, and denote by $\mathcal{F}$ the algebra generated by taking the union over all possible (double cones) $\mathcal{O}$. Actually, for simplicity, and in order to avoid domain problems as much as possible, we shall consider the algebra $\mathcal{A}$ generated by bounded functions of localized Wightman fields: $\mathcal{A} = \cup_\mathcal{O} \mathcal{A}(\mathcal{O})$, with $\mathcal{A}(\mathcal{O})$ the algebra of bounded operators localized in $\mathcal{O}$. As usual, locality, translation covariance

$$A(\mathbf{x}, x_0) = U(\mathbf{x}, x_0)\, A\, U(\mathbf{x}, x_0)^{-1}, \quad \forall A \in \mathcal{A},$$

relativistic spectral condition, and cyclicity of the vacuum with respect to $\mathcal{A}$, are assumed.

The mass gap condition is replaced by the following:

Massless state condition. There are states $\Phi \neq \Psi_0$ satisfying

$$(H - |\mathbf{P}|)\, \Phi = 0. \tag{6.3.3}$$

$\mathcal{P}_1$ will denote the projection on the corresponding subspace.

[15] D. Buchholz, Comm. Math. Phys. **42**, 269 (1975); *ibid.* **52**, 147 (1977) and Lectures at the XVth Winter School of Theoretical Physics in Karpacz, 1978, Acta Universitatis Wratislaviensis No 519, p. 189.

3.2 One-body problem

As in the Haag–Ruelle theory, the first step is the solution of the one-body problem. Given an operator $A \in \mathcal{A}(\mathcal{O})$, with $< A >_0 = 0$ and a distributional solution $g(\mathbf{x}, t)$ of the free wave equation, corresponding to the initial data $g(\mathbf{x}, 0) = 0$, $\dot{g}(\mathbf{x}, 0) = 2\pi\, \delta(\mathbf{x})$, i.e.,

$$g(\mathbf{x}, x_0) = \pi^{-1}\varepsilon(x_0)\, \delta(x_0^2 - |\mathbf{x}|^2),$$

we consider the (formal) expression

$$A_g(t) = \int_{x_0=t} d^3x\, g(\mathbf{x}, x_0)\, \overset{\leftrightarrow}{\partial_0}\, A(\mathbf{x}, x_0).$$

For simplicity, in the sequel, we shall consider only the first of the two terms on the right-hand side of the above equation, namely

$$A(t) \equiv -\pi^{-1}\int d^3x\, \varepsilon(t)\, \delta(t^2 - |\mathbf{x}|^2)\, \partial_t A(\mathbf{x}, t) = -2|t|\int d\omega(\mathbf{e})\, \dot{A}(t\mathbf{e}, t), \qquad (6.3.4)$$

where a radial integration has been performed, $d\omega(\mathbf{e})$ is the normalized invariant measure on the unit sphere $S^2 = \{\mathbf{e}, \mathbf{e}^2 = 1\}$, and the right hand side can be given the meaning of a Bochner integral (i.e., by norm-dominated convergence, $\int f d\mu = \lim_{n\to\infty}\int f_n d\mu$, if $\lim_{n\to\infty}\int ||f - f_n||d\mu = 0$).

By applying $A(t)$ to the vacuum we obtain

$$iA(t)\Psi_0 = 2|t|\int d\omega(\mathbf{e})[H,\, A(t\mathbf{e}, t)]\Psi_0 = 2|t|\int d\omega(\mathbf{e})\, e^{it(H-\mathbf{P}\cdot\mathbf{e})}HA\Psi_0$$

$$= i\, \mathrm{sign}\, t\, (e^{it(H-|\mathbf{P}|)} - e^{it(H+|\mathbf{P}|)})\, |\mathbf{P}|^{-1}\, H\, A\, \Psi_0.$$

In the spectral representation of the right-hand side, the exponentials have rapid oscillations for large t, and, by the Riemann–Lebesgue lemma, the only surviving contributions are those from the components of $|\mathbf{P}|^{-1}\, H\, A\, \Psi_0$ which lie in the kernels of $(H \pm |\mathbf{P}|)$. Since $A_t \Psi_0$ has no vacuum contribution, by eq. (6.3.3) and the spectral condition, one has

$$\mathrm{weak} - \lim_{t\to\infty} A(t)\Psi_0 = \mathcal{P}_1|\mathbf{P}|^{-1}HA\Psi_0 = \mathcal{P}_1 A\Psi_0.$$

A strong limit can be obtained by performing a suitable time average,

$$\bar{A}(t) \equiv T(t)^{-1}\int_t^{T(t)} dt'\, A(t'),$$

where $T(t)$ is a slowly increasing function, e.g., $T(t) = \log(1 + t^2)$. Then, by the mean ergodic theorem,[16] it follows that

$$\mathrm{strong} - \lim_{t\to\infty} \bar{A}(t)\Psi_0 = \mathcal{P}_1 A\Psi_0. \qquad (6.3.5)$$

The time average is crucial for obtaining the strong asymptotic limit.

---

[16] See, e.g., M. Reed and B. Simon, *Methods of Modern Mathematical Physics, Vol. 1* , Academic Press 1972, Theor. II.11.

3.3 Asymptotic limit

The first result is the existence of asymptotic fields A^{out} on a dense set.

Theorem 3.1 *Given a local operator $A \in \mathcal{A}(\mathcal{O})$, with the property $\mathcal{P}_1 A \Psi_0 \neq 0$, one can construct a corresponding* **asymptotic** *(densely defined)* **operator** A^{out} *such that*

$$A^{out}\Psi_0 = strong - \lim_{t \to \infty} \bar{A}(t)\Psi_0 = \mathcal{P}_1 A \Psi_0. \tag{6.3.6}$$

Furthermore, if $\mathcal{O}_+$ denotes the future tangent of $\mathcal{O}$, $\forall B \in \mathcal{A}(\mathcal{O}_1)$, $\mathcal{O}_1 \subset \mathcal{O}_+$, one has

$$[A^{out}, B] = 0, \tag{6.3.7}$$

(localization property).

Proof. By Huyghens' principle, if A is localized in $\mathcal{O}$, then the operator $A(t)$ defined in eq. (6.3.4) is localized in the region $\{\mathcal{O} + (t\mathbf{e}, t); |\mathbf{e}| = 1\}$. Hence, the operator $\bar{A}(t)$ is localized in

$$\mathcal{O}_{l.c} \equiv \cup_{\tau > t} \{\mathcal{O} + (\tau\mathbf{e}, \tau); |\mathbf{e}| = 1\}$$

and for t large enough $\mathcal{O}_{l.c}$ is spacelike separated from any bounded region $\mathcal{O}_1 \subset \mathcal{O}_+$. Then, by locality $\forall B \in \mathcal{A}(\mathcal{O}_1)$, $\mathcal{O}_1 \subset \mathcal{O}_+$, $\bar{A}(t)$ commutes with B, and eq. (6.3.5) gives

$$strong - \lim_{t \to \infty} \bar{A}(t)\, B\,\Psi_0 = B \mathcal{P}_1 A \Psi_0. \tag{6.3.8}$$

Thus, the strong limit exists on the domain $\mathcal{D}_+ \equiv \mathcal{A}(\mathcal{O}_+)\,\Psi_0$, which is dense by the Reeh–Schlieder theorem; this defines $\mathcal{A}^{out}$.

In order to control the limit of strings of operators $\bar{A}_i$ one needs a weak form of Ruelle's cluster property adapted to the massless case, namely, the following relations:

$$\|\bar{A}_1(t)\dots\bar{A}_n(t)\Psi_0\| \leq C, \quad \text{uniformly in } t, \tag{6.3.9}$$

$$\lim_{t \to \infty} (\Psi_0, A_1(t)\dots A_n(t)\Psi_0) = \sum (\Psi_0, A_{i_1}\mathcal{P}_1 A_{i_2}\Psi_0)\dots(\Psi_0, A_{i_{n-1}}\mathcal{P}_1 A_{i_n}\Psi_0) \tag{6.3.10}$$

for $n = $ even and zero otherwise, where the sum is over all partitions of $(1,\dots n)$ into ordered pairs.[17]

Theorem 3.2 *For a collection of local operators $A_i \in \mathcal{A}(\mathcal{O}_i)$, $i = 1,\dots n$, the following asymptotic limits exist:*

$$weak - \lim_{t \to \infty} \bar{A}_1(t)\dots\bar{A}_n(t)\Psi_0 = \Psi^{out}(A_1,\dots A_n) \tag{6.3.11}$$

and if $A \in \mathcal{A}(\mathcal{O})$, $A_i \in \mathcal{A}(\mathcal{O}_+)$, then Ψ^{out} is in the domain of the operator A^{out} defined by Theorem 3.1 and

$$A^{out}\Psi^{out}(A_1,\dots A_n) = \Psi^{out}(A, A_1,\dots A_n). \tag{6.3.12}$$

---

[17] For the proof, see D. Buchholz, Comm. Math. Phys. **52**, 147 (1977), Appendix.

Moreover, one has

$$(\Psi^{out}(A_n, \ldots A_1), \; \Psi^{out}(A_{n+1}, \ldots A_k)) =$$

$$\sum (\Psi_0, \, A_{i_1} P_1 A_{i_2} \Psi_0) \ldots (\Psi_0, \, A_{i_{k-1}} P_1 A_{i_k} \Psi_0) \qquad (6.3.13)$$

for $k = $ even and zero otherwise, where the sum is over all partitions of $(1, \ldots k)$ into ordered pairs (free field factorization of the correlation functions). This implies that the convergence in eq. (6.3.11) is actually strong.

Proof. Following Buchholz, the proof of eq. (6.3.11) is obtained by induction. By eq. (6.3.8), it holds for $n = 1$. Moreover, $\forall B \in \mathcal{A}(\mathcal{O}_1)$, $\mathcal{O}_1 \subset \mathcal{O}_+$, by eqs. (6.3.8) and (6.3.9) one has

$$\lim_{t \to \infty} ((\bar{A}_1(t) - A^{out}) B \Psi_0, \, \bar{A}_2(t) \ldots \bar{A}_n(t) \Psi_0) = 0$$

and therefore, if eq. (6.3.11) holds for $n - 1$, one has

$$\lim_{t \to \infty} (B \Psi_0, \, \bar{A}_1(t) \ldots \bar{A}_n(t) \Psi_0) = \lim_{t \to \infty} (A_1^{out} B \Psi_0, \, \bar{A}_2(t) \ldots \bar{A}_n(t) \Psi_0) =$$

$$= (A^{out} B \Psi_0, \, \Psi^{out}(A_2, \ldots A_n)).$$

Thus, one has convergence of the scalar products with a dense set, and, since by eq. (6.3.9) the sequence is uniformly bounded in t, this implies weak convergence for n.

The above equation shows that eq. (6.3.12) holds for the adjoint $A^{out \, *} \supset A^{out}$ (since $A = A^*$ implies that A^{out} is hermitian), and, since the vectors $\Psi^{out}(A_1, \ldots A_n)$ are in the domain of A^{out}, as a consequence of eq. (6.3.10),[18] eq. (6.3.12) holds for A^{out}. Thus, for any collection $A_i \in \mathcal{A}(\mathcal{O}_i)$, $\mathcal{O}_{i-1} \subset \mathcal{O}_{i+}$, $i = 1, \ldots n$, one has

$$\Psi^{out}(A_m, \ldots A_1) = A_m^{out} \ldots A_1^{out} \Psi_0 \qquad (6.3.14)$$

and the scalar product of these vectors is governed by eq. (6.3.10). The extension to vectors $\Psi^{out}(A_m, \ldots A_1)$ arising from arbitrary local operators $A_1, \ldots A_m$ is obtained by continuity.

Finally, one has

$$\text{weak} - \lim_{t \to \infty} ||\bar{A}_1(t) \ldots \bar{A}_n(t) \Psi_0|| = \lim_{t \to \infty} (\Psi_0, \, \Psi^{out}(A_n \ldots A_1 A_1 \ldots A_n)) =$$

$$(\Psi^{out}(A_1, \ldots A_n), \, \Psi^{out}(A_1, \ldots A_n)) = ||\text{weak} - \lim_{t \to \infty} \bar{A}_1(t) \ldots \bar{A}_n(t) \Psi_0||,$$

so that the sequence actually converges strongly.

The same type of arguments may be used for proving the existence of the asymptotic limits $t \to -\infty$, which define the asymptotic fields A^{in}.

Both A^{out} and A^{in} are defined on a dense domain of the Hilbert space (by the cyclicity of the vacuum with respect to the local algebra $\mathcal{A}$), without the need of asymptotic completeness. The combined effect of locality and Huyghens' principle

---

[18] For the detailed argument, see Buchholz's lectures.

makes the asymptotic limit much simpler than in the massive case; in particular, no information is needed on the energy–momentum spectrum above the light cone.

Buchholz analysis shows that the absence of a mass gap and/or long-range correlations (reflected in a cluster property decay like $1/r^2$) do not preclude the existence of the S-matrix for massless particles.

It is important to stress that Buchholz theory applies to the vacuum sector $\mathcal{H}_0$ of quantum electrodynamics, defined by the cyclicity of the vacuum with respect to the local observable algebra (where all the assumptions are expected to hold, even if asymptotic completeness fails because of the electron–positron pair creation). In this case one obtains a proof of the existence of the asymptotic electromagnetic algebras $\mathcal{A}^{as}(F_{\mu\nu})$, $as = in/out$, and the corresponding S-matrix.

4 Remarks on the infrared problem

The physical relevance of the infrared problem first appeared in quantum electrodynamics (QED), where the scattering processes of charged particles, computed at a given perturbative order n, diverge when the infrared cutoff, typically a fictitious photon mass μ, is removed. The perturbative solution of this problem is to let $\mu \to 0$ only after summing over all the processes, at the given order, which involve emission of photons with total energy less than the resolution ΔE of the experimental apparatus, briefly called *soft photons*.[19] The physical motivations for such a prescription are very strong, and in fact from a practical point of view this has been considered as the solution of the problem.

However, careful examination reveals that such a prescription hides non-trivial conceptual problems. First, as explicit perturbative second-order calculations show, the infrared-divergent terms in the amplitudes are proportional to $(\alpha/\pi)\log(m/\mu)$, m the electron mass, so that for $\mu \to 0$ the expansion parameter is not merely α. As a matter of fact, the cancellation of the infrared divergences, after summing over the transition probabilities corresponding to emission of photons with energy less than ΔE, is very peculiar, and it may seem paradoxical to derive cancellations by summing positive probabilities. Actually, the cancellation at a given order is achieved by keeping only the terms up to the given order of α in the transition probabilities of the various processes, and therefore by neglecting terms of higher orders in α which actually becomes very large in the limit $\mu \to 0$, because of the factors $\log(m/\mu)$. The neglect of such "higher orders" terms violates positivity of the transition probabilities when $\mu \to 0$ and makes cancellations possible.

As an illustrative example we consider the electron scattering by the Coulomb potential.[20] At the lowest order, the amplitude $R_0(p_1 \to p_2)$ for such a process is proportional to the elementary vertex $e\gamma_\mu$, where e is the electron charge. Radiative corrections of order α arising from diagrams with no emission of photons add the following term to the vertex (in the limit $q^2 << m^2$, $q \equiv p_2 - p_1$, m the electron mass):

[19] J. Schwinger, Phys. Rev. **76**, 790 (1949); J. M. Jauch and F. Rohrlich, *The Theory of Photons and Electrons*, 2nd exp. ed., 2nd corr. print, Springer 1980.

[20] For the details of the calculations, see, e.g., J. M. Jauch and F. Rohrlich, *The Theory of Photons and Electrons*, Addison-Wesley 1959, Sect. 15.2.

$$-\frac{\alpha}{3\pi}\left[\frac{q^2}{m^2}\left(\log\frac{m}{\mu}-\frac{3}{8}\right)e\gamma_\mu+\frac{3}{4m}\sigma_{\mu\nu}q^\nu\right]+O(q^2).$$

Thus, the corresponding transition probability contains the following infrared divergent terms:

$$\left[1-\frac{2\alpha}{3\pi}\frac{q^2}{m^2}\left(\log\frac{m}{\mu}-\frac{1}{5}\right)\right]|R_0|^2+O(\alpha^2).$$

On the other hand, the amplitude for electron scattering with the emission of a soft photon of momentum k and polarization $\varepsilon(\mathbf{k})$, with $k_0 \leq \Delta E << m$ (ΔE measured in the reference frame in which the electron is at rest), is proportional to the following modified vertex:

$$\frac{e}{\sqrt{2(2\pi)^3 k_0}}\left[\frac{p_2\cdot\varepsilon(\mathbf{k})}{p_2\cdot k}-\frac{p_1\cdot\varepsilon(\mathbf{k})}{p_1\cdot k}\right]e\gamma_\mu$$

and the corresponding transition probability (the photon is not observed) contains the following infrared-divergent terms (for q^2 small)

$$\frac{2\alpha}{3\pi}\frac{q^2}{m^2}\left[\log\frac{2\Delta E}{\mu}-\frac{5}{6}\right]|R_0|^2+O(\alpha^2).$$

Thus, the sum of the two transition probabilities, neglecting terms of order α^2 (which are actually divergent in the limit $\mu \to 0$), has a finite limit $\mu \to 0$, since the above infrared-divergent terms add up to

$$|R_0|^2\left(1-\frac{2\alpha}{3\pi}\frac{q^2}{m^2}\log\frac{m}{2\Delta E}\right). \tag{6.4.1}$$

As remarked above, such a cancellation crucially relies on neglecting the (infrared-divergent) terms of order α^2; in a certain sense, the prescription is to take the limit α small before the limit $\mu \to 0$. The violation of positivity implied by such a prescription appears clearly in the resulting differential cross-section,

$$\frac{d\sigma}{d\Omega}=\frac{d\sigma^M}{d\Omega}(1-\delta(\Delta E)), \tag{6.4.2}$$

where the index M denotes the Mott cross-section, corresponding to the vertex $e\gamma_\mu$, and in the non-relativistic limit $(v/c)^2 << 1$,

$$\delta(\Delta E)=\frac{8\alpha}{3\pi}\left(\log\frac{m}{2\Delta E}+\frac{19}{30}\right)(v/c)^2\sin^2(\theta/2),\quad \sin^2\theta\equiv q^2/4m^2.$$

As already noted by Schwinger, an improvement of the energy resolution with a decrease of ΔE leads to negative cross-sections. To solve this paradox Schwinger conjectured that an exponentiation $e^{-\delta}$ should take place, instead of the lowest-order term $1-\delta$. This corresponds to the sum over all possible emissions of soft photons accompanying the given process, and implies the *vanishing of the cross section if no soft photon is emitted*. The conceptual basis of such a mechanism is the

Bloch–Nordsieck model and the consequent analysis of the infrared divergences (see Chapter 2, Section 3.2).

A systematic analysis of the summation over soft photon emission has been achieved by Yiennie, Fratschi, and Suura. A simplified analysis with a conceptual improvement was presented by Weinberg and we refer to his excellent account for a comprehensive treatment.[21] We briefly recall the basic steps in the Weinberg argument.

The virtual photon radiative corrections to the $\alpha \to \beta$ transition amplitude $M_{\alpha\beta}$ due to *any number* of virtual photons of momentum $|\mathbf{q}| \geq \lambda$, λ a small infrared cutoff, lead to an amplitude denoted by $M_{\beta\alpha}^{\lambda}$, whereas if one considers only virtual photons of momentum greater than Λ, briefly called hard photons, one obtains the amplitude $M_{\beta\alpha}^{\Lambda}$. One of the basic steps in the Weinberg analysis is the relation between the two amplitudes,

$$M_{\beta\alpha}^{\lambda} = M_{\beta\alpha}^{\Lambda} \exp[(2\pi)^{-4}\tfrac{1}{2} \sum_{n\,m} e_n e_m \, \eta_n \, \eta_m J_{n\,m}],$$

$$J_{n\,m} \equiv -i(p_n \cdot p_m) \int_{\lambda \leq |\mathbf{q}| \leq \Lambda} d^4q[(q^2 - i\varepsilon)(p_n \cdot q - i\varepsilon\eta_n)(-p_m \cdot q - i\varepsilon\eta_m]^{-1},$$

where e_n, p_n, e_m, p_m denote the charges and momenta of the nth and mth particle lines joined by a virtual photon line of momentum q, and $\eta_n = \pm 1$ for particles in the final/initial state β/α, respectively. Since

$$\mathrm{Re}\, J_{n\,m} = \frac{2\pi^2}{\beta_{n\,m}} \ln\left(\frac{1 + \beta_{n\,m}}{1 - \beta_{n\,m}}\right) \ln\frac{\Lambda}{\lambda}$$

one obtains the following relation between the corresponding transition rates:

$$\Gamma_{\beta\alpha}^{\lambda} = \left(\frac{\lambda}{\Lambda}\right)^{A_{\beta\alpha}} \Gamma_{\beta\alpha}^{\Lambda}, \tag{6.4.3}$$

$$A_{\beta\alpha} \equiv -\frac{1}{8\pi^2} \sum_{n\,m} \frac{e_n e_m \eta_n \eta_m}{\beta_{n\,m}} \ln\left(\frac{1 + \beta_{n\,m}}{1 - \beta_{n\,m}}\right) > 0.$$

Thus, the rate vanishes in the limit $\lambda \to 0$. This reflects the physical fact that if the process $\alpha \to \beta$ involves charged particles, it is always accompanied by the emission of undetectable soft photons, and therefore what is measured is an inclusive rate with unobserved soft photons.

Now, the amplitude $M_{\beta\alpha}^{\mu_1 \cdots \mu_N}(q_1, .. q_n)$ for the emission of N soft photons with momenta and polarizations $q_1, \ldots q_N$, $\mu_1, \ldots \mu_N$ in the process $\alpha \to \beta$, factorizes in the limit $q \to 0$,

$$M_{\beta\alpha}^{\mu_1 \cdots \mu_N}(q_1, .. q_n) \to_{q \to 0} M_{\beta\alpha} \prod_{r=1}^{N} \left(\sum_n \frac{\eta_n e_n p_n^{\mu_r}}{p_n \cdot q_r - i\varepsilon\eta_n}\right),$$

---

[21] D. Yiennie, S. Frautschi and H. Suura, Ann. Phys. **13**, 379 (1961); S. Weinberg, Phys. Rev. **B140**, 516 (1965); *The Quantum Theory of Fields, Vol. I*, Cambridge University Press 1995, Chap. 13.

and therefore the corresponding rate for the process $\alpha \to \beta$ with emission of N soft photons of energy $\omega_1, \ldots \omega_N$, $\lambda \leq \omega_r \leq E$, $\sum_r \omega_r \leq E$ is

$$\Gamma^\lambda_{\beta\alpha}(E) = \Gamma^\lambda_{\beta\alpha} \sum_{N=0}^{\infty} A^N_{\beta\alpha} \int_{\lambda \leq \omega_r \leq E;\, \sum_r \omega_r \leq E} \prod_{r=1}^{N} d\omega_r / \omega_r.$$

The rate for the emission of an arbitrary number of soft photons with the above energy restrictions is obtained by summing over N, and in the limit $\lambda << E$ one obtains

$$\Gamma^\lambda_{\beta\alpha}(E) \to F(E; A_{\beta\alpha}) \left(\frac{E}{\lambda}\right)^{A_{\beta\alpha}} \Gamma^\lambda_{\beta\alpha}, \tag{6.4.4}$$

where $F(E, A_{\beta\alpha})$ is a function of order 1. The combination of this equation with eq. (6.4.3) gives the cancellation of the infrared divergences in the limit $\lambda \to 0$:

$$\Gamma^\lambda_{\beta\alpha} \to F(E, A_{\beta\alpha}) \left(\frac{E}{\Lambda}\right)^{A_{\beta\alpha}} \Gamma^\Lambda_{\beta\alpha}. \tag{6.4.5}$$

The energy E has the physical meaning of the total energy carried by the undetected soft photons, and Λ defines the hard (detectable) photons.

A crucial feature of the above analysis is that soft and hard photons are treated very differently with respect to an expansion in powers of α. The lesson from the above formula is that a perturbative expansion is allowed for the amplitude involving only hard photons, but a summmation over all contributions of soft photons, i.e., to all orders of α, is needed in order to cancel the infrared divergences. This implies that the perturbative expansion crucially depends on the parameter E strictly related to the apparatus sensibility ΔE; i.e., apparatus with different sensibilities define different perturbative expansions, and therefore different theories. A natural question is whether there is an E-independent theory or S-matrix, so that the rates of the processes observed with an apparatus sensibility E are obtained, according to the standard quantum-mechanical rules, by summing over the rates with final unobserved soft photons. As in Schwinger's conjectured formula, according to eq. (6.4.5), there is no non-trivial extrapolation to an E-independent theory, since the rates for processes $\alpha \to \beta$ involving charged particles vanish in the limit $E \to 0$, corresponding to finer and finer apparatus sensibility.

Furthermore, the above cancellation of infrared divergences is obtained by a crucial interplay between the summation over soft photon emission, physically motivated by the apparatus sensibility, and the physically less compelling summation to all orders over the radiative corrections due to virtual soft photons.

A significant step in the direction of a ΔE-independent theory was made by Chung and Kibble[22] in the framework of the perturbative theory.

The basic input is the realization that scattered charged particles are always accompanied by a cloud of soft photons described by a coherent function, as clearly displayed by the Bloch–Nordsieck (BN) model; therefore asymptotic charged states are

[22] V. Chung, Phys. Rev. **B140**, 1110 (1965); T. W. Kibble, Phys. Rev. **173**, 1527 (1968); **174**, 1882 (1968); **175**, 1624 (1968).

not described by free Fock states, but rather by states which are coherent states with respect to the asymptotic electromagnetic field. For a charged particle of momentum p_μ the coherent function is characterized by the infrared singularity $p_\mu/p \cdot k$, in the region of small photon momentum k, and otherwise dependent on the preparation or detection of the charged (scattering) state. The charged scattering states are therefore *non-Fock coherent states* of the electromagnetic radiation. Such a characteristic feature of the charged (scattering) states, advocated by Bloch and Nordsieck on the basis of their model, will hereafter be referred to as the *Bloch–Nordsieck ansatz*.

As in the BN model, the removal of the infrared cutoff gives vanishing S-matrix elements between free Fock charged states (i.e., states with no accompanying electromagnetic radiadion), but finite matrix elements between non-Fock coherent states. This follows from the (perturbative) factorization of the infrared singularities in transition amplitudes; in fact, the infrared-divergent part due to soft photon emission or absorption in S-matrix elements is of the form

$$\exp e\, \mathbf{a}^*\Big(\frac{\mathbf{p}^{out}}{p^{out}\cdot k} - \frac{\mathbf{p}^{in}}{p^{in}\cdot k}\Big),$$

and depends only on the initial and final momenta, as in the BN model. The important contributions by Chung and Kibble is to have shown that the perturbative expansion of the S-matrix based on the BN ansatz is infrared-convergent.

The indication by the BN model is further supported by other soluble models such as the Blanchard–Pauli–Fierz model.[23] It is also at the basis of the Kulish–Faddeev approach to quantum electrodynamics, where the infrared coherent factors are related to the asymptotic dynamics in the charged sectors, which is argued to differ from the free dynamics as in the Dollard treatment of Coulomb scattering.[24]

The conceptual open problems are i) the non-perturbative proof of the Bloch–Nordsieck ansatz, ii) the unwanted superselection rule of the momentum of the charged particles, or of the velocity of the associated Lienardt–Wiechert potentials, since such quantum numbers label the inequivalent non-Fock representation defined by the coherent functions, iii) the compatibility of the BN coherent functions and transversality condition for the (asymptotic) photons (Zwanziger paradox), iv) the implications on Lorentz invariance.

For a general overview of these non-perturbative problems and their possible solution, see the Erice lectures.[25]

For the non-perturbative analysis of the infrared problem in QED a crucial input is the Buchholz proof of the existence of the asymptotic limit of the photon field

[23] P. Blanchard, Comm. Math. Phys. **15**, 156 (1959).

[24] P. P. Kulish and L. D. Faddeev, Theor. Matem. Fiz. **4**, 153 (1970).

[25] G. Morchio and F. Strocchi, Infrared Problem, Higgs Phenomenon and Long Range Interactions, in *Fundamental Problems of Gauge Field Theory*, G. Velo and A.S. Wightman eds. Plenum 1986, p. 301. In particular for point i) and the solution of Zwanziger paradox see G. Morchio and F. Strocchi, Nucl. Phys. **B211**, 471 (1983); **B232**, 547 (1984); for the general approach to the non-perturbative solution of the infrared problem in QED see J. Fröhlich, G. Morchio and F. Strocchi, Ann. Phys. **119**, 241 (1979); for the breaking of the Lorentz group see J. Fröhlich, G. Morchio and F. Strocchi, Phys. Lett. **B89**, 61 (1979), D. Buchholz, Phys. Lett. **B174**, 331 (1986).

(see Section 6.3). Once the asymptotic electromagnetic algebra is given, the problem is to characterize its representations in the (scattering) charged sectors. Contrary to the massive case, such representations need not to be equivalent to the Fock representations, since the particle number is not bounded by the (free) Hamiltonian (see Chapter 1, Section 5) and, in fact, in the charged sectors the long range of the electric field expectations requires a non-Fock representation.[26]

---

[26] See the Erice lectures quoted above.

7
Quantization of gauge field theories

1 Physical counterpart of gauge symmetry

The extraordinary success of the standard model indicates that elementary particle interactions are described by gauge field theories. Furthermore, by the mechanism of asymptotic freedom, gauge field theories may escape the triviality theorem which affects the φ^4 theory (and probably also Yukawa-type field theories) in 3+1 dimensions. Gauge field theories exhibit experimentally established properties, which appear as very distinctive features with respect to the standard quantum field theories, like spontaneous symmetry breaking with an energy gap (Higgs mechanism) in apparent contradiction with Goldstone theorem, quark confinement and linearly rising potential in contrast with the cluster property, axial current anomaly, asymptotic freedom, etc.[1]

It is then natural to try to understand such departures from standard quantum field theory in terms of general properties, independently of the specific model. The original characterization by Yang and Mills is that quantum numbers or charges associated to gauge transformations (briefly, *gauge charges*) have only a local meaning, so that their relative identification at different spacetime points (governed by their commutation relations) is physically meaningless. Such a property does not have a direct experimental verification, since, as a consequence of confinement and symmetry breaking, the observed physical states do not carry (non-abelian) gauge charges. Furthermore, by definition, gauge transformations reduce to the identity on the observables, so that they can be defined only by introducing non-observable fields. In fact, the standard formulation of gauge field theories is done in terms of *non-observable* fields which transform non-trivially under the group $\mathcal{G}$ of *local gauge transformations* (also called gauge transformations of second kind), and the Hamiltonian (or the Lagrangian) is invariant under such transformations. The pointwise invariance of the observables under gauge transformations raises the conceptual question of which

---

[1] For the general structure and properties of gauge field theories, see S. Weinberg, *The Quantum Theory of Fields*, Vol. II, Cambridge University Press 1996. For the lack of locality and the violation of cluster property, see F. Strocchi, Phys. Rev. **D17**, 2010 (1978); for a non-perturbative discussion of the evasion of the Goldstone theorem in gauge theories, see F. Strocchi, Comm. Math. Phys. **56**, 57 (1977); G. Morchio and F. Strocchi, Removal of the infrared cutoff, seizing of the vacuum and symmetry breaking in many body and in gauge theories, invited talk at the *IX Int. Cong. on Mathematical Physics*, B. Simon *et al.* (eds.), Hilger 1989, p. 490 and references therein; for a non-perturbative proof of the Higgs phenomenon, see G. Morchio and F. Strocchi, Jour. Phys. A: Math. Theor. **40**, 3173 (2007).

structural property distinguishes the *observable algebras* of gauge field theories, and which is the meaning of gauge symmetry breaking.

As discussed in Chapter 4, Section 2, for the permutation group, the role of (unbroken) gauge symmetries appears to be that of providing a classification of the (inequivalent) representations of the observable algebra, not through the gauge charges, but through the invariants which characterize the irreducible representations of the gauge group. It is still unexplained why only states corresponding to one-dimensional representations of gauge groups (which include the permutation group) occur in elementary particles. The role and the physical interpretation of the gauge charges is one of the issues for understanding the physical basis of gauge invariance and therefore of gauge theories.

In order to formulate gauge invariance, we start by recalling that in classical field theory the invariance of the Lagrangian or of the Hamiltonian under a (n-dimensional) Lie group G can be checked by considering the infinitesimal variation of the fields φ_i, $i = 1, \ldots d$, (for simplicity we consider internal symmetries, i.e., not affecting the spacetime points, take G compact, and include the coupling constants in the generators):

$$\delta\varphi_i(x) = i\varepsilon_a t^a_{ij}\, \varphi_j(x) \equiv i(\varepsilon t\, \varphi)_i(x), \tag{7.1.1}$$

where $a = 1, \ldots n$, $i = 1, \ldots d$, summation over repeated indices is understood, ε are the infinitesimal group parameters, and t is the (d-dimensional) hermitian matrix representation of the generators of the group G, provided by the fields φ_i.[2]

In view of the local gauge extension of the above infinitesimal transformations, it is convenient to distinguish the so-called matter fields, which transform according to eq. (7.1.1), and the vector fields A^a_μ, which transform according to the adjoint representation of G,

$$\delta A^a_\nu(x) = i\varepsilon_c T^c_{ab} A^b_\nu(x) \equiv i(\varepsilon T A_\nu)^a(x), \quad T^c_{ab} = if^a_{cb} = -if^c_{ab}, \tag{7.1.2}$$

f being the Lie algebra (real) structure constants.

The *local gauge group $\mathcal{G}$ associated with G* (also briefly called the G *local gauge group*, whereas G is called the *global group* or *charge group*) is the infinite-dimensional group obtained by letting the group parameters be regular functions $\varepsilon(x)$ of the spacetime points, typically $\varepsilon \in \mathcal{D}(\mathbf{R}^4)^n$ or $\in \mathcal{S}(\mathbf{R}^4)^n$. The corresponding infinitesimal transformations of the fields are

$$\delta^\varepsilon \varphi_i(x) = i\varepsilon_a(x) t^a_{ij}\, \varphi_j(x) \equiv i(\varepsilon(x)\, t\, \varphi)_i(x), \tag{7.1.3}$$

$$\delta^\varepsilon A^b_\nu(x) = i\varepsilon_a(x) T^a_{bc} A^c_\nu(x) + \partial_\nu \varepsilon^b(x) \equiv i(\varepsilon(x)\, T A_\nu)^b(x) + \partial_\nu \varepsilon^b(x). \tag{7.1.4}$$

---

[2] For the basic elements of group theory and group representations, see, e.g., Y. Choquet-Bruhat and C. DeWitt-Morette, *Analysis, Manifolds and Physics*, Part I, North-Holland 1991, esp. Chap. III, Sect. D.

The need of the inhomogeneous term in eq. (7.1.4) is well known in the (abelian) $U(1)$ case, which corresponds to electrodynamics; its presence in the general case can be argued on the basis of the geometry of connection 1-forms.[3]

It is very important to keep separate the Lie algebra $L(G)$ of G and the infinite-dimensional Lie algebra corresponding to $\mathcal{G}$, briefly denoted by $L(\mathcal{G})$. In our opinion, both from a mathematical and from a physical point of view, it is useful to insist on such a distinction. In fact, as we shall see, trivial representations of $L(\mathcal{G})$ need not to be trivial representations of $L(G)$. For the same reasons, calling G the global gauge subgroup of the group of local gauge transformations may be a misleading glossary, since the structures of $L(G)$ and of $L(\mathcal{G})$ are mathematically very different.[4]

We may now investigate the physical consequences of *local gauge invariance*; namely, the invariance under local gauge transformations.

A naive extrapolation from the first Noether theorem (for the finite-dimensional case) might suggest that the invariance of the Lagrangian under the infinite-dimensional gauge group $\mathcal{G}$ would lead to an infinite set of conservation laws. Actually, one rather obtains a stronger form of the local conservation laws which follow from the invariance under the *global group* G. This is essentially the content of the *second Noether theorem*, the proof of which is briefly sketched.

The invariance of the Lagrangian $\mathcal{L}(\varphi, \partial\varphi, A, \partial A)$ under the infinitesimal transformations of eqs. (7.1.3) and (7.1.4) gives ($\delta\partial_\mu\varphi = \partial_\mu\delta\varphi$ and summation over repeated indices is understood)

$$i\left(\frac{\delta\mathcal{L}}{\delta\varphi_i}(t^a\varphi)_i + \frac{\delta\mathcal{L}}{\delta\partial_\mu\varphi_i}(t^a\partial_\mu\varphi)_i + \frac{\delta\mathcal{L}}{\delta A_\nu^b}(T^aA_\nu)^b + \frac{\delta\mathcal{L}}{\delta\partial_\mu A_\nu^b}(T^a\partial_\mu A_\nu)^b\right)\varepsilon^a +$$

$$\left(i\frac{\delta\mathcal{L}}{\delta\partial_\mu\varphi_i}(t^a\varphi)_i + \frac{\delta\mathcal{L}}{\delta A_\mu^a} + i\frac{\delta\mathcal{L}}{\delta\partial_\mu A_\nu^b}(T^aA_\nu)^b\right)\partial_\mu\varepsilon^a + \frac{\delta\mathcal{L}}{\delta\partial_\mu A_\nu^a}\partial_\mu\partial_\nu\varepsilon^a = 0.$$

Since $\varepsilon^a(x)$ are arbitrary functions, the coefficients of $\varepsilon, \partial_\mu\varepsilon$ and the $\mu \leftrightarrow \nu$ symmetrized coefficient of $\partial_\mu\partial_\nu\varepsilon$ must vanish. Hence, from the latter condition,

$$G_b^{\mu\,\nu} \equiv -\frac{\delta\mathcal{L}}{\delta\partial_\mu A_\nu^b} = -G_b^{\nu\,\mu} \tag{7.1.5}$$

is an antisymmetric tensor.

---

[3] See, e.g., the above-quoted book by Y. Choquet-Bruhat and C. DeWitt-Morette, Vol. I, Chap. V, Sect. B, Vol. II, Chap. V *bis*.

[4] Also, the name of gauge transformations of first kind should be avoided for G; its use is probably related to the case of $U(1)$ transformations which are improperly given this name even in theories with no local gauge invariance (this is standard in the discussion of the applications of the first Noether theorem; see, e.g., N. N. Bogoliubov and D. V. Shirkov, *Introduction to the Theory of Quantized Fields*, Interscience 1959, Chap. I). The basic difference between the global group G and the gauge group $\mathcal{G}$ is at the basis of the construction of the Dirac gauge-invariant *charged* field: P. A. M. Dirac, Canad. J. Phys. **33**, 650 (1955); E. D' Emilio and M. Mintchev, Fortsch. Physik, **32**, 473 (1984); O. Steinmann, *Perturbative Quantum Electrodynamics and Axiomatic Field Theory*, Springer 2000.

Moreover, for the currents associated with the invariance under the global group G (by the first Noether theorem)

$$J^{a\,\mu} \equiv -i\frac{\delta\mathcal{L}}{\delta\partial_\mu\varphi_i}(t^a\varphi)_i - i\frac{\delta\mathcal{L}}{\delta\partial_\mu A^b_\nu}(T^a A_\nu)^b \equiv j^{a\,\mu}(\varphi) + j^{a\,\mu}(A), \qquad (7.1.6)$$

from the vanishing of the coefficient of $\partial_\mu\varepsilon$, one obtains

$$J^b_\mu = \frac{\delta\mathcal{L}}{\delta A^\mu_b} = \partial^\nu G^b_{\mu\nu} + E[A]^b_\mu, \qquad (7.1.7)$$

where $E[A]$ is the Euler–Lagrangian operator for the field A, which vanishes as a consequence of the corresponding Euler–Lagrangian equations

$$E[A]^b_\mu \equiv \partial_\nu\frac{\delta\mathcal{L}}{\delta\partial_\nu A^b_\mu} - \frac{\delta\mathcal{L}}{\delta A^b_\mu} = 0. \qquad (7.1.8)$$

By similarly introducing the Euler–Lagrangian operator for the matter fields φ_i

$$E[\varphi]_i \equiv \partial_\mu\frac{\delta\mathcal{L}}{\delta\partial_\mu\varphi_i} - \frac{\delta\mathcal{L}}{\delta\varphi_i} = 0, \qquad (7.1.9)$$

from the vanishing of the coefficient of ε, one obtains

$$\partial^\mu J^b_\mu + E[\varphi]\,i\,t^b\,\varphi + E[A]\,i\,T^b\,A = 0. \qquad (7.1.10)$$

This equation reflects the invariance under the global group G; it gives the local conservation law for the current J^b_μ,

$$\partial^\mu J^b_\mu(x) = 0,$$

and, *at the classical level*, identifies the corresponding charge

$$Q^b = \int d^3x\, J^b_0(\mathbf{x},0) \qquad (7.1.11)$$

as the generator of the transformations with the group parameter ε pointing in the b-direction:

$$[Q^b, \varphi_i] = -\delta^b\varphi_i, \quad [Q^b, A^\gamma_\mu] = -\delta^b A^\gamma_\mu, \qquad (7.1.12)$$

where the square bracket denotes the Poisson bracket.

Eqs. (7.1.5), (7.1.6), and (7.1.7) encode the invariance under the local gauge group $\mathcal{G}$, and can be taken as a characterization of such an invariance property

$$J^b_\mu = \partial^\nu G^b_{\mu\nu}, \quad G^b_{\mu\nu} = -G^b_{\nu\mu}. \qquad (7.1.13)$$

Such a characteristic equation will be called the *local Gauss law*, since the Gauss theorem represents the integrated form of it. In general, a charge given by eq. (7.1.11), with J_μ a current which obeys a Gauss law, will be called a *Gauss charge*; thus, the second Noether theorems says that *the charges defined by the global group G are Gauss charges*.

The current continuity equation is trivially implied by eq. (7.1.13) without recourse to the evolution equation for the matter (charge carrying) fields φ_i. From a geometrical point of view, the local Gauss law says that the currents associated with the global group G define differential forms which are δ-boundaries or coexact,[5] and therefore the conservation of the corresponding Gauss charges has a geometrical or kinematical meaning.

In conclusion, the invariance under the infinite-dimensional local gauge group $\mathcal{G}$ yields a stronger form of the current conservation laws associated with the global group G (rather than infinite conservation laws).

By the above remarks, the validity of a local Gauss law appears to have a more direct physical meaning than a gauge symmetry, which is non-trivial only on non-observable fields and therefore cannot be defined without them. It is therefore tempting to regard the validity of local Gauss laws as the basic characteristic feature of gauge quantum field theories, and to consider local gauge invariance merely as a useful recipe for writing down Lagrangian functions which automatically lead to the validity of local Gauss laws.

Actually, as we shall discuss below, for the canonical formulation of gauge field theories one must add a gauge fixing term in the Lagrangian (irrelevant for the physical implications), and this can be done even at the expense of totally breaking the gauge invariance of the Lagrangian (as, e.g., in the so-called unitary gauge). Thus, the gauge invariance of the Lagrangian is not so crucial from a physical point of view, whereas so is the validity of local Gauss laws, which hold on the physical states independently of the gauge-fixing. In fact, the local Gauss law must be taken as a firm requirement for the renormalization conditions on the physical states (as it is the validity of the Ward identities), whereas other consequences of the classical second Noether theorem (like eqs. (7.1.11) and (7.1.12)) do not survive quantization (see Section 5.3).

The relevance of the local Gauss law is that it encodes a general property largely independent of the specific Lagrangian model, and actually, as we shall show below, most of the peculiar features of gauge quantum field theories, with respect to standard quantum field theories, like the evasion of the Goldstone theorem (Higgs mechanism), the lack of locality and the violation of the cluster property by charge-carrying fields (linearly rising quark potential and confinement), the superselection of Gauss charges, etc., can be understood as consequences of the validity of local Gauss laws at the quantum level.[6]

Eqs. (7.1.13) are obviously satisfied in the case of the standard gauge-invariant Lagrangians constructed on the basis of the minimal coupling prescription

---

[5] G. De Rham, *Differential Manifolds*, Springer 1984.

[6] The recognition of local Gauss laws as the basic characteristic features of gauge field theories has been argued and stressed in view of the *quantum* theories in F. Strocchi and A. S. Wightman, Jour. Math. Phys. **15**, 2198 (1974); F. Strocchi, Gauss law in local quantum field theory, in *Field Theory, Quantization and Statistical Physics*, D. Reidel 1981, pp. 227–36; F. Strocchi, *Elements of Quantum Mechanics of Infinite Systems*, World Scientific 1985, Part C, Chap. II, and later reproposed, without mentioning the above references, by D. L. Karatas and K. L. Kowalski, Am. J. Phys. **58**, 123 (1990); H. A. Al-Kuwari and M.O. Taha, *ibid.* **59**, 363 (1990); N. Nakanishi and I. Ojima, *Covariant Operator Formalism of Gauge Theories and Quantum Gravity*, World Scientific 1990; K. Brading and H. R. Brown, arXiv:hep-th/0009058 v1.

$\partial_\mu \varphi_i \to \partial_\mu \varphi_i - i t^a_{ij} A^a_\mu \varphi_j \equiv (D_\mu \varphi)_i$ (called the *covariant derivative*), and in fact in the abelian case coincide with the Maxwell equations; however, they hold also if one adds other (non-minimal) gauge-invariant terms in the Lagrangian.

The standard Yang–Mills Lagrangian, essentially selected by the minimal coupling prescription, has the form

$$\mathcal{L} = -\tfrac{1}{4} F_{a\,\mu\nu} F_a^{\mu\nu} + \mathcal{L}(\varphi, D_\mu \varphi), \tag{7.1.14}$$

where

$$F^a_{\mu\nu} \equiv \partial_\mu A^a_\nu - \partial_\nu A^a_\mu + i T^a_{bc} A^b_\mu A^c_\nu = \partial_\mu A^a_\nu - \partial_\nu A^a_\mu + f^a_{bc} A^b_\mu A^c_\nu, \tag{7.1.15}$$

and $\mathcal{L}(\varphi, D_\mu \varphi)$ is obtained from the matter field Lagrangian $\mathcal{L}(\varphi, \partial_\mu \varphi)$, without gauge-coupling, through the minimal coupling replacement $\partial_\mu \to D_\mu$.

For example, for a complex scalar field φ with self-interaction potential $U(\varphi)$ one has

$$\mathcal{L}(\varphi, D_\mu \varphi) = (D_\mu \varphi)^* D^\mu \varphi - U(\varphi), \tag{7.1.16}$$

and for a Dirac field,

$$\mathcal{L}(\psi, D_\mu \psi) = \tfrac{1}{2} \overline{\psi} (i\gamma_\mu D^\mu - m)\psi + \quad \text{hermitian conj.} \tag{7.1.17}$$

The definition (7.1.15) and the Jacobi identity satisfied by the matrices T of the adjoint representation, $f^d_{ab} f^e_{dg} + f^d_{ga} f^e_{db} + f^d_{bg} f^e_{da} = 0$, imply the *Bianchi identities*:

$$D_\lambda F^a_{\mu\nu} + D_\mu F^a_{\nu\lambda} + D_\nu F^a_{\lambda\mu} = 0. \tag{7.1.18}$$

It may useful to remark that the current j^a_μ appearing in the Yang–Mills equations: $(D^\nu F_{\mu\nu})^a = j^a_\mu$ corresponds to $j_\mu(\varphi)^a$ of eq. (7.1.6) and does not provide the generator of the global gauge transformations, which requires the sum $j_\mu(\varphi)^a + j_\mu(A)^a = J^a_\mu$.

Since the so-called "gauge principle" is a statement about charges having only a local meaning (as in the original Yang-Mills motivation), the analysis of their properties at the quantum level becomes a basic conceptual issue. Therefore, we are led to discuss the physical meaning and properties of the *Gauss charges* associated to a quantum gauge theory and of the related "charged" states.

According to the standard wisdom, two phenomena may occur:
i) the global gauge group is spontaneously broken, and correspondingly one has *broken Gauss charges*; this characterizes the Higgs mechanism in the electroweak theory (see Section 6 below),
ii) the global gauge group is unbroken, as in quantum electrodynamics and in quantum chromodynamics (QCD), and one has *unbroken Gauss charges*.

Quite generally, only group invariant functions of the Gauss charges may define observable quantum numbers; in particular, in the QCD case, only color-invariant quantum numbers may have a physical meaning (the generators of the color group define non-abelian superselection rules and cannot define observable quantum numbers). Thus, a careful discussion of the properties of Gauss charges is a preliminary step for understanding the phenomena of *screening and confinement*, at a non-perturbative level.

As we shall see, the fields which carry a Gauss charge cannot be local and therefore one of the basic properties of the standard quantum fields fails; the corresponding physical consequence is that states carrying a "Gauss" charge cannot be local. From a more technical point of view, such a conflict between the Gauss law and locality leads to various strategies for the quantization of gauge theories.

Essentially, there are two possible choices:

I) One insists on the Gauss law as an operator equation and adopts a quantization leading to a non-local field algebra. The prototype is the Coulomb gauge quantization of quantum electrodynamics, where the electron field (and more general any charged field) is non-local. This leads to very important properties and mechanisms of the corresponding quantum field theory, such as the non-locality of the charged states (Section 3.4), the possible failure of the cluster property, the evasion of the Goldstone theorem (Section 6.2), and so on.

II) One adopts a quantization which leads to a local field algebra, at the expense of weakening the Gauss law to a subsidiary condition of the physical states; the prototype is the Feynman gauge quantization of quantum electrodynamics. The technical advantages are many, since all the standard technical tools of quantum field theory are available; this is the class of renormalizable gauges in which most of the perturbative calculations have been performed. However, the inner product vector space defined by the vacuum expectation values of the (local) field algebra cannot satisfy positivity. This provides a mechanism for avoiding the unwanted (massless) Goldstone bosons in the physical spectrum (Section 6.1). The need of selecting the Hilbert space of physical states by a subsidiary condition will also provide a mechanism for the solution of the $U(1)$ problem in QCD (Chapter 8, Section 2.4).

In our opinion, a careful analysis of the non-locality of the charged states and of the general properties of the different quantizations is crucial for a mathematical and non-perturbative understanding of the important physical phenomena predicted by gauge quantum field theories.

2 Gauss law and locality

We shall prove that the validity of Gauss laws has very strong consequences for the quantization of gauge theories and leads to crucial differences with respect to standard quantum field theories.

We start by discussing the abelian case, where the Gauss law reduces to the Maxwell equation $j_0(x) = \operatorname{div} \mathbf{E}(x)$, and therefore, by the Gauss theorem, establishes a tight link between the local properties of the solutions and their behavior at infinity. In fact, at the classical level the charge of a solution of the electrodynamics equations can be computed either by integrating the charge density, i.e., a local function of the charge carrying fields, or by computing the flux of the electric field at space infinity. In the quantum case, this implies that the charge carrying fields cannot be local with respect to the (local) electric field.[7] The argument is simple, but it requires to carefully take care of some delicate points.

<hr>

[7] R. Ferrari, L. E. Picasso, and F. Strocchi, Comm. Math. Phys. **35**, 25 (1974).

The relevant issue is the quantum version of *the relation between the algebraic derivation* δF defined by a global, i.e., space- and time-independent, symmetry group β^λ, $\lambda \in \mathbf{R}$,

$$\delta F \equiv d\beta^\lambda(F))/d\lambda|_{\lambda=0}, \tag{7.2.1}$$

and the charge density of the associated conserved Noether current j_μ; namely, the quantum analog of eqs. (7.1.11) and (7.1.12).

For this purpose, the first point is that the naive integral of the charge density does not carry through in the quantum case, both for the ultraviolet singularities and for the lack of convergence of the infinite integral. Since the current operator is a distribution, which in general does not have a restriction at sharp time (Chapter 4, Section 6.1), one may consider a *regularization of the charge integral* by restricting it to a sphere of radius R and by smearing it in time. The strategy is then to take a suitable limit $R \to \infty$ and prove that it is independent of the time smearing, as a consequence of the current continuity equation.

The standard choice is to consider the following regularization:

$$Q_R = \int d^3x \, dt \, f_R(\mathbf{x})\alpha(t) \, j_0(\mathbf{x},t) \equiv j_0(f_R\alpha), \tag{7.2.2}$$

with $f_R(\mathbf{x}) = f(|\mathbf{x}|/R) \in \mathcal{D}(\mathbf{R})$, $f(x) = 1$ for $|x| < 1$, $f(x) = 0$ for $|x| > 1 + \varepsilon$, $\varepsilon > 0$, $\alpha \in \mathcal{D}(\mathbf{R})$, $\operatorname{supp}\alpha \subset [-\delta, \delta]$, $\int dt \, \alpha(t) = \tilde{\alpha}(0) = 1$.[8]

Quite generally, the smearing in eq. (7.2.2) guarantees that the integral is well defined, but not the convergence of Q_R when $R \to \infty$. In fact, Q_R does not converge weakly, not even on the vacuum; in order to obtain weak convergence on the local states, an improved smearing[9] is necessary, as discussed in Section 5.3 below.

However, for the generation of the field transformations, all that is needed is the *convergence of the commutators of* Q_R with the field algebra in the limit $R \to \infty$. For that purpose, a crucial role is played by *locality*. In fact, if a field F is *local* with respect to j_μ, the distribution $[\,j_0(\mathbf{x},\alpha), F\,]$ has compact support in $\mathbf{x}$, since, for R sufficiently large, the spacetime points $(\mathbf{x},t)$ with $|\mathbf{x}| \geq R$, $|t| \leq \delta$ become spacelike with respect to the localization region of F, so that they give vanishing contribution to commutators. Thus, the limit of $[\,j_0(f_R\alpha), F\,]$ exists, and is actually reached for finite (sufficiently large) values of R. Clearly, the existence of the limit persists if F is a quasi-local field, i.e., a polynomial of local field operators smeared with test functions of fast decrease.

Furthermore, for *local* fields the $R \to \infty$ limit of the commutator is *independent of the smearing function* $\alpha \in \mathcal{D}(\mathbf{R})$, with the normalization condition $\tilde{\alpha}(0) = 1$. In fact, if α_1, α_2 are two such functions, one has

$$\alpha_1(t) - \alpha_2(t) = \partial_t\beta(t), \quad \beta(t) \equiv \int_{-\infty}^{t} ds \, [\alpha_1(s) - \alpha_2(s)] \in \mathcal{D}(\mathbf{R});$$

[8] For a discussion of the generation of symmetry transformations by charges, see, e.g., F. Strocchi, *Symmetry Breaking*, 2nd ed., Springer 2008, Sect. 15.2.

[9] M. Requardt, Comm. Math. Phys. **50**, 259 (1976); G. Morchio and F. Strocchi, J. Math. Phys. **44**, 5569 (2003).

by the current continuity equation $j_0(f_R(\alpha_1 - \alpha_2)) = j_0(f_R \partial_t \beta) = j_i(\partial_i f_R \beta)$, and, since supp $\partial_i f_R \, \beta \subset \{\mathbf{x}, t; |\mathbf{x}| \geq R, \, t \in \mathrm{supp}\beta\}$,

$$\lim_{R \to \infty} [\, j_0(f_R(\alpha_1 - \alpha_2)), \, F\,] = 0.$$

This argument shows that the current continuity equation implies that

$$\lim_{R \to \infty} (d/dt)[\, j_0(f_R, t), \, F\,] = 0 \qquad (7.2.3)$$

(as a distribution in t), i.e., the commutator is independent of t for R large enough, and the role of smearing with α is that of regularizing the equal-time commutator without infinite renormalization constants.

The independence of α may be regarded as the quantum version of the time independence of the charge, as given by the Noether theorem at the classical level; clearly, such a property is crucial for the possibility of generating the global *time-independent* symmetry group β^λ.

Furthermore, for a local field algebra, due to the time independence given by eq. (7.2.3), the relation between the algebraic derivation, eq. (7.2.1), and the (integral of the) density of the associated conserved Noether current, *at the level of commutators*, can be decided on the basis of equal-time (anti)commutation relations. Typically, for a field operator $\varphi(\mathbf{x}, t)$, one has ($g \in \mathcal{D}(\mathbf{R}^3)$)

$$i \lim_{R \to \infty} [\, Q_R, \, \varphi(g, t)\,] = i \lim_{R \to \infty} [\, j_0(f_R, t), \varphi(g, t)\,] = \delta \, \varphi(g, t), \qquad (7.2.4)$$

where the first equality crucially requires the time-independence, i.e., the *relative locality between the field operator φ and j_μ*, and the second is decided by the equal-time canonical (anti)commutators. In general, the control of such a relative locality is a genuine dynamical problem; however, the equal-time commutators between a field φ and j_μ are in general local, and therefore the relative locality between φ and j_μ holds at unequal times if the time evolution of φ and j_μ is local.

In conclusion, the *Noether relation between the infinitesimal symmetry transformation and the integral of the density* of the associated conserved current extends to the quantum case *at the level of commutators*, if the field algebra satisfy locality and the charge integral is suitably regularized.

The above picture must be revisited for the case of quantum electrodynamics (QED), since the validity of the Gauss law for the electric current requires that the *field algebra $\mathcal{F}$, generated by the charged fields and the electromagnetic field, cannot satisfy locality*. In fact, one has:

Proposition 2.1   *If a field φ is local with respect to the electric field* $\mathbf{E}$*, the Gauss law gives* $\lim_{R \to \infty}[\, j_0(f_R \alpha), \, \varphi\,] = 0$.

Proof. In fact, by the local Gauss law one has

$$\lim_{R \to \infty} [\, j_0(f_R \alpha), \varphi(y)\,] = - \int d^3x \, dt \, \nabla_i f_R(\mathbf{x}) \alpha(t) [\, E_i(\mathbf{x}, t), \varphi(y)\,] = 0,$$

since $\mathrm{supp}\,\nabla_i f_R \alpha \subset \{R \leq |\mathbf{x}| \leq R(1+\varepsilon),\ |t| \leq \delta\}$ becomes spacelike with respect to any bounded spacetime region, for R large enough.[10]

Thus, if a field is charged, in the sense that it transforms non-trivially under the charge group, i.e., $\delta\varphi \neq 0$, it cannot be local with respect to F_{0i}, because otherwise eq. (7.2.4) and Proposition 2.1 imply $\delta\varphi(y) = i\lim_{R\to\infty}[\,j_0(f_R\alpha),\varphi(y)\,] = 0$.

The non-locality of the charged fields may appear as a mere gauge artifact with no physical relevance, since charged fields are not observable fields and, as discussed in Chapter 3, Section 1, for them locality is not a must. This is clearly displayed by the Coulomb gauge in QED, where

$$\mathrm{div}\,\mathbf{A} = 0, \quad A_0(\mathbf{x}, t) = \int d^3y\,(4\pi|\mathbf{x} - \mathbf{y}|)^{-1}\,j_0(\mathbf{y}, t),$$

and the canonical quantization of the canonical variables $A_i,\ \partial_0 A_i, \psi, \bar\psi$ gives, for the electric field $E_i = \partial_0 A_i - \partial_i A_0$ (without taking care of renormalization constants),

$$[\,E_i(\mathbf{x}, t),\ \psi(\mathbf{y}, t)\,] = -e\,\partial_i\,(4\pi\,|\mathbf{x} - \mathbf{y}|)^{-1}\,\psi(\mathbf{y}, t). \tag{7.2.5}$$

This indicates a failure of commutativity at spacelike points, with a Coulomb tail delocalization, as needed in order to account for a non-vanishing integral of the charge density commutator.[11]

The non-locality of the charged fields implies that the charged states cannot be in the closure of $\mathcal{F}_0\Psi_0$, where $\mathcal{F}_0$ denotes the local (and therefore chargeless) field algebra.[12] Thus, the non-local charged fields of the Coulomb gauge play the role of generating from the vacuum the non-local charged states. The local Gauss law therefore has the important physical consequence that the charged states cannot be local, in the sense discussed in Chapter 3, Section 1.

This can be directly seen by noticing that for any given bounded region $\mathcal{O}$, by the local Gauss law Q_R belongs to the observable algebra $\mathcal{A}(\mathcal{O}')$, for R sufficiently large, and therefore, if the state ω is localized in $\mathcal{O}$, according to eq. (3.1.4), $\omega(Q_R) = (\Psi_0, Q_R\,\Psi_0) = 0$, i.e., ω has zero charge.

In the language of algebraic quantum field theory, the charged fields define charged *non-local morphisms* ρ of the observable algebra, (typically by means of the charged "unitary" operators U constructed in terms of them, $\rho(A) = U^*AU$, as discussed in Chapter 3, Section 1), since the corresponding charged states ω_ρ:

$$\omega_\rho(A) = (\Psi_0, \rho(A)\,\Psi_0),$$

cannot be localized, as a consequence of the local Gauss law.[13]

---

[10] The vanishing of the commutator persists also for more general time smearing $\alpha_R(x_0)$, with $\mathrm{supp}\,\alpha_R \subset \{-R(1-\varepsilon) < x_0 < R(1-\varepsilon)\}$ (see Section 5.3 below).

[11] For a careful discussion, which takes into account the UV singularities of the equal-time commutators and the need of renormalization, see K. Symanzik, Lectures on Lagrangian Field Theory, DESY report T-71/1, p. 110. Moreover, an improved time-smearing is necessary for generating β^λ by the "integral" of j_0, as discussed in Section 5.3.

[12] For the detailed argument, see R. Ferrari, L. E. Picasso, and F. Strocchi, Nuovo Cim. **39 A**, 1 (1977), and Section 3.4 below.

[13] S. Doplicher, R. Haag, and J. Roberts, Comm. Math. Phys. **23**, 199 (1971).

Thus, the general properties of the charged fields are not confined to non-physical parts of the theory, as sometimes stated,[14] since they are relevant for the description of the physical charged states and of their expectations (see Section 3.4 below).

A further relevant consequence of the non-locality of the charged fields is that the relation between the derivation defined by the $U(1)$ charge group and the (suitably regularized) integral of the charge density j_0 has to be discussed anew; actually, as we shall see below (see Section 5.3), *the classical Noether relation given by eqs. (7.1.11), (7.1.12) does not always carry through in the quantum case.*[15]

Since locality has played a crucial role in the analysis of the general properties of QFT discussed in Chapters 3–6, it should not be a surprise that the correlation functions of the non-local field algebra, needed for the description of the states on the observables of a gauge quantum field theory, exhibit very different properties with respect to those of a standard QFT. As we shall see, the charge superselection rule, the Higgs mechanism, and the failure of the cluster property may be understood as consequences of the local Gauss law.[16]

3   Local gauge quantization of QED

3.1   Weak Gauss law

The lack of locality is a technical drawback, especially for the perturbative expansion, and in fact, since the early days of perturbative QED, the strategy of keeping locality, at the expense of weakening the local Gauss law, has been adopted. In fact, local quantizations of gauge field theories, *in primis* of QED, are the most convenient (if not the exclusive) ones for the discussion and proof of renormalization, and for this reason are extensively used in the perturbative expansion.

However, the perturbative expansion in terms of Feynman diagrams, e.g., in the Feynman gauge, somewhat hides the problem of the identification of the physical states and the quantum-mechanical structure, except for asymptotic times, i.e., for the S-matrix elements. As a matter of fact, in QFT textbooks little attention is paid to the quantum-mechanical interpretation of local gauge quantizations, except for the free case (see Section 8.2 below). This may lead to the superficial impression that apart from a more complicated book-keeping of indices, the structure is similar to that of the scalar QFT model.

The strategy of the local quantizations of QED, whose prototype is the *Feynman–Gupta–Bleuler (FGB)* gauge, is to approach the dynamical problem in terms of a field algebra $\mathcal{F}$ generated by fields which obey local equal-time (anti)commutation relation and hyperbolic evolution equations, so that $\mathcal{F}$ may be taken as a *local field algebra*. Then, given the (e.g., perturbative) solution of the dynamics in terms of local fields,

---

[14] See, e.g., C. A. Hurst, Math. Rev. MR2053089.

[15] G. Morchio and F. Strocchi, Jour. Phys. A: Math. Theor. **40**, 3173 (2007).

[16] For a look at the general properties of gauge quantum field theories, in connection with the local Gauss law, see F. Strocchi, *Selected Topics on the General Properties of Quantum Field Theory*, World Scientific 1993, Chaps. VI, VII, and references therein to the previous papers.

the physical results are obtained by considering the state vectors which satisfy the Gauss law constraint.

In the *FGB quantization of QED*, the field algebra $\mathcal{F}$ is generated by the electron field ψ and the four independent components A_μ of the vector potential. The field equations are the Dirac equation with minimal electromagnetic coupling and the following equations for A_μ:

$$\Box A_\mu = -j_\mu, \quad \text{i.e.,} \quad \partial^\nu F_{\mu\nu} = j_\mu + \partial_\mu \partial^\nu A_\nu. \tag{7.3.1}$$

The equations of motion are similar to that of a Yukawa theory of fermions interacting with four (independent) fields A_μ, and the perturbative expansion indicates that the vacuum expectations of the field algebra $\mathcal{F}$ generated by ψ and A_μ satisfy locality, covariance, and the spectral condition. However, the Gauss law is violated by the "longitudinal" field $\mathcal{L}_\mu \equiv \partial_\mu \partial^\nu A_\nu$ and, as in the free case, one must re-establish it at the level of physical expectations.

By the same argument of Chapter 3, Section 3, the vacuum correlation functions of $\mathcal{F}$ define a vector space $\mathcal{D}_0 = \mathcal{F}\Psi_0$ of local vectors and (with the standard hermiticity properties of A_μ, ψ) a hermitian inner product $< ., . >$ on $\mathcal{D}_0$. Clearly, not all the vectors of $\mathcal{D}_0$ correspond to physical states. The vectors $\Psi \in \mathcal{D}_0$, which have a physical interpretation, in terms of expectations given by $< ., . >$, must clearly satisfy the *subsidiary condition*

$$< \Psi, \partial^\nu A_\nu \Psi >= 0, \quad \text{i.e.,} \quad < \Psi, (j_\mu - \partial^\nu F_{\mu\nu}) \Psi >= 0, \tag{7.3.2}$$

in order to exclude violations of the Maxwell equations by physical expectations (*weak Gauss law*).

In the *FGB quantization*, the non-linear condition (7.3.2) is replaced by the *linear Gupta–Bleuler (GB) condition*, which characterizes the physical vectors Ψ:

$$\partial A^- \Psi = 0, \quad \partial A \equiv \partial^\mu A_\mu. \tag{7.3.3}$$

In fact, since ∂A obeys a free field equation and A_μ is required to be $<,>$-hermitian, a splitting into its positive and negative energy parts is possible:

$$\partial A = \partial A^+ + \partial A^-, \quad \partial A^+ = (\partial A^-)^* \tag{7.3.4}$$

(* denotes the $<,>$-adjoint), and it is easy to see that eq. (7.3.3) implies eq. (7.3.2). The (Fermi) subsidiary condition $\partial A \, \Psi = 0$ would imply $\partial A \, \Psi_0 = 0$, and if locality holds, by the Reeh–Schlieder theorem $\partial A = 0$; hence, $j_\mu = \partial^\nu F_{\mu\nu}$, in conflict with local charged fields.

The GB condition states that there are no non-transverse photons in the physical states, and the vacuum vector satisfies it by the spectral condition. The distinguished linear subspace of $\mathcal{D}_0$, consisting of physical vectors, will be denoted by $\mathcal{D}_0'$ and $\mathcal{D}_0'' \equiv \{\Psi \in \mathcal{D}_0'; < \Psi, \Psi >= 0\}$.

In contrast with the free case (see Section 8.2), a control of the consistency of the quantum-mechanical structure associated with the FGB quantization is lacking in the interacting case and one needs further arguments for the physical acceptance of such a quantization. The relevant questions are a better understanding of the state content of $\mathcal{D}_0'$ and its relation with the algebra of observables.

3.2 Subsidiary condition and gauge invariance

In order to better understand the role of the field $\partial A(x)$ and its relation with the observables, we start by arguing that ∂A has simple commutation relations with the elements of the field algebra $\mathcal{F}$.

In fact, ∂A satisfies the free wave equation, as a consequence of the current continuity equation, and therefore so do the commutators $[\partial A(x), A_\mu(y)]$ and $[\partial A(x), \psi(y)]$. This implies that they are determined by the initial data, i.e., by the corresponding equal-time commutators, which cannot involve infinite renormalization constants because, by the Green-function formula for the solution of the generalized Cauchy problem, they are expressible in terms of the (distributionally) well defined unequal-time commutators. This means that such commutators have well-defined restrictions at equal times; canonical quantization[17] and the equation $\partial_0 \partial A = \Delta A_0 - j_0 - \partial_0 \partial_i A_i$, implied by eq. (7.3.1), give:

$$[\partial A(x), A_0(y)]_{x_0=y_0} = i\,\delta(\mathbf{x} - \mathbf{y}), \quad [\partial A(x), A_i(y)]_{x_0=y_0} = 0, \tag{7.3.5}$$

$$[\partial_0\partial A(x), A_0(y)]_{x_0=y_0} = 0, \quad [\partial_0\partial A(x), A_i(y)]_{x_0=y_0} = i\,\partial_i\delta(\mathbf{x} - \mathbf{y}), \tag{7.3.6}$$

$$[\partial A(x), \psi(y)]_{x_0=y_0} = 0, \quad [\partial_0\partial A(x), \psi(y)]_{x_0=y_0} = e\,\delta(\mathbf{x} - \mathbf{y})\,\psi(y). \tag{7.3.7}$$

The stability of the above equations under renormalization (without the occurrence of renormalization constants) has been discussed in detail by Symanzik,[18] and the constant e in eq. (7.3.7) is the (finite) renormalized charge.

By the Green-function formula, the above equal-time commutators imply the following unequal-time commutators:

$$[\partial A(x), A_\mu(y)] = -i\partial_\mu\, D(x - y), \quad D(x) \equiv \Delta(x; m^2 = 0); \tag{7.3.8}$$

$$[\partial A(x), \psi(y)] = -e\, D(x - y)\,\psi(y). \tag{7.3.9}$$

The equations of motion of the FGB quantization are no longer invariant under local gauge transformations with arbitrary parameters $\varepsilon(x) \in \mathcal{D}(\mathbf{R}^4)$, or $\in \mathcal{S}(\mathbf{R}^4)$, but only under those parametrized by functions $\varepsilon(x)$ which satisfy $\square\,\varepsilon(x) = 0$. As regularity and locality conditions for the gauge functions $\varepsilon(x)$, one may restrict the initial data to $\varepsilon(\mathbf{x},0)$, $\partial_0\varepsilon(\mathbf{x},0) \in \mathcal{D}(\mathbf{R}^3)$ or $\in \mathcal{S}(\mathbf{R}^3)$.

Proposition 3.1 *The commutators (7.3.8) and (7.3.9) imply that the (c- number)* **local gauge transformations** *on $\mathcal{F}$, with parameters $\varepsilon(x)$, satisfying*

$$\square\varepsilon(x) = 0, \quad \varepsilon(\mathbf{x},0),\ \partial_0\varepsilon(\mathbf{x},0) \in \mathcal{D}(\mathbf{R}^3)\ \ or\ \in \mathcal{S}(\mathbf{R}^3), \tag{7.3.10}$$

---

[17] In the canonical quantization the fields A_μ and ψ are treated as independent variables, and

$$[\dot{A}_\mu(x), A_\nu(y)]_{x_0=y_0} = ig_{\mu\nu}\,\delta(\mathbf{x} - \mathbf{y}).$$

[18] K. Symanzik, Lectures on Lagrangian Field Theory, DESY report T-71/1, Sect. 5. The absence of infinite renormalization constants in eqs. (7.3.5) and (7.3.6) follows also from the fact that the spectral measure $d\sigma(m^2)$ which appears in the two-point function $< \partial A(x)\, A_\mu(y) >$ is proportional to $\delta(m^2)$, since ∂A obeys the free wave equation, and therefore the corresponding renormalization constant $Z = (\int d\sigma(m^2))^{-1}$ is finite.

are generated by the field $\partial A(x)$ in the following sense: $\forall F \in \mathcal{F}$

$$\lim_{R \to \infty} -i \int d^4x \, [\, f_R(\mathbf{x}) \alpha(x_0) \, (\varepsilon(x) \stackrel{\leftrightarrow}{\partial_0} \partial A(x)), \, F\,] = \delta^\varepsilon F. \tag{7.3.11}$$

Proof. It is enough to prove eq. (7.3.11) for the fields $F = \psi$, A_μ, since they generate $\mathcal{F}$. Then, one has (with $\stackrel{\leftrightarrow}{\partial_0}$ defined as in eq. (6.1.2))

$$\lim_{R \to \infty} \partial_0 \int d^3x [\, f_R(\mathbf{x}) \, \varepsilon(x) \stackrel{\leftrightarrow}{\partial_0} \partial A(x), \, F\,] = 0,$$

because i) both $\varepsilon(x)$ and $\partial A(x)$ satisfy the (free) wave equation, ii) for each given x_0, by hyperbolicity with initial data (7.3.10), $\varepsilon(\mathbf{x}, x_0)$ and $\dot\varepsilon(\mathbf{x}, x_0)$ vanish sufficiently rapidly at space infinity, so that the (local) commutator of $f_R \varepsilon \Delta \partial A - f_R \Delta \varepsilon \partial A = f_R \partial_i (\varepsilon \partial_i \partial A - \partial_i \varepsilon \partial A)$ with F integrates out to zero, for $R \to \infty$. Hence, the integral

$$\int d^3x [\, f_R(\mathbf{x}) \, \varepsilon(x) \stackrel{\leftrightarrow}{\partial_0} \partial A(x), \, F\,]$$

may be computed by taking $x_0 = 0$ and eq. (7.3.11) easily follows from the commutators (7.3.8) and (7.3.9) and the Green-function formula:

$$\varepsilon(y) = \int_{x_0=0} d^3x \, [\, \partial_0 \varepsilon(x) \, D(x-y) - \varepsilon(x) \, \partial_0 D(x-y) \,]. \tag{7.3.12}$$

The possibility of defining local gauge transformations is due to the fact that the FGB quantization of QED does not completely fix the gauge (as it is instead the case of the Coulomb gauge).

This allows the identification of the *-algebra $\mathcal{F}_{obs}$ of local observable field operators as the subalgebra $\mathcal{F}_{gi}$ of $\mathcal{F}$, which is pointwise-invariant under the local gauge transformations of eq. (7.3.11) (*gauge-invariant subalgebra*). Equivalently, $\mathcal{F}_{gi}$ is the subalgebra of fields which commute with the field $\partial A(x)$. In fact, in eq. (7.3.11) one can omit the α-smearing, fix the time $x_0 = 0$, and vary $\varepsilon(\mathbf{x}, 0)$, $\dot\varepsilon(\mathbf{x}, 0)$ arbitrarily.[19]

We may now better understand the physical meaning of the subsidiary condition and the physical content of $\mathcal{D}'_0$. To this purpose we denote by $\mathcal{F}'$ the (local) *-subalgebra of $\mathcal{F}$ with the property that $\partial A^- F \Psi_0 = 0$, $\forall F \in \mathcal{F}'$.

Proposition 3.2 *In the FGB quantization, a local vector, i.e., of the form $\Psi = F \Psi_0$, with $F \in \mathcal{F}$, satisfies the GB-subsidiary condition iff F commutes with ∂A^-; then $\mathcal{F}'$ is pointwise invariant under the local gauge transformations of eq. (7.3.11) and $\mathcal{F}' = \mathcal{F}_{gi}$.*

Proof. Clearly, $\mathcal{F}_{gi} \Psi_0 \in \mathcal{D}'_0$, because if F commutes with $\partial A(f)$, $\forall f \in \mathcal{S}(\mathbf{R}^4)$, it commutes also with $\partial A^\pm$, since, for test functions $g_\pm$ with $\operatorname{supp} \tilde{g}_\pm \subset \overline{V}_\pm$, $\partial A^\pm(g_\pm) = \partial A(g_\pm)$. Thus, $\mathcal{F}_{gi} \subseteq \mathcal{F}'$.

For the converse, we note that the commutator $[\, \partial A^-(x), F\,]$, $F \in \mathcal{F}$, is a local operator, as can be easily checked for the fields $F = \psi$, A_μ, which generate $\mathcal{F}$. In fact,

---

[19] In the literature, this gauge invariance is sometimes called strict gauge invariance; see K. Symanzik, Lectures on Lagrangian QFT, 1971, Sect. 6.2; F. Strocchi and A. S. Wightman, Jour. Math. Phys. **15**, 2198 (1974), pp. 2216–19.

$\mathrm{supp}_y\, D^-(x - y)\, f(y) \subset \mathrm{supp}_y\, f(y)$, and therefore locality follows from eqs. (7.3.8) and (7.3.9). Hence, by the Reeh–Schlieder theorem,

$$[\partial A^-(x), F]\Psi_0 = \partial A^-(x)\, F\, \Psi_0 = 0, \quad \forall F \in \mathcal{F}'$$

implies $[\partial A^-(x), F] = 0, \forall F \in \mathcal{F}'$.

Moreover, by the stability of $\mathcal{F}'$ under the adjoint, also $[F^*, \partial A^-] = 0$. This implies

$$[F, \partial A^+] = -[F^*, \partial A^-]^* = 0,$$

so that F commutes with ∂A, i.e., $F \in \mathcal{F}_{gi}$.

This suggests that in the FGB gauge the local physical states might be either identified by the GB-subsidiary condition or equivalently as the vectors of $\mathcal{F}_{gi}\,\Psi_0$, and, as we shall see, this provides a possible useful strategy for the construction of the physical charged states (see Section 3.4 below).

Quite generally, in the following, a *local quantization* of electrodynamics is defined by

i) a *local field algebra* $\mathcal{F}$, generated by the fields ψ, A_μ, transforming covariantly under the Poincaré group,

ii) a group of *local gauge transformations* of $\mathcal{F}$, which (defines automorphisms of $\mathcal{F}$ and) identifies its pointwise-invariant subalgebra $\mathcal{F}_{gi}$,

iii) a *Poincaré-invariant vacuum functional* $\omega_0(\mathcal{F}) =< \mathcal{F} >_0$, which satisfies the spectral condition and, on the gauge-invariant subalgebra $\mathcal{F}_{gi}$, the positivity condition

$$\omega_0(B^* B) \geq 0, \quad \forall B \in \mathcal{F}_{gi},$$

iv) the *weak Gauss law*,

$$j_\mu = \partial^\nu F_{\mu\nu} - \mathcal{L}_\mu,$$

$$< \Psi, (\partial^\nu F_{\mu\nu} - j_\mu)\Phi > = < \Psi, \mathcal{L}_\mu \Phi > = 0,$$

for all vectors Ψ, Φ which describe physical states.

In general, Ψ_0 will denote the representative vector of ω_0, and $\mathcal{D}_0' \subset \mathcal{D}_0 \equiv \mathcal{F}\Psi_0$ the subspace of local vectors satisfying the weak Gauss law. In the FGB quantization, on which we focus in the following, $\mathcal{L}_\mu = \partial_\mu\, \partial A$.

The positivity condition on the vacuum expectations of the local observable field algebra has the same motivations and justification as the positivity condition in standard QFT. Due to Proposition 3.2, it implies the *semipositivity of the inner product* $< .\,, . >$ *on* $\mathcal{D}_0'$ in the FGB realization.

3.3 Indefinite metric and Hilbert–Krein structure

The weakening of the Gauss law, necessary for a local quantization of QED, has the inevitable consequence of *violation of positivity* by the vacuum expectations of the local field algebra, so that one of the basic properties of standard quantum field theory fails.

Proposition 3.3   *In the local FGB quantization of QED the inner product defined on $\mathcal{D}_0$ by the vacuum correlation functions of the field algebra $\mathcal{F}$ cannot be positive semidefinite.*

Proof.   If a local operator F commutes with $\partial A(x)$, $\forall x \in \mathbf{R}^4$, it also commutes with $\partial A^-(x)$, and therefore $\partial A^- F \Psi_0 = 0$. Thus, since the fields $F_{\mu\nu}$ and $j_\mu = \partial^\nu F_{\mu\nu} + \partial_\mu \partial A$ commute with ∂A (see eqs. (7.3.8) and (7.3.9)), they map physical vectors into physical vectors and, by eq. (7.3.2),

$$< (j - \partial F)\Psi_0, \, (j - \partial F)\Psi_0 >= 0. \tag{7.3.13}$$

If semipositivity holds, eq. (7.3.13) implies

$$< \Psi, \, (j - \partial F)\Psi_0 >= 0, \qquad \forall \Psi \in \mathcal{D}_0.$$

Then, by the same argument of the Reeh–Schlieder theorem, one has that $\forall A, B \in \mathcal{F}$, $< A\Psi_0, (j - \partial F)\, B\, \Psi_0 >= 0$, and

$$< A\Psi_0, \, [\, (j - \partial F), \, \psi\,]\, B\, \Psi_0 >= 0,$$

which is incompatible with the existence of local charged fields (with non-trivial correlation functions). Hence, the Wightman functions of the local field algebra $\mathcal{F}$ cannot satisfy the positivity condition.

This is a general feature of local quantizations of QED: *locality is incompatible with positivity*. In conclusion, in the quantization of QED one may either have positivity and non-locality (the prototype being the Coulomb gauge) or locality and non-positivity (the prototype being the Feynman–Gupta–Bleuler gauge).

By the above proposition, in contrast with the standard case, the vector space $\mathcal{D}_0$ defined by the Wightman functions of the local field algebra $\mathcal{F}$ does not have the structure of a pre-Hilbert space (possibly modulo null subspaces). The lack of positivity allows for a failure of the cluster property (see footnote 1).

For the interpretation of the Wightman functions in terms of Hilbert space expectations of field operators, one needs a Hilbert topology to be associated with $\mathcal{D}_0$. The existence of a Hilbert space structure is proved in the free case (see Section 8.2), and must be required in the interacting case.[20]

W5′ (Hilbert space structure condition) The vacuum expectations of $\mathcal{F}$ satisfy the following regularity condition (in the notation of eq. (3.3.9)): there are Hilbert products $(.\,,\,.)_n$ on $\mathcal{S}(\mathbf{R}^n)$, $\forall n$, such that for any pair f_n, f_m of test functions

$$|\mathcal{W}(f_n^* \times f_m)| \leq (f_n, \, f_n)_n \, (f_m, \, f_m)_m.$$

The Hilbert space structure condition can be regarded as a substitute of the positivity condition W5, and its fulfilment is guaranteed by general regularity conditions (see the references in the footnote 20). It implies that there is a Hilbert product $(.\,,\,.)$ on D_0, and the vacuum expectation values of the fields can be written in the form

[20] G. Morchio and F. Strocchi, Ann. Inst. H. Poincaré, **A33**, 251 (1980); F. Strocchi, *Selected Topics on the General Properties of Quantum Field Theory*, World Scientific 1993, Sects. 6.3, 6.4, and Appendix 1.

$$< \Psi_0, \, A^* B \Psi_0 >=< A \Psi_0, \, B \Psi_0 >= (A \Psi_0, \, \eta \, B \Psi_0),$$

with $\eta^* = \eta$, $\eta^2 = 1$, $\eta \Psi_0 = \Psi_0$, called the *metric operator*.

The closure of $\mathcal{D}_0$ with respect to the scalar product $(.,.)$ yields a Hilbert space $\mathcal{K}$ with a metric operator, also called a *Hilbert–Krein space*, and by an adaptation of the reconstruction theorem one gets that the fields are operator valued-distributions in $\mathcal{K}$.

This structure is simply realized in the free quantization of the electromagnetic potential, where it is easy to see that the indefinite product $< .,.>$ is semipositive on $\mathcal{D}_0'$ and allows for a construction of the Hilbert space of local physical states.[21] In the interacting case, the semipositivity on $\mathcal{D}_0'$ is not obvious; it can be argued on the basis of the perturbative expansion or, more generally, as a consequence of the positivity of the vacuum functional on the gauge invariant field subalgebra, due to Proposition 3.2.

3.4 Charged states

There is "experimental" evidence that in QED the $U(1)$ global group is not broken, i.e., all the correlation functions of the field algebra $\mathcal{F}$ are invariant under infinitesimal $U(1)$ transformations (eqs. (7.1.3) and (7.1.4) with $\varepsilon = $ const), so that the generator Q of the $U(1)$ transformations on $\mathcal{F}$ exists, and $< \delta F >_0 = i < [Q, F] >_0 = 0$, $Q \Psi_0 = 0$. For a local field algebra $\mathcal{F}$, the relation between Q and $Q_R = j_0(f_R \alpha)$, given by eq. (7.2.4) for the vacuum expectations

$$< \delta F >_0 = i \lim_{R \to \infty} < [Q_R, F] >_0 = 0, \tag{7.3.14}$$

may be extended to the action on the local states by

$$Q F \Psi_0 = [Q, F] \Psi_0 = \lim_{R \to \infty} [Q_R, F] \Psi_0, \quad \forall F \in \mathcal{F}. \tag{7.3.15}$$

As shown by Swieca,[22] as a consequence of eq. (7.3.14) such a definition is consistent, i.e., $A \Psi_0 = B \Psi_0$ implies $\lim_{R \to \infty} [Q_R, A] \Psi_0 = \lim_{R \to \infty} [Q_R, B] \Psi_0$. In fact, $(A - B) \Psi_0 = 0$ implies, $\forall C \in \mathcal{F}$,

$$< C \Psi_0, \, [Q_R, (A - B)] \Psi_0 >=< \Psi_0, \, [Q_R, C^* (A - B)] \Psi_0 >$$

and the right-hand side vanishes in the limit $R \to \infty$, by eq. (7.3.14). By a similar argument the operator defined by the r.h.s of eq. (7.3.15) is hermitian on the local vectors and the compactness of the charge group yields the existence of a dense domain of analytic vectors.

It is worthwhile to remark that eq. (7.3.15) does not imply that

$$Q F \Psi_0 = \lim_{R \to \infty} Q_R F \Psi_0, \quad \forall F \in \mathcal{F}$$

[21] See, e.g., Section 8.2 below; A. S. Wightman and L. Gårding, Arkiv f. Fysik, **28**, 129 (1964); F. Strocchi, *Selected Topics on the General Properties of Quantum Field Theory*, World Scientific 1993, Sect. 7.1.

[22] J. Swieca, Goldstone's theorem and related topics, in *Cargèse Lectures in Physics*, Vol. 4, D. Kastler (ed.), Gordon and Breach 1970; for the mechanism of symmetry breaking, see F. Strocchi, *Symmetry Breaking*, 2nd ed., Springer 2008.

(not even as a weak limit with respect to the local states), unless by an improved smearing one gets the weak convergence of $Q_R \Psi_0$ (Section 5.3, below).

The vector space $\mathcal{D}_0$ contains charged vectors, e.g., those obtained by applying charged fields to the vacuum, but none of them belongs to $\mathcal{D}_0'$.[23] In particular, in the FGB gauge $\psi(f)\Psi_0$ has charge $-e$, but it is not a physical vector; in fact, by eq. (7.3.9), $\widetilde{\partial A}^{\,-}(k)\,\psi(g)\,\Psi_0 = [\,\widetilde{\partial A}^{\,-}(k), \psi(g)]\Psi_0 = -e(2\pi)^{-3}\theta(-k_0)\delta(k^2)\int d^4p\,\tilde g(p)\,\tilde\psi(k-p)\,\Psi_0 \neq 0$.

Proposition 3.4 *In a local quantization of electrodynamics, the matrix elements of the charge Q on local physical vectors vanish:*

$$< \Psi, Q\,\Phi > = 0, \quad \forall\, \Psi,\ \Phi \in \mathcal{D}_0'.$$

Proof. By the same argument used for the proof of Proposition 2.1, putting $\partial F_R \equiv \partial^i F_{0i}(f_R \alpha)$, one has

$$\lim_{R\to\infty}\,[\,\partial F_R,\, F\,] = 0, \quad \forall F \in \mathcal{F}.$$

Now, vectors of $\mathcal{D}_0'$ are of the form $\Psi = A\Psi_0$, $\Phi = B\Psi_0$, $A,\ B \in \mathcal{F}$, and putting $j_R \equiv j_0(f_R\alpha)$, by eq. (7.3.15), one has

$$< \Psi, Q\Phi > = \lim_{R\to\infty} < \Psi, [\,Q_R,\, B\,]\Psi_0 > =$$

$$= \lim_{R\to\infty} < \Psi, [\,j_R - \partial F_R,\, B\,]\Psi_0 > .$$

By locality, the limit is reached for finite (large) values of R, and by the weak Gauss law, eq. (7.3.2), the above commutator reduces to $< \Psi, B\,(j_R - \partial F_R)\,\Psi_0 > \equiv W(0,0)$. Such a matrix element is related to the three-point function

$$W(y - x, z - y) \equiv < A^*(x)\,B(y)\,(j_R - \partial F_R)(z) >_0,$$

where $A(x) \equiv U(x)A\,U^{-1}(x)$ and similarly for the other operators. By the spectral condition, W is the boundary value of a function $W(\zeta_1, \zeta_2)$ analytic in the tube $\mathcal{T}_2$. For ζ_1, ζ_2 real, and ζ_2 sufficiently large spacelike, $B(y)$ and $(j_R - \partial F_R)(z)$ commute, and W vanishes by the weak Gauss law. Hence, by the edge of the wedge theorem $W = 0$ everywhere, $W(0,0) = 0$ and $< \Psi, Q\,\Phi > = 0$.

The result agrees with the general argument of Section 2 that physical charged states cannot be local in DHR sense. The widespread belief that a vector $\psi(f)\Psi_0$ will become a physical state by going on the mass-shell is not correct, for several reasons. In particular, one does not get a state which satisfies the GB-subsidiary condition. In fact, as we shall see below, in order to satisfy the subsidiary condition an infrared dressing must be performed.[24]

The construction of physical charged states is one of the basic problems of gauge field theories. It is deeply related to the solution of the infrared problem in QED,

[23] R. Ferrari, L. E. Picasso, and F. Strocchi, Nuovo Cim. **39A**, 1 (1977); see the related Proposition 3.4 below.

[24] For the discussion of this problem, see G. Morchio and F. Strocchi, Ann. Phys. **168**, 27 (1986), esp. Sect. 4.

since a physical charged particle must be accompanied by its radiation field, i.e., by a "cloud" of soft photons. Moreover, the possibility of constructing color-charged states is at the root of the confinement problem.

The advantage of the FGB quantization, or more generally of local quantizations of QED, is that one can dispose of local charged vectors for the construction of non-local physical charged states; this means that one should look for i) solutions Ψ of the GB subsidiary condition, in some Hilbert–Krein closure of $\mathcal{D}_0$, carrying a non-zero charge, i.e., $< \Psi, Q\,\Psi > \neq 0$, ii) satisfying a (inner product) positivity condition $< \Psi, \Psi >> 0$, and iii) having positive finite energy.

In contrast with the free case, properties ii) and iii) are not obviously guaranteed by the GB condition; also, the property of non-zero charge may be difficult to verify, because in general Q_R does not converge on non-local states. Thus, the construction of charged physical states does not appear easy.[25]

A possible alternative to the strategy of the GB subsidiary condition, as a condition for the physical states, is to replace it (or the weak Gauss law condition iv of Section 3.2) by the following one, which looks more effective in view of the non-locality of the charged physical states:

iv′) *the physical states are obtained by applying gauge-invariant (not necessarily local) operators to the vacuum.*

Propositions 3.2 and 3.4 indicate that for the construction of charged physical states one must use *non-local* gauge-invariant operators F. The weak Gauss law is automatically satisfied, since gauge invariance implies that F commutes with ∂A (Proposition 3.1) and the fulfilment of the GB condition follows from $\partial A^- \Psi_0 = 0$. Furthermore, the *positivity property ii)* amounts to the positivity of the vacuum functional on the algebra of local gauge-invariant fields and on its non-local limits. Furthermore, correlation functions of gauge-invariant fields satisfy the *cluster property*.

For the construction of non-local limits of local fields, one needs a (possibly Hilbert–Krein) topology, beyond that defined by the inner product given by the Wightman functions of the local field algebra. This point is usually overlooked in the literature on the construction of physical charged states,[26] but it plays a crucial role and cannot be dismissed as a mathematical subtlety. The Hilbert–Krein topology should actually be

[25] Furthermore, the fulfilment of the above conditions requires that such states cannot transform covariantly under the Lorentz group and cannot have a sharp mass (*infraparticle structure*): J. Fröhlich, G. Morchio, and F. Strocchi, Phys. Lett. **89B**, 61 (1979); G. Morchio, and F. Strocchi, Nucl. Phys. **B211**, 471 (1983): *ibid* **232**, 547 (1984); for a general review, see G. Morchio and F. Strocchi, Infrared Problem, Higgs Phenomenon and Long Range Interactions, Erice Lectures in *Fundamental Problems of Gauge Field Theory*, G. Velo and A. S. Wightman eds. Plenum 1986, p. 301; for the inevitable infraparticle structure, see D. Buchholz, Phys. Lett. **B174**, 31 (1986), which will be reviewed in Section 7 below.

Even the fulfilment of condition iii) may be non-trivial and crucially related to the severity of the infrared singularities; in fact, the confinement of charged particles in 1+1 and in 2+1 dimensions and of massless charged particles in 3+1 dimensions may be understood as the impossibility of fulfilling the finite-energy condition: G. Morchio, and F. Strocchi, Ann. Phys. **172**, 267 (1986); F. Acerbi, G. Morchio, and F. Strocchi, Jour. Math. Phys. **34**, 899 (1993).

[26] See e.g., E. D'Emilio and M. Mintchev, Fortschr. Physik, **32**, 473 (1984); N. N. Bogoiubov, A. A. Logunov, A. I. Oksak, and I. T. Todorov, *General Principles of Quantum Field Theory*, Kluwer Acad. 1990, Chap. 10.

selected by the condition that it yields the required physical charged states as limits of local vectors.[27]

One might be led to believe that it is impossible to obtain charged gauge-invariant fields, by the prejudice that the global group is a subgroup of the gauge group, and therefore the invariance under the latter implies the invariance under the former, i.e., zero charge. A counter-example to this prejudice is the Dirac–Symanzik–Steinmann field,[28] formally given by

$$\psi_{gi}(x) = e^{-i\,e\,[(-\Delta)^{-1}\partial_i A^i](x)}\,\psi(x), \tag{7.3.16}$$

where A^i, ψ are the (renormalized) fields which describe the vector potential and the Dirac field, respectively, in the FGB quantization of QED (and e is the renormalized charge).

It is not difficult to see that ψ_{gi} commutes with ∂A, and therefore is invariant under the local gauge transformations of eqs. (7.3.10) and (7.3.11):

$$[\partial A(x),\,\psi_{gi}(y)] = e(-\Delta^{-1}\partial_i\partial^i D)(x-y)\psi_{gi}(y)+$$
$$-e\,D(x-y)\psi_{gi}(y) = 0,$$

but it does not commute with the charge Q, since under the (global) charge group $\psi(x) \to e^{i\alpha}\psi(x)$, $A_i(x) \to A_i(x)$.[29]

Eq. (7.3.16) together with

$$A_\mu(x) \to A_\mu(x) - \partial_\mu[(-\Delta)^{-1}\partial_i A^i](x) \equiv A_\mu^{gi}(x) \tag{7.3.17}$$

correspond to the classical gauge transformation from the FGB gauge to the Coulomb gauge, and the fields ψ_{gi}, A_i^{gi} can be identified with the electron field ψ_C and the vector potential A_{iC} in that gauge. In fact, by using $\partial_0 A_i = \partial_i A_0 + F_{0i}$, one obtains

$$\partial_i A_{gi}^i = 0, \qquad A_0^{gi} = -(-\Delta)^{-1}\,\partial^i F_{0i} = \Delta^{-1} j_0^C, \qquad j_0^C \equiv j_0 + \partial_0\partial A, \tag{7.3.18}$$

which characterize the Coulomb gauge (see Symanzik, Lectures 1971, eqs. (6.30), (6.35), and (6.38)). The equations (7.3.8) imply the gauge invariance of A_μ^{gi}: $[\partial A,\,A_\mu^{gi}(x)] = 0.$

---

[27] For a discussion of these points, see G. Morchio and F. Strocchi, Nucl. Phys. **B211**, 471 (1983): Errata ibid **232**, 547 (1984); Jour. Math. Phys. **44**, 5569 (2003); for a general review see G. Morchio, and F. Strocchi, Erice Lectures 1986, Sect. 2.3, and F. Strocchi, *Selected Topics on the General Properties of QFT*, World Scientific 1993, Part II, Chap. VI. For the conditions on the Hilbert–Krein topology, see G. Morchio and F. Strocchi, Representations of *-algebras in indefinite inner product spaces, in *Stochastic Processes, Physics and Geometry: New Interplays. II*, Canad. Math. Soc. Conference Proceedings Vol. 29, 2000, p. 491, esp. Sect. 4.

[28] P. A. M. Dirac, Canad. J. Phys. **33**, 650 (1955); K. Symanzik, Lectures 1971; O. Steinmann, *Perturbative Quantum Electrodynamics and Axiomatic Field Theory*, Springer 2000.

[29] Note, however, that the transformations under the global group are not given by (the limits of) $[Q_R,\,\psi_{gi}]$, with $Q_R = j_0(f_R\alpha)$, since ψ_{gi} is a non-local field, and in fact a divergent renormalization constant appears in the equal-time commutator of j_0 and ψ_{gi}; see K. Symanzik, Lectures 1971, eqs. (5.92a), (6.42c). Thus, the check that ψ_{gi} carries a non-zero charge cannot invoke the (naive) equal-time canonical commutation relations (as done in the literature). This point further clarifies the basic difference between charge transformations and local gauge transformations, since the generation of the latter by ∂A, eq. (7.3.11), does not have the problems of the generation of the first, because the fall-off in space of $\varepsilon(x)$, eq. (7.3.10), allows the extension to the non-local field ψ_{gi}.

The positivity of the Coulomb gauge implies the positivity of the vacuum correlation functions of the non-local gauge invariant algebra generated by the fields $\psi_{gi} = \psi_C$ and A_μ^{gi}. Furthermore, the complete description of the charged states in terms of $\psi_{gi} = \psi_C$, as given by the Coulomb gauge, supports the characterization iv') of the physical states, proposed above as an alternative to the GB condition. The Coulomb gauge also supports the fulfilment of the cluster property by the correlation functions of the gauge-invariant fields.

The general lesson from the Dirac–Symanzik–Steinmann field, for the construction of a gauge-invariant field starting from the local charged field $\psi(x)$, is the need of a "dressing" factor $U(x)$, which commutes with the $U(1)$ charge group (e.g., a function of A_μ non-involving the charged field ψ), and does not commute with ∂A, in order to compensate the local gauge transformation of ψ; i.e., $U(x)$ should satisfy

$$[Q, U(y)] = 0, \quad [\partial A(x), U(y)] = eD(x - y)U(y). \tag{7.3.19}$$

There is no unique solution of eq. (7.3.19), and therefore there is not a unique way of constructing gauge invariant charged fields; ψ_{gi} defined by eq. (7.3.16) is a special case of the more general formula

$$\Psi(x, f) = \exp\left(-ie \int dy\, f^\nu(x - y)\, A^\nu(y)\right) \psi(x), \tag{7.3.20}$$

with f a real function satisfying $\partial_\mu f^\mu(x) = \delta(x)$. The invariance of $\Psi(x, f)$ under local gauge transformations follows, due to the fall-off in space of $\varepsilon(x)$ which allows an integration by parts with no boundary terms.[30] Eq. (7.3.16) corresponds to the choice

$$f_0 = 0, \quad f_i(x) = (4\pi)^{-1}\partial_i|\mathbf{x}|^{-1}.$$

The construction of the physical charged states (through the gauge-invariant charged fields) is a crucial step for the non-perturbative solution of the infrared problem. In particular eq. (7.3.20) shows that an infrared dressing (given by the exponential of the vector potential) is necessary for obtaining physical states through the Dirac field ψ in the FGB formulation; the on-shell restriction, which is actually precluded by the infraparticle structure (see Section 7 below), does not work.

4   Local gauge quantization of the Yang–Mills theory

The quantization of the Yang–Mills (YM) theory remained an unsolved problem until the proof of renormalization, which was achieved by using a local field algebra and a weak Gauss law, as in the electrodynamical case. However, the way the Gauss law has to be weakened is more tricky than in the abelian case, where this is obtained by the addition of the gauge fixing term $-\tfrac{1}{2}\alpha(\partial A)^2$ to the gauge-invariant Lagrangian, $\mathcal{L} = \mathcal{L}_{inv}$ ($\alpha = 1$ corresponds to the FGB gauge).

---

[30] The mathematical status of the gauge-invariant charged fields defined above has been discussed by Steinmann in perturbation theory: O. Steinmann, *Perturbative QED and Axiomatic Field Theory*, Springer 2000. However, a modification of the Dirac factor is needed for removing infrared divergences of the scalar products of local and physical charged states: G. Morchio and F. Strocchi, Jour. Math. Phys. **44**, 5569 (2003).

In the non-abelian case, if one tries to fix the gauge in a similar way by the addition of the term $\mathcal{L}_{GF} \equiv -\frac{1}{2}\alpha(\partial A^a)^2$, one meets non-trivial difficulties. In this case, an argument similar to the second Noether theorem for $\mathcal{L} = \mathcal{L}_{inv} + \mathcal{L}_{GF}$, gives the following equations:

1) the invariance under the charge group identifies the associated conserved current

$$J^{a\,\mu} \equiv -i\frac{\delta\mathcal{L}}{\delta\partial_\mu\varphi_i}(t^a\varphi)_i - i\frac{\delta\mathcal{L}}{\delta\partial_\mu A_\nu^b}(T^a A_\nu)^b, \quad \partial^\mu J_\mu^a(x) = 0, \tag{7.4.1}$$

2) the transformation under the local gauge group

$$\delta^\varepsilon\mathcal{L} = \delta^\varepsilon\mathcal{L}_{GF} = -\alpha\, iT_{bc}^a A_\nu^c\,\partial A^b\,\partial^\nu\varepsilon^a - \alpha\, g^{\mu\nu}\,\partial A^b\,\partial_\mu\partial_\nu\varepsilon^b$$

gives the analogs of eqs. (7.1.5) and (7.1.7),

$$F_b^{\mu\nu} \equiv -\frac{\delta\mathcal{L}}{\delta\partial_\mu A_\nu^b} - \alpha\, g^{\mu\nu}\partial A_b = -F_b^{\nu\,\mu}, \tag{7.4.2}$$

$$J_\mu^b = \partial^\nu F_{\mu\nu}^b - \alpha(D_\mu\,\partial A)^b, \quad (D_\mu A_\nu)^a \equiv \partial_\mu A_\nu^a - iT_{bc}^a A_\mu^c A_\nu^b. \tag{7.4.3}$$

Hence, the deviation from the Gauss law is given by

$$\partial^\nu F_{\mu\nu}^a = J_\mu^a + \alpha\, D_\mu\partial A^a, \tag{7.4.4}$$

and since $\partial^\mu J_\mu^a = 0$, one has

$$\partial^\mu D_\mu\partial A^a = 0. \tag{7.4.5}$$

It is worth remarking that if $\mathcal{L}_{inv}$ is the gauge-invariant Lagrangian (7.1.14), the tensor $F_{\mu\nu}^a$ defined in eq. (7.4.2) coincides with the $F_{\mu\nu}^a$ of eq. (7.1.15); however, since the gauge-fixing term contains the charged field ∂A^a, the conserved current J_μ^a, which generates the global group transformations, becomes modified. In fact, one has

$$J_\mu^a(\alpha) = J_\mu^a(\alpha = 0) + i\alpha(T^a A_\mu)^b\,\partial A_b.$$

Sometimes, in the literature the weak Gauss law is written in the form

$$J_\mu^a(\alpha = 0) = \partial^\nu F_{\mu\nu}^a - \alpha\partial_\mu\partial A^a \tag{7.4.6}$$

where neither $J_\mu^a(\alpha = 0)$ nor $\partial_\mu\partial A^a$ satisfies a continuity equation. In our opinion, the above equation is somewhat misleading, because the Gauss law is the relation between the *conserved* current associated to the global group and the divergence of an antisymmetric tensor, and therefore the deviation from the Gauss law is not given by $\alpha\partial_\mu\partial A^a$, but by the conserved current $(D_\mu\partial A)^a$.

The weak Gauss law requires that the physical vectors Ψ satisfy

$$< \Psi, (J_\mu^a - \partial^\nu F_{\mu\nu}^a)\,\Psi > = 0 = < \Psi, D_\mu\partial A^a\Psi >, \tag{7.4.7}$$

but the explicit realization and control of this condition appear substantially more difficult than in the abelian case, because:

i) since ∂A^a is not a free field, it is not obvious to find a subsidiary condition which linearizes eq. (7.4.7) and is obviously satisfied by the vacuum and by the vectors $A\Psi_0$, with A any gauge-invariant field,

ii) contrary to the abelian case the deviation from the Gauss law is not given by an operator which generates the local gauge transformations,

iii) actually, since $\delta^\varepsilon \mathcal{L}_{GF} \neq 0$, for non-constant ε, there cannot be a conserved current which generates them. On the other hand, the crucial role played by local gauge transformations for the proof of renormalization in the abelian case (through the corresponding Ward identities) suggests that one should look for a substitute for them.

A solution of these problems is provided by the BRS quantization, which introduces two additional sets of anticommuting (local) hermitian fields $c^a(x)$, $\bar{c}^a(x)$, called *Faddeev–Popov ghosts*, and a modified gauge-fixing term, conveniently written in the Nakanishi–Lautrup formalism,[31]

$$\mathcal{L}_{GF} = -\partial^\mu B^a A^a_\mu + \tfrac{1}{2}\xi\, B^a B^a - i\partial^\mu \bar{c}^a \,(D_\mu c)^a, \tag{7.4.8}$$

where B^a is the Nakanishi–Lautrup field (for its use in the quantization of QED, see Section 8.2), the variation with respect to which gives

$$\partial A^a + \xi B^a = 0.$$

This choice leads to the following important features.

i) The functional integral associated with the Lagrangian $\mathcal{L}_{inv} + \mathcal{L}_{GF}$ does not suffer the ill-definiteness arising from the infinite contribution of gauge-equivalent configurations.

ii) The *role of the local gauge transformations of the abelian case is taken by the (local) BRS (also called BRST) transformations* (see below). They are parametrized by a parameter θ, which anticommutes with $c^a, \bar{c}^a$ and with all fermionic matter fields, and have the following infinitesimal form: for any element F of the local field algebra $\mathcal{F}$, $\delta^\theta F = \theta s F$, with

$$s\psi = it^a c^a\,\psi, \quad sA^a_\mu = (D_\mu c)^a, \quad sB^a = 0, \quad sc^a = -\tfrac{1}{2}f^a_{bc}c^b c^c, \quad s\bar{c}^a = iB^a. \tag{7.4.9}$$

Note that $s(AB) = s(A)B + \varepsilon(A)\,A\,s(B)$, $\varepsilon(A) = \pm$, according as A is a bosonic/fermionic operator.

Such transformations are unambiguously defined in the classical case, but in the quantum case require the definition of products of fields at the same point, and therefore involve a regularization. By exploiting the antisymmetry of the structure constants, the Jacobi identity, and the anticommutativity of the ghost fields, it is not difficult to see that such transformations are nilpotent, i.e.,[32]

$$ssF = 0, \tag{7.4.10}$$

and that they reduce to a gauge transformation with parameter $\theta c^a(x)$ on the matter, and gauge fields (eq. (7.1.4) can be written as $\delta^\varepsilon A^b_\nu = (D_\nu \varepsilon)^b$).

iii) The action is invariant under (local) BRS transformations, since the gauge-invariant part is BRS-invariant (by the above remark) and (it is easy to see that)

---

[31] For a more detailed account, see N. Nakanishi and I. Ojima, *Covariant Operator Formalism of Gauge Theories and Quantum Gravity*, World Scientific 1990; S. Weinberg, *The Quantum Theory of Fields*, Vol. II, Cambridge University Press 1996, Chap. 15.

[32] For the details of the proof, modulo regularization, see, e.g., S. Weinberg, 1996, Vol. II, pp. 29–30.

the gauge-fixing term can be written as

$$\mathcal{L}_{GF} = is[\partial^\mu \bar{c}^a A_\mu^a - \tfrac{1}{2}\xi \bar{c}^a B^a], \tag{7.4.11}$$

so that its invariance follows trivially from the nilpotency of s.

iv) The ghosts fields contribute to both the conserved current J_μ^a associated with the Yang–Mills charge group (not broken by $\mathcal{L}_{GF}$) and the conserved current J_μ^B associated with the BRS invariance:

$$J_\mu^a = j_\mu^a(\varphi) + f^a_{bc}[A^{b\,\nu} F^c_{\nu\,\mu} + A_\mu^b B^c - i\bar{c}^b (D_\mu c)^c + i\partial_\mu \bar{c}^b c^c],$$

$$J_\mu^B = B^a(D_\mu c)^a - \partial_\mu B^a c^a + \tfrac{1}{2}i\partial_\mu \bar{c}^a f^a_{bc} c^b c^c - \partial^\nu (F_{\mu\nu}{}^a c^a). \tag{7.4.12}$$

The *modified Gauss law* reads

$$\partial^\nu F_{\mu\nu}{}^a = J_\mu^a - \partial_\mu B^a - f^a_{bc} A_\mu^b B^c + if^a_{bc}\bar{c}^b (D_\mu c)^c. \tag{7.4.13}$$

The BRS charge Q_B is assumed to be unbroken, i.e., $Q_B \Psi_0 = 0$, and, as in Section 3.4, on the local vectors $((Q_B)_R \equiv J_0^B(f_R\,\alpha))$,

$$Q_B F \Psi_0 \equiv \lim_{R\to\infty} [(Q_B)_R, F]_{\mp} \Psi_0 = -isF\Psi_0, \quad \forall F \in \mathcal{F},$$

where the commutator $(-)$ or the anticommutator $(+)$ has to be taken according as F is a bosonic or a fermionic operator, respectively (then the last term on the right-hand side of eq. (7.4.12) does not contribute by locality). By eq. (7.4.10), $Q_B^2 = 0$.

v) The realization of the BRS symmetry leads to the Slavnov–Taylor identities which allow for a control and proof of the perturbative renormalization, as the Ward identities do in the abelian case.

In conclusion, the BRS quantization provides a local quantization which is at least perturbatively under control, just as the FGB quantization does for the abelian case.

Once the dynamical problem is formulated in a tractable way, it remains to discuss the physical interpretation; namely, the selection of the physical vectors. For this purpose, it is easy to check, by using eqs. (7.4.9), that the *modified Gauss law*, eq. (7.4.13), can be written in the form[33]

$$\partial^\nu F_{\mu\nu}{}^a = J_\mu^a - \{Q_B, (D_\mu \bar{c})^a\}, \tag{7.4.14}$$

which clearly implies that the vectors Ψ satisfying the *BRS subsidiary condition*

$$Q_B \Psi = 0 \tag{7.4.15}$$

satisfy the weak Gauss law,

$$< \Psi, (\partial^\nu F_{\mu\nu}^a - J_\mu^a) \Psi >= 0. \tag{7.4.16}$$

Conversely, the condition that on the vector space of physical vectors Ψ the inner product should be independent of the gauge fixing requires (due to the nilpotency of s) that $Q_B \Psi = 0$.[34]

[33] I. Ojima, Nucl. Phys. **B143**, 340 (1978).

[34] For more details and relevant references, see N. Nakanishi and I. Ojima, *Covariant Operator Formalism*, 1990, and S. Weinberg, *The Quantum Theory of Fields*, Vol. II, 1996.

In the following, $\mathcal{D}_0' \subseteq \mathcal{D}_0 \equiv \mathcal{F}\,\Psi_0$ denotes the subspace of local physical vectors selected by the BRS subsidiary condition (7.4.15) and $\mathcal{D}_0'' \equiv \{\Psi \in \mathcal{D}_0';\ <\Psi,\Psi>=0\}$.

As in the abelian case, the unbroken YM charges Q^a are defined on local states by

$$Q^a\,F\,\Psi_0 = \lim_{R\to\infty}[Q_R^a, F]\Psi_0, \quad \forall F \in \mathcal{F}. \tag{7.4.17}$$

Proposition 4.1   *The subspace $\mathcal{D}_0'$ of local physical vectors has the following properties:*

i) $\forall \Psi \in \mathcal{D}_0'$

$$<\Psi, Q^a\,\Psi>=0,$$

ii) $\forall F \in \mathcal{F}$, $F\,\Psi_0 \in \mathcal{D}_0'$, *iff*

$$[Q_B, F]_{\mp} = 0,$$

($\mp$ for bosonic/fermionic F), i.e., iff F is BRS invariant,

iii) as in the abelian case

$$(\partial^\nu F_{\mu\nu}{}^a - J_\mu^a)\,\mathcal{D}_0' \subseteq \mathcal{D}_0''. \tag{7.4.18}$$

More generally, for any BRS-invariant (possibly non-local) operator A, such that $Q_B\,A\,\Psi_0 = 0$, putting $\Psi \equiv (\partial^\nu F_{\mu\nu}{}^a - J_\mu^a)\,A\,\Psi_0$ one has

$$Q_B\,\Psi = 0, \quad <\Psi,\Psi>=0. \tag{7.4.19}$$

As in the abelian case, the inner product defined on $\mathcal{D}_0$ by the Wightman functions of $\mathcal{F}$ cannot be semidefinite.

Proof. i) The proof is the same as for Proposition 3.4.

ii) Clearly, $[Q_B, F]_{\mp} = 0$ implies $Q_B\,F\Psi_0 = 0$, since $Q_B\,\Psi_0 = 0$. Conversely, $Q_B\,F\,\Psi_0 = 0$ implies $[Q_B, F]_{\mp}\,\Psi_0 = 0$ and, since $[Q_B, F]_{\mp}$ is a local operator, by the Reeh–Schlieder theorem it vanishes.

iii) If $Q_B\,A\Psi_0 = 0$, one has

$$Q_B\{Q_B, (D_\mu\bar{c})^a\}\,A\Psi_0 = Q_B^2\,(D_\mu\bar{c})^a\,A\Psi_0 = 0,$$

and eqs. (7.4.19) follow easily.

As in the abelian case, the indefiniteness follows from locality and the vanishing of the two-point function of $\partial^\nu F_{\mu\nu}{}^a - J_\mu^a$.

As in the abelian case, the selection of the physical vectors by means of a subsidiary condition and its explicit realization was historically argued on the basis of the perturbative expansion, i.e., for local vectors. This was regarded as sufficient for the quantum mechanical interpretation of the theory, until it was realized (see Propositions 3.3 and 4.1) that the physical charged vectors cannot belong to $\mathcal{D}_0$, and should be obtained as suitable limits of local vectors.

Thus, strictly speaking, the GB subsidiary condition (7.3.3) and the BRS subsidiary condition (7.4.15) are modeled for the identification of the (physical) vacuum sector and it is not obvious what condition should select the non-local limits of local vectors which describe physical (charged) states. The extrapolation of the GB and

BRS subsidiary conditions from the vacuum sector to the charged sectors has in fact been criticized in the literature.[35]

As mentioned before, the construction of physical charged states as limits of local vectors is a very delicate and still debated problem, especially in the non-abelian case. Actually, the *a priori* characterization of the charged states is a prerequisite for the discussion of their existence and properties, in particular for the confinement problem, and more generally for the physical interpretation of the local (renormalizable) quantizations of gauge field theories.

The BRS transformations play a role very close to that of the local gauge transformations in the abelian case, and therefore, in analogy with the abelian case, it is tempting to replace the BRS subsidiary condition, eq. (7.4.15), by the requirement that the physical states are obtained by applying BRS-invariant operators to the vacuum. Thanks to ii) of Proposition 4.1, for local states this characterization coincides with the BRS subsidiary condition (both yielding the physical states of the vacuum sector), but for charged states it may offer a simpler recipe for exploring their possible construction. It is worthwhile to remark that a BRS-invariant operator is not necessarily invariant under local gauge transformations, nor necessarily invariant under the YM charge group, generated by J^a_μ, but it is invariant modulo operators with vanishing matrix elements on the physical vectors, i.e., it is weakly invariant.[36]

In this framework, the confinement of YM charges amounts to the impossibility of constructing BRS-invariant operators A, which have a finite commutator with the energy–momentum P_μ (so that they carry a finite energy–momentum) and are not color singlets:

$$< A\,\Psi_0, \sum_a (Q^a)^2\, A\,\Psi_0 > \neq 0.$$

It is worth remarking that Q_B does not commute with the YM charges, but it commutes with the invariant $\sum_a (Q^a)^2$.

It is worthwhile mentioning that another way to obtain the BRS symmetry— actually the original one—is through the functional integral quantization, by the work of Faddeev and Popov and of Slavnov.[37] In this approach, the Faddeev–Popov fields arise in the representation of the determinant $\det F_g$, where F_g is the gauge-fixing function, as a path integral. Such a determinant is needed in order to guarantee the gauge invariance of the correlation functions of gauge-invariant operators.[38]

---

[35] For the QED case, see D. Zwanziger, Phys. Rev. **D14**, 2570 (1976) and references therein; for a way out of Zwanziger obstructions, see G. Morchio and F. Strocchi, Erice Lectures 1985, in *Fundamental Problems of Gauge Field Theory*, 1986. For the discussion of the infrared dressing needed for obtaining a physical charged state out of a local charged state, see G. Morchio and F. Strocchi, Ann. Phys. **168**, 27 (1986), for the QED case.

[36] E.g., B^a is not invariant under local gauge transformations, and does not commute with the YM charges Q^b, since its canonical momentum is $\pi^a_B \equiv \delta\mathcal{L}/\delta\dot{B}^a = -A^a_0$, and the other canonical moment are $\pi^a_c = i\dot{\bar{c}}^a$, $\pi^a_{\bar{c}} = -i(D_0 c)^a$, $\pi_{A^a_k} = F^a_{0\,k}$. For details, see N. Nakanishi and I. Ojima, *Covariant Operator Formalism.*, 1990. For weak gauge invariance, see F. Strocchi and A.S. Wightman, Jour. Math. Phys. **15**, 2198 (1974).

[37] For a review, see L. D. Faddeev and A. A. Slavnov, *Gauge Fields: Introduction to Quantum Theory*, Addison-Wesley 1991.

[38] For an account of this strategy, see S. Weinberg, *The Quantum Theory of Fields*, Vol. II, 1996 and references therein.

5   Gauss law and charge superselection rule

One of the Dirac–Von Neumann axioms of quantum mechanics is that the states of a quantum-mechanical system are described by vectors of a Hilbert space $\mathcal{H}$, and that every vector describes a state, equivalently all projections, and therefore all (bounded) self-adjoint operators, represent observables (briefly $\mathcal{A}_{obs} = \mathcal{B}(\mathcal{H})$).

It was later realized[39] that typically (but not exclusively) for systems with infinite degrees of freedom, the physical states may belong to a direct sum of irreducible representations of the observable algebra, and therefore one cannot measure coherent superpositions of vectors belonging to inequivalent representations of the observable algebra. This means that if $\mathcal{H} = \oplus_j \mathcal{H}_j$, each $\mathcal{H}_j$ carrying an irreducible representation of $\mathcal{A}_{obs}$, a linear combination $\alpha \Psi_1 + \beta \Psi_2$ of vectors Ψ_1, Ψ_2 belonging to different $\mathcal{H}_j$ is not a physically realizable pure state, and it rather describes a mixture with density matrix $|\alpha|^2 \Psi_1 \otimes \Psi_1 + |\beta|^2 \Psi_2 \otimes \Psi_2$.

The impossibility of measuring such relative phases of a linear combination of vectors is equivalent to the existence of operators Q, called *superselected charges*, which commute with all the observables (and have a discrete spectrum if the Hilbert space is separable).[40]

In fact, the relative phase in a linear combination of eigenvectors $Q \Psi_i = q_i \Psi_i$, $i = 1, 2$, $q_1 \neq q_2$, is measurable iff there is an observable A with non-vanishing matrix element

$$(\Psi_1,\, A\, \Psi_2) = (q_1 - q_2)^{-1}\, (\Psi_1,\, [\, Q,\, A\,]\, \Psi_2).$$

Thus, a charge Q is superselected iff it does not only commute with the Hamiltonian (so that it defines a conserved quantum number and selection rules), but commutes with *all* the observables.

Wick, Wightman, and Wigner (WWW) proved that rotation and time-reversal invariance imply that the operator $Q_F = (-1)^{2J} = (-1)^F$, where J is the angular momentum and F is the fermion number modulo 2, is a superselected charge (*univalence superselection rule*, also called *fermion–boson superselection rule*). It was later shown that only rotational invariance was needed for the proof.[41] WWW also suggested that the electric charge and possibly the baryon and lepton number define superselected charges.

---

[39] G. C. Wick, A. S. Wightman, and E. P. Wigner, Phys. Rev. **88**, 101 (1952). For a discussion of the physical principles of QM, see F. Strocchi, Eur. Phys. J. Plus, **127**, 12 (2012).

[40] The superselected charges are often called *gauge charges*, and the group generated by them *global gauge group*, but this should not dim the crucial distinction from the group of local gauge transformations, of which the global gauge transformations should not be regarded as a subgroup. Under general localization properties of the states (see eq. (3.1.4)), it has been proved that the superselected group is compact (S. Doplicher and J. Roberts, Comm. Math. Phys. **131**, 51 (1990); S. Doplicher, in *Proceedings of the International Congress of Mathematicians*, Kyoto 1990, Springer 1991, p. 1319).

[41] G. C. Hegerdeldt, K. Kraus, and E. P. Wigner, Jour. Math. Phys. **9**, 2029 (1968); for a nice review of the superselection rules, see A. S. Wightman, Nuovo Cim. **110 B**, 751 (1995).

The superselection rule for the electric charge was later questioned and debated.[42] Actually, as we shall show below, it can be proved to be a consequence of the Gauss law and locality of the observables.[43]

The proof may be dismissed as trivial by arguing that observables must be gauge invariant and that gauge-invariance implies zero charge, but, as stressed before, such an argument is not correct, since the latter implication is contradicted by the Dirac–Symanzik–Steinmann field operator, showing that gauge invariant operators need not commute with the electric charge.[44]

The proof relies on the relation between the generator of the $U(1)$ charge group and the "integral" of the electric current density j_0, which obeys a Gauss law. As we shall see below, such a relation, which is usually taken for granted, becomes problematic if the charged fields are not (sufficiently) local with respect to the electric field, and in fact the discussion takes different paths in local and non-local gauges. Actually, in the Coulomb gauge, in order to generate the (time-independent) $U(1)$ transformations by j_0, one should adopt a spacelike time average in the integral of the current density (see Section 5.3).

5.1   Gauss charges in local gauges

As discussed in Section 2, in local gauges the action of the generator Q of the $U(1)$ global charge group on the local fields can be expressed in terms of the (suitably regularized) integral of the current density $j_0(f_R\,\alpha)$. This allows us to compute the commutator $[Q, A]$, with A a (local) observable field, in terms of $j_0(f_R\,\alpha)$; then *locality and (weak) Gauss law* give the superselection of Q.[45]

The proof in local gauges, typically in the FGB gauge, makes clear that local gauge invariance implies invariance under the (charge) global group for *local* operators, but not in general.

Theorem 5.1   *In the local quantization of QED, as a consequence of Einstein causality and the Gauss law, all observables commute with the charge Q defined on the local states by*

$$Q\,F\Psi_0 = \lim_{R\to\infty}\,[\,j_0(f_R\alpha), F\,]\,\Psi_0, \quad \forall F \in \mathcal{F}.$$

Proof. The proof exploits locality of the observables (required by Einstein causality), (weak) Gauss law, and invariance of the observables under local gauge transformations. In fact, for any *local* operator A (see Section 2),

$$[Q, A] = \lim_{R\to\infty}[Q_R, A] = \lim_{R\to\infty}[Q_R - \partial^i F_{0i}(f_R\alpha), A],$$

[42] R. Mirman, Phys. Rev. **186**, 1380 (1969); Y. Aharonov and L. Susskind, Phys. Rev. **155**, 1428 (1967); D. Kershaw and C. H. Woo, Phys. Rev. Lett. **33**, 918 (1974); G. C. Wick, A. S. Wightman, and E. P. Wigner, Phys. Rev. **D1**, 3267 (1970).

[43] F. Strocchi and A. S. Wightman, Jour. Math. Phys. **15**, 2198 (1974).

[44] This incorrect logic invalidates the proof claimed in L.V. Prokhorov, Lett. Math. Phys. **19**, 245 (1990), which completely misses the role of locality and does not even mention the previous proof.

[45] F. Strocchi and A. S. Wightman, Jour. Math. Phys. **15**, 2198 (1974).

where $Q_R \equiv j_0(f_R \alpha)$. Furthermore, by the (weak) Gauss law

$$Q_R - \partial^i F_{0i}(f_R \alpha) = -\partial_0 \, \partial A(f_R \alpha)$$

and if A is invariant under local gauge transformations, by Proposition 3.1 it commutes with ∂A:

$$[\partial A(x), A] = 0.$$

Hence, for any observable A, one has $[Q, A] = 0$.

A similar result can be proved for YM charges, in local gauges.

Proposition 5.2 *In a local quantization of YM theory, the unbroken YM charges (if any), defined on the local vectors by eq. (7.4.17), have vanishing commutator with any (local) observable A on the physical vectors, i.e.,*

$$< \Psi, [Q^a, A]\Phi > = \lim_{R \to \infty} < \Psi, [Q_R^a, A] \, \Phi >= 0, \tag{7.5.1}$$

$\forall \Psi, \Phi$ *satisfying the BRS supplementary condition.*

Proof. By locality of the observables

$$\lim_{R \to \infty} [\partial^i F_{0\,i}^a(f_R \alpha), A] = 0$$

and, by eq. (7.4.14), for any pair of physical states Ψ, Φ, i.e., satisfying eq. (7.4.15),

$$< \Psi, (\partial^\nu F_{\mu\nu}{}^a - J_\mu^a) \, \Phi >= 0.$$

Now, by Proposition 4.1, any (local) observable A (which by the superselection of Q_F is a bosonic field), satisfies $Q_B A \Psi_0 = 0$ and therefore commutes with Q_B, so that $Q_B \Psi = 0$ implies $Q_B A \Psi = 0$. Then, by eq. (7.4.19),

$$\lim_{R \to \infty} < \Psi, [Q_R^a, A] \, \Phi >= \lim_{R \to \infty} < \Psi, [Q_R^a - \partial^i F_{0i}^a(f_R \alpha), A] \, \Phi >= 0.$$

This proof shows that the observables are not only invariant under the (infinitesimal) local gauge transformations of $\mathcal{G}$, but, in matrix elements on the physical states, they are also invariant under the (infinitesimal) transformations of the global group G; i.e., for any observable A and for any physical state Ψ,

$$< \Psi, \delta^a A \, \Psi >\equiv i \lim_{R \to \infty} < \Psi, [Q_R^a, A] \, \Psi >= 0. \tag{7.5.2}$$

The unbroken YM charges Q^a defined by eq. (7.4.17) generate a group G_c, which is in general *non-abelian* and plays the role of a *color group*, commuting with the observables.

It is then natural to enquire about the physical implications of the existence of superselected (possibly non-commuting) charges. The answer will also shed light on the role of gauge groups, which by definition are not seen by the observables, but may have the role of classifying their representations.

5.2 Superselected charges and physical states

As we have seen for the non-abelian group of permutations (see Chapter 4, Section 2), the non-abelian color group G_c has the following implications on the representations of the observable algebra $\mathcal{A}$: pure states ω on $\mathcal{A}$ can be described in terms of vectors Ψ, which transform under an irreducible representation $U(g)$ of G_c, as

$$\omega(A) = \mathrm{Av}_{g \in G_c} < U(g)\Psi,\, AU(g)\,\Psi >,$$

where $\mathrm{Av}_{g \in G_c}$ denotes the average over the group G_c. Hence, ω is described by a colorless density matrix in the Hilbert space which is the carrier of irreducible representations of G_c. The extension of such states to the YM charges, by the above formula, gives $\omega(Q^a) = 0$.

E.g., for $G_c = SU(2)$ the charge which classifies the irreducible representations of $\mathcal{A}$ and (therefore) defines a superselection rule is the (gauge-invariant) Casimir operator $Q^2 = \sum_a (Q^a)^2$, which can be easily shown to commute with the observable algebra. Hence, the *problem of confinement* becomes the problem of non-existence of states ω on the observables such that $\omega(Q^2) \neq 0$, even if $\omega(Q^a) = 0$.

Thus, the important (if not the exclusive) role of the gauge group is that of providing, through the invariants of its representations, the superselected quantum numbers or charges, which classify the irreducible representations of the observable algebra $\mathcal{A}$. The unobservable charged fields play the role of intertwiners (or charge raising/lowering operators) between the vacuum representation (vacuum sector) and the charged sectors. This justifies the introduction of the enlarged algebra $\mathcal{F}$ generated by the observables and the charged fields (*field algebra*), since $\mathcal{F}$ is the carrier of the irreducible representations of the gauge group and provides the corresponding invariants.

As shown by Doplicher, Haag, and Roberts,[46] all the *local* pure states on the observable algebra $\mathcal{A}$ (i.e., those defined by local morphisms on $\mathcal{A}$) can be obtained in this way in terms of a *compact* gauge group. The classification of the representations of $\mathcal{A}$ defined by non-local morphisms, corresponding to the electrically charged states or to the possible colored states, is still a deep open problem.

5.3 Electric charge, current, and photon mass

In this section we will discuss general properties of the charge in QED. Since important physical issues are involved, the discussion shall be done in the physical Coulomb gauge, where the Maxwell equations hold as operator equations and, in each charge sector, all the states have a physical interpretation.

Moreover, since the global $U(1)$ gauge group is not broken, and therefore implementable by a one-parameter group of unitary operators, its generator is well defined and unambiguously identifiable with the electric charge Q^{el}, the latter being defined by

---

[46] For a review of their results, see R. Haag, *Local Quantum Physics*, Springer 1996, Chap. IV; we refer to this book also for an extensive discussion of charges, global gauge groups, and exchange symmetry in terms of observable algebras.

$$[Q^{el}, \psi_C] = -e\psi_C, \quad Q^{el}\Psi_0 = 0.$$

This implies that Coulomb fields of definite electric charge q, when applied to the vacuum, yield *physical states of electric charge q*. As we have seen, this property is not shared by the local fields of the FGB gauge.

The first issue is the *relation between the electric charge Q^{el} and the electric current j_μ*. On the basis of the wisdom of standard quantum field theory described by a local field algebra, it is usually taken for granted that the relation between Q^{el} and the integral of j_0, given by the Noether theorem in classical field theory, keeps holding in quantum electrodynamics. We have already remarked that such a relation holds at the level of commutators with local fields, and that a suitable regularization is needed for the integral of the charge density. The non-locality of the Coulomb gauge poses the problem anew, and, as we shall see, the naive picture requires a substantial revisitation.

Another issue is the proof that Q^{el} is not only unbroken but also *superselected*.

A third issue is a possible *explanation of the photon being massless*. A folklore statement is that local gauge invariance constrains the photon mass to zero. The basis of such a statement is the perturbative approach to QED, where at the tree level a mass term $\mu^2 A_\nu A^\nu$ is forbidden by local gauge invariance (which, however, gets broken by the gauge-fixing term) and this property is not spoiled by the radiative corrections, as a consequence of the Ward identities corresponding to unbroken $U(1)$ global gauge symmetry. However, in the early 1960s Schwinger provided a counter-example: in the so-called Schwinger model, i.e., QED in 1+1 spacetime dimensions, the $U(1)$ global gauge symmetry is unbroken and the "photon" field $F_{\mu\nu}$ is massive.[47]

Later, Swieca[48] pointed out that if the $U(1)$ global gauge group is unbroken *and* there are physical charged states, there cannot be a gap in the energy–momentum spectrum of the two-point function of $F_{\mu\nu}$.

All these conditions appear to be essentials: if the $U(1)$ is broken, the perturbative expansion around a mean field symmetry breaking parameter gives a mass to the corresponding vector boson (*Higgs mechanism*); moreover, as displayed by the Schwinger model, the confinement of physical charged states allows for a mass gap in the vector boson spectrum.

We shall give a general non-perturbative argument in the Coulomb gauge showing that i) the electric charge Q^{el} can be expressed in terms of a suitably smeared integral of the charge density iff the field $F_{\mu\nu}$ has a massless contribution, i.e., if there are massless photons, ii) *if there are massless photons the $U(1)$ global gauge symmetry cannot be broken*.

We start by discussing the possibility of generating the $U(1)$ transformations by a suitably regularized integral of the charge density j_0^C, a crucial issue, not only in the

[47] J. Schwinger, Phys. Rev. **128**, 2425 (1962); J. Schwinger, Trieste Lectures 1962, in *Theoretical Physics*, I.A.E.A. Vienna 1963, p. 89; J. H. Lowenstein and J. A. Swieca, Ann. Phys. **68**, 172 (1971); G. Morchio, D. Pierotti, and F. Strocchi, Ann. Phys. **188**, 217 (1988).

[48] J. Swieca, Phys. Rev. **D13**, 312 (1976); his assumption about the momentum-space behavior of form factors of the *non-local* charged states has been proved by D. Buchholz and K. Fredenhagen, Nucl. Phys. **B154**, 226 (1979), however, by using Poincaré covariance, which cannot hold in charged sectors (see Section 7).

unbroken case (where it explains the properties of the electric charge, in particular its superselection), but also for a non-perturbative control (and proof) of the Higgs mechanism (see Section 6.2).

For this purpose we denote by $\mathcal{F}_C$ the field algebra generated by the Coulomb gauge fields ψ_C, A_μ^C, and assume that its vacuum correlation functions are well defined.[49]

The following proposition shows that the non-locality of the charged fields implies the α-dependence (and therefore the time-dependence) of the commutators $[\, j_0^C(f_R\,\alpha),\, F\,]$, $F \in \mathcal{F}_C$, in the limit $R \to \infty$, in contrast to the time-independence of the $U(1)$ transformations.

Proposition 5.3 *In the Coulomb gauge, $\forall \Psi$, $\Phi \in \mathcal{F}_C \Psi_0$, the limits*

$$\lim_{R \to \infty} (\Psi, [\, j_0^C(f_R\,\alpha),\, \psi_C(y)\,]\,\Phi) \qquad (7.5.3)$$

exist but are α-dependent, and therefore the time-independent $U(1)$ global gauge group cannot be generated by such integrals of the charge density.

Proof. The proof exploits the relation between the fields of the Coulomb gauge and the FGB fields, eqs. (7.3.16) and (7.3.17),

$$\psi_C(x) = e^{-i\,e\,[(-\Delta)^{-1}\partial_i A^i](x)}\,\psi(x),$$

$$A_\mu^C = A_\mu(x) - \partial_\mu[(-\Delta)^{-1}\partial_i A^i](x), \quad j_0^C = j_0 + \partial_0\partial A.$$

In particular, the above relation between ψ_C and ψ provides an explicit control of the non-locality of the Coulomb charged fields (otherwise difficult to obtain).[50] Clearly, $(\Psi, (j_\mu^C - j_\mu)\Phi) = 0$, $\forall \Psi, \Phi \in \mathcal{F}_C\,\Psi_0$, since $\partial A \mathcal{F}_C \Psi_0 = \mathcal{F}_C \partial A \Psi_0$ and, by the gauge invariance of ψ_C, $[\, j_0(f_R, x_0), \psi_C(y)\,] = [\, j_0^C(f_R, x_0), \psi_C(y)\,]$. The convergence and time-dependence of the charge density commutators $[\, j_0(f_R, x_0), \psi_C(y)\,] = [\, j_0^C(f_R, x_0),\, \psi_C(y)\,] = [\, F_{0i}(\partial_i f_R, x_0), \psi_C(y)\,]$ are governed by the large (spacelike) distance behavior of the commutator $[\, F_{0\,j}(\mathbf{x}, x_0),\, \psi_C(y)\,]$. By the locality of the FGB charged field ψ, $\lim_{R \to \infty} [\, F_{0i}(\partial_i f_R, x_0), \psi(y)\,] = 0$, and therefore, by the above expression of ψ_C in terms of ψ, one has to analyse the large spacelike distance behavior of the commutator $[\, F_{0j}(x),\, \exp(-ie(-\Delta^{-1}\partial_i A^i)(y))\,]$. Now, by the locality of the Feynman–Gupta–Bleuler fields the commutator $[\, F_{\mu\nu}(x + \mathbf{a}),\, A_j(\mathbf{z}, y_0)\,]$ is a local operator which vanishes whenever $(\mathbf{x} + \mathbf{a} - \mathbf{z})^2 > (x_0 - y_0)^2$; therefore, it has a compact support in the variable $\mathbf{x} + \mathbf{a} - \mathbf{z}$, and the convolution with $\partial_z^j|\mathbf{y} - \mathbf{z}|^{-1}$ decreases at least as $|\mathbf{a}|^{-2}$, for $|\mathbf{a}| \to \infty$. By the same reasons, the commutator of

$$C_{\mu\nu}(x + \mathbf{a}, y) \equiv \int d^3z\, \partial_z^j|\mathbf{y} - \mathbf{z}|^{-1}[\, F_{\mu\nu}(x + \mathbf{a}),\, A_j(\mathbf{z}, y_0)\,]$$

[49] For the necessary UV regularization see, for a perturbative control, O. Steinmann, *Perturbative Quantum Electrodynamics and Axiomatic Field Theory*, Springer 2000, and for a general control which exploits the properties of the FGB gauge, D. Buchholz, S. Doplicher, G. Morchio, J. E. Roberts, and F. Strocchi, Ann. Phys. **290**, 53 (2001).

[50] G. Morchio and F. Strocchi, Jour. Phys. A: Math. Theor. **40**, 3173 (2007); F. Strocchi, *Symmetry Breaking*, 2nd ed., Springer 2008.

with $(-\Delta^{-1}\partial^j A_j)(y)$ decreases at least as $|\mathbf{a}|^{-4}$; the same decrease holds for the commutator with each term of the expansion of the exponential, as well as with ψ_C and A_i^C. Thus, for $|\mathbf{a}| \to \infty$,

$$[F_{\mu\nu}(x+\mathbf{a}), \, e^{-ie(-\Delta^{-1}\partial_i A^i)(y)}]\,\psi(y) \sim$$

$$\sim \int d^3z\, \partial_z^j \frac{1}{|\mathbf{y}-\mathbf{z}|}[F_{\mu\nu}(x+\mathbf{a}), \, A_j(\mathbf{z},y_0)]\,\psi_C(y) + O(|\mathbf{a}|^{-4}),$$

and in the matrix elements between Coulomb states $C_{\mu\nu}(x+\mathbf{a},y)$ may be pushed to the extreme left, apart from terms descreasing as $|\mathbf{a}|^{-4}$. Furthermore, by eqs. (7.3.8), $C_{\mu\nu}(x+\mathbf{a},y)$ commutes with ∂A; i.e., it is a gauge-invariant field. Then, the cluster property satisfied by the gauge-invariant fields (see Section 3.4) implies that the vacuum insertion gives the leading contribution, and one obtains, for $|\mathbf{x}| \to \infty$,[51]

$$[F_{\mu\nu}(x), \, \psi_C(y)] \sim \frac{ie}{4\pi} \int d^3z\, \partial_z^j \frac{1}{|\mathbf{y}-\mathbf{z}|} < [F_{\mu\nu}(x), \, A_j(\mathbf{z},y_0)] > \psi_C(y), \qquad (7.5.4)$$

the corrections being at least $O(|\mathbf{x}|^{-4})$. This and all the following equations are understood to hold in matrix elements between *Coulomb states* $\Psi, \Phi \in \mathcal{F}_C \Psi_0$. The large R limit of the charge-density commutator appearing in eq. (7.5.3) is easily obtained from eq. (7.5.4). In fact, since

$$< [F_{\mu\nu}(x), \, A_j(z)] > = (\partial_\nu g_{\mu j} - \partial_\mu g_{\nu j})F(x-z),$$

where $F(x) \equiv F^+(x) - F^+(-x)$ is the invariant commutator function of $F_{\mu\nu}$ (see, e.g., eq. (7.5.10) below), one has, for $R \to \infty$,

$$[j_0(f_R, x_0), \psi_C(y)] \sim -ie\partial_0 \int d^3x\, f_R(\mathbf{x})F(x-y)\psi_C(y). \qquad (7.5.5)$$

By locality, $F(x)$ vanishes for $|\mathbf{x}|$ large enough, so that the limit $R \to \infty$ of eq. (7.5.5) exists, i.e., the commutator of the charge density is integrable in $\mathbf{x}$. Thus, the non-locality of the charged Coulomb fields does not preclude the existence of the commutators of the integral of the charge density.

On the other hand, in all correlation functions of the Coulomb field algebra, one has

$$\lim_{R\to\infty} [j_0(f_R, x_0), \, \psi_C(y)] = -e \int d\rho(m^2)\, \cos(m(x_0-y_0))\,\psi_C(y), \qquad (7.5.6)$$

---

[51] Quite generally, for a field algebra satisfying asymptotic abelianess and with a center invariant under translations, one has

$$\text{w} - \lim_{|\mathbf{a}|\to\infty} A_{\mathbf{a}}B\,\Psi_0 = < A > B\Psi_0.$$

(For a simple proof, see F. Strocchi, *Symmetry Breaking*, 2nd ed. Springer 2008, p. 102.) For a more complete proof of the behavior (7.5.4), which takes into account the need of a UV regularization of eq. (7.3.16) and exploits the locality of the charged fields in the FGB gauge and the cluster property, see D. Buchholz, S. Doplicher, G. Morchio, J. E. Roberts, and F. Strocchi, Ann. Phys. **290**, 53 (2001).

and the r.h.s is independent of time iff $d\rho(m^2) = \lambda\delta(m^2)$, i.e., if $F_{\mu\nu}$ is a free field. This implies the α-dependence of the charge-density commutator, eq. (7.5.3), in the interacting case.[52]

The time dependence of $[\,j_0(f_R, x_0), \psi_C(y)\,]$ can also be derived by the general estimate of the spacelike large distance behavior of the commutator of $\mathbf{j}(\mathbf{x}, x_0)$, given by eq. (7.5.4), i.e.,

$$[j_i(x), \psi_C(y)] \sim i(e/4\pi) \int d^3z\, \partial_z^i |\mathbf{z} - \mathbf{y}|^{-1} \partial_0^2 F(\mathbf{x} - \mathbf{z}, x_0 - y_0)\, \psi_C(y),$$

$$\lim_{R\to\infty} [\,\dot{Q}_R(x_0), \psi_C(y)\,] = \lim_{R\to\infty} [\,\mathrm{div}\,\mathbf{j}(f_R, x_0), \psi_C(y)\,] \neq 0.$$

The time dependence of the charge-density commutators invalidates the non-renormalization arguments for conserved (local) currents, and equal-time restrictions may fail to exist. In fact, an infinite renormalization constant appears in equal-time commutators (see Symanzik Lecture Notes, eq. (6.42c)),

$$[\,j_0(x), \psi_C(y)\,]_{x_0=y_0} = -e(Z_3)^{-1}\,\delta(\mathbf{x} - \mathbf{y})\,\psi_C(y) \tag{7.5.7}$$

(all fields being renormalized fields and e the renormalized charge). For such a phenomenon, the vacuum polarization due to loops of charged fields plays a crucial role, so that the semiclassical approximation does not provide relevant information about the time dependence of the charge commutators, and in fact the phenomenon does not appear in the classical theory, where there are finite-energy localized solutions with non-zero charge and localized current j_μ, only the electric field being a Coulomb delocalized function of j_0.[53]

Thus, as shown by the above proposition, if the field algebra is not local, the heuristic argument, by which if the symmetry commutes with time translation (equivalently if the current continuity equation holds), then the generating charge commutes with the Hamiltonian (and is therefore independent of time), is not correct.

Even if the equal-time commutators, in particular $[\,\mathbf{j}, \varphi\,]$, have a sufficient localization, the time evolution may induce a delocalization leading to a failure of eq. (7.2.3). For these reasons, no reliable information can be inferred from the equal-time commutators, and the check of the basic assumptions of the Goldstone theorem becomes interlaced with the dynamical problem, as it happens for non-relativistic systems.

---

[52] G. Morchio and F. Strocchi, Jour. Math. Phys. **44**, 5569 (2003), Appendix. The same conclusions hold if instead of eq. (7.3.16) one uses the regularized version of Buchholz *et al.*, since eqs. (7.5.4)–(7.5.6) are changed only by a convolution with a test function $h(y_0) \in \mathcal{D}(\mathbf{R})$.

The time dependence of the charge commutators in the presence of Coulomb interactions has been argued in the literature on the basis of simple models or of the perturbative approach: G. S. Guralnik, C. R. Hagen, and T. W. Kibble, Phys. Rev. Lett. **13**, 585 (1964); T. W. Kibble, Phys. Rev. **155**, 1554 (1966), G. S. Guralnik, C. R. Hagen, and T. W. Kibble, Broken symmetries and the Goldstone Theorem, in *Advances in Particle Physics*, Vol. 2, R. L. Cool and R. E. Marshak (eds.), Interscience 1968, p. 567. The above exploitation of the relation between the Coulomb fields and the FGB fields has allowed a general non-perturbative proof.

[53] D. Buchholz, S. Doplicher, G. Morchio, J. E. Roberts, and F. Strocchi, Ann. Phys. **290**, 53 (2001).

In the unbroken case, a relation between the electric charge Q^{el}, defined in the Coulomb gauge as the generator of the $U(1)$ charge group, and the associated Noether current can be established by taking a suitable (spacelike) time average in the smeared integral of j_0:

$$Q_{R\delta} \equiv j_0(f_R \alpha_{\delta R}), \quad \alpha_{\delta R}(x_0) \equiv \alpha(x_0/\delta R)/\delta\, R, \qquad (7.5.8)$$

with $0 < \delta < 1$, $\alpha \in \mathcal{D}(\mathbf{R})$, $\operatorname{supp}\alpha \subset [-\varepsilon, \varepsilon]$, $\varepsilon << 1$.[54]

Proposition 5.4 *In the Coulomb gauge, the (time-independent) $U(1)$ charge transformations*

$$\delta\psi_C(x) = -e\psi_C(x), \quad \delta A_\mu^C(x) = 0.$$

are generated by the current j_μ^C

$$\delta F = i \lim_{\delta \to 0} \lim_{R \to \infty} [Q_{R\delta}, F], \quad \forall F \in \mathcal{F}_C, \qquad (7.5.9)$$

iff the spectral measure $d\rho$, which characterizes the two-point function of the electromagnetic field

$$< F_{\mu\nu}(\tfrac{1}{2}x)F_{\lambda\sigma}(-\tfrac{1}{2}x) >_0 = i\, d_{\mu\nu\lambda\sigma} \int d\rho(m^2)\Delta^+(x; m^2) \equiv d_{\mu\nu\lambda\sigma}F^+(x), \qquad (7.5.10)$$

$d_{\mu\nu\lambda\sigma} \equiv (g_{\nu\sigma}\partial_\mu\partial_\lambda + g_{\mu\lambda}\partial_\nu\partial_\sigma - g_{\nu\lambda}\partial_\mu\partial_\sigma - g_{\mu\sigma}\partial_\nu\partial_\lambda)$, *contains a $\delta(m^2)$; i.e., there are* **massless photons.**

Quite generally, as a consequence of the Gauss law, denoting by Ψ_0 the vacuum vector, one has

$$strong - \lim_{R \to \infty} j_0(f_R\, \alpha_{\delta R})\, \Psi_0 = 0, \qquad (7.5.11)$$

so that if there are massless photons, the $U(1)$ group, being generated by $j_0(f_R\, \alpha_{\delta R})$, is unbroken; furthermore, in this case, one can express the electric charge Q^{el} as an integral of the charge density j_0 not only in the commutators with charged fields, but also in the matrix elements of Coulomb charged states $\Phi, \Psi \in \mathcal{F}_C\, \Psi_0$:

$$(\Phi, Q^{el}\, \Psi) = \lim_{\delta \to 0} \lim_{R \to \infty} (\Phi, j_0(f_R\, \alpha_{\delta R})\, \Psi) \qquad (7.5.12)$$

Proof. For the proof of eq. (7.5.9) it is enough to discuss the commutators with the fields ψ_C, A_i^C which generate $\mathcal{F}_C$.

1. For $F = \psi_C$, the time-smearing of eq. (7.5.5) with $\alpha_{\delta R}(x_0)$ gives

$$[j_0(f_R\alpha_{\delta R}), \psi_C(y)] \sim$$

$$\sim e \int d\rho(m^2)\, d^3q\, \tilde{f}(\mathbf{q})\, \operatorname{Re}[e^{-i\omega(\mathbf{q},m)y_0}\, \tilde{\alpha}(\delta R\, \omega_R(\mathbf{q}, m))]\, \psi_C(y),$$

where $\omega_R(\mathbf{q}, m) \equiv (\mathbf{q}^2 R^{-2} + m^2)^{\frac{1}{2}}$. Then, since α is of fast decrease, $\tilde{\alpha}(\delta R\, \omega_R(\mathbf{q}, m)) \to 0$ if $m \neq 0$, and, by the dominated convergence theorem, the

---

[54] G. Morchio and F. Strocchi, Jour. Math. Phys. **44**, 5569 (2003); G. Morchio and F. Strocchi, Jour. Phys. A: Math. Theor. **40**, 3173 (2007).

right-hand side vanishes if the $d\rho(m^2)$ measure of the point $m^2 = 0$ is zero, i.e., if there is no $\delta(m^2)$ contribution to $d\rho$.

For the $m^2 = 0$ contribution, one has that $e^{-i|\mathbf{q}|y_0/R} \to 1$ when $R \to \infty$, and $\tilde{\alpha}(\delta|\mathbf{q}|) \to \tilde{\alpha}(0) = 1$ when $\delta \to 0$, moreover, $\int d^3q \tilde{f}(\mathbf{q}) = f(0) = 1$. Then, if the point $m^2 = 0$ has measure λ, the limit $R \to \infty$ of the above commutator is $\lambda e \varphi_C(y)$; finally the renormalization condition of the asymptotic electromagnetic field gives $\lambda = 1$.

2. For $F = A_i^C = A_i - \partial_i(-\Delta^{-1}\partial_j A^j)$, we remark that the commutator $\lim_{R\to\infty}[\partial_j F_{0\,j}(f_R, x_0), A_i]$ vanishes by the locality of the FGB fields. Moreover, the second term

$$[F_{0\,j}(\mathbf{x}, x_0), \partial_i(-\Delta^{-1}\partial^k A_k)(\mathbf{y}, y_0)]$$

contains an additional space derivative with respect to the commutator $C_{0\,j}$, defined in the proof of Proposition 5.3, so that it decreases at least as $|\mathbf{x}|^{-3}$; therefore, it gives a vanishing contribution, after smearing with f_R, in the limit $R \to \infty$.

For the proof of eq. (7.5.11) we note that by the Gauss law,

$$< j_0(x)\, j_0(y) >_0 = \Delta\square F^+(x - y),$$

so that, putting $d\nu(m^2) \equiv m^2 d\rho(m^2)$, $Q_{RT} \equiv j_0(f_R\alpha_T)$, $T = \delta R$, one has

$$\|Q_{RT}\,\Psi_0\|^2 = \int d\nu(m^2)\,\mathrm{Re}\,\Big[\int d^3q\,|\tilde{\alpha}(T\,\omega_R(\mathbf{q}, m)|^2\,\frac{|\mathbf{q}\tilde{f}(\mathbf{q})|^2}{2\omega_R(\mathbf{q}, m)}\Big]R.$$

Since α is of fast decrease, $\forall N \in \mathbf{N}$,

$$|\tilde{\alpha}(T\,\omega(\mathbf{q}, m))|^2 \le C_N\,(1 + T^2 m^2)^{-N},$$

and, since $d\rho$ is tempered, $\exists M \in \mathbf{N}$ such that $(1 + m^2)^{-M}\,d\rho(m^2) \equiv d\rho'(m^2)$ is a finite measure. Then, by taking $N = M + 2$, one has, for $T > 1$,

$$\|Q_{R,T}\,\Psi_0\|^2 \le C_N \int m^2 d\rho'(m^2)\,R\,[(1 + T^2\,m^2)^{-2}C/m]$$

$$\le C'(R/T) \int d\rho'(m^2)Tm/(1 + T^2m^2)^2 \equiv (R/T)G(T).$$

By the dominated convergence theorem, $G(T) \to 0$ as $T \to \infty$, since the integrand function is bounded and converges to zero pointwise. Then, for $T(R) = \delta R$ one has

$$\lim_{R\to\infty}\,Q_{RT}\Psi_0 = 0. \tag{7.5.13}$$

Eq. (7.5.12) follows from eq. (7.5.9) and eq. (7.5.11).

The above proposition settles the problem of *the relation between the $U(1)$ generator, Q^{el}, and the (suitably smeared) integral of the charge density of the corresponding conserved current*, in a very stringent way: such a relation can be established *iff there are massless vector bosons*, and in this case *the $U(1)$ symmetry is unbroken*.

Furthermore, in the unbroken QED case, *the electric charge on the* physical *Coulomb states can be expressed as the integral of the charge density j_0*, and by exploiting the Gauss law this provides a proof of the *electric charge superselection rule*.[55]

Theorem 5.5 *(Electric charge superselection rule) Any observable A commutes with the electric charge Q^{el} on the (Coulomb) charged states*

$$(\Phi_C, [Q^{el}, A] \Psi_C) = \lim_{\delta \to 0,\, R \to \infty} (\Phi_C, [j_0(f_R \alpha_{\delta R}), A] \Psi_C) = 0. \qquad (7.5.14)$$

Proof. The proof follows from Gauss law $j_0(f_R \alpha_{\delta R}) = F_{0\,i}(\partial_i f_R, \alpha_{\delta R})$ and the locality of the observables (relative to the observable field $F_{\mu\nu}$), since for any given compact region $\mathcal{O}$ in spacetime, the points $(\mathbf{x}, \delta R z_0)$, $|\mathbf{x}| > R$, $z_0 \in \operatorname{supp} \alpha \subset [-\varepsilon, \varepsilon]$, $|\delta| < 1$, $|\varepsilon| << 1$, become spacelike with respect to $\mathcal{O}$, for R large enough.

6 Gauss law and Higgs mechanism

The Higgs mechanism relative to the breaking of the global group G plays a crucial role in the standard model of elementary particle physics. The standard discussion of this mechanism is based on the perturbative expansion, and in particular, the evasion of the Goldstone theorem is checked at the tree level with the disappearance of the massless Goldstone bosons and with the vector bosons becoming massive.[56]

This is clearly displayed by the Higgs–Kibble (abelian) model of a (complex) scalar field φ interacting with a real gauge field A_μ, defined by the following Lagrangian ($D_\mu = \partial_\mu - ieA_\mu$):

$$\mathcal{L} = -\tfrac{1}{4}F_{\mu\nu}^{\,2} + |D_\mu \varphi|^2 - U(\rho); \quad \rho(x) \equiv |\varphi(x)|. \qquad (7.6.1)$$

$\mathcal{L}$ is invariant under the $U(1)$ charge group: $\beta^\lambda(\varphi) = e^{i\lambda}\varphi$, $\beta^\lambda(A_\mu) = A_\mu$, and under the corresponding local gauge transformations,

$$\beta^\Lambda(\varphi(x)) = e^{i\Lambda(x)}\varphi(x), \quad \beta^\Lambda(A_\mu(x)) = A_\mu(x) + e^{-1}\partial_\mu \Lambda(x).$$

At the classical level one may argue that by a local gauge transformation

$$\varphi(x) = e^{i\theta(x)}\rho(x) \to \rho(x), \ A_\mu(x) \to A_\mu(x) - e^{-1}\partial_\mu \theta(x) \equiv W_\mu(x)$$

[55] The proof would not be possible without curing the α-dependence of $j_0(f_R \alpha)$, in the limit $R \to \infty$, by the improved smearing (7.5.8), as discussed in G. Morchio and F. Strocchi, J. Math. Phys. **44**, 5569 (2003); Jour. Phys. A: Math. Theor. **40**, 3173 (2007).

It is worth remarking that Theorem 5.1 establishes the superselection of the $U(1)$ generator Q, as defined on the local fields of the FGB quantization (equivalently as defined on the local states), whereas Q^{el} is the electric charge defined on the physical states.

[56] P. W. Higgs, Phys. Lett. **12**, 132 (1964); Phys. Rev. Lett. **13**, 508 (1964); Phys. Rev. **145**, 1156 (1966); Spontaneous symmetry breaking, in *Phenomenology of Particle Physics at High Energy: Proc. 14th Scottish Univ. Summer School in physics 1973*, R. L. Crawford and R. Jennings (eds.), Academic Press 1974, p. 529; G. S. Guralnik, C. R. Hagen, and T. W. Kibble, Phys. Rev. Lett. **13**, 585 (1964); Broken symmetries and the Goldstone theorem, in *Advances in Particle Physics*, Vol. 2, R. L. Cool and R. E. Marshak (eds.), Interscience 1968, p. 567; T. W. Kibble, Broken Symmetries, in *Proc. Oxford Int. Conf. on Elementary Particles*, 1965, Oxford University Press 1966, p. 19; Phys. Rev. **155**, 1554 (1966); F. Englert and R. Brout, Phys. Rev. Lett. **13**, 321 (1964).

one may eliminate the field θ from the Lagrangian, which becomes

$$\mathcal{L} = -\tfrac{1}{4}F_{\mu\nu}{}^2 + \tfrac{1}{2}e^2\rho^2\,W_\mu^2 + \tfrac{1}{2}(\partial_\mu\rho)^2 - U(\rho). \tag{7.6.2}$$

If the (classical) potential U has a non-trivial (absolute) minimum $\rho = \bar{\rho}$, one can consider a semiclassical approximation based on the expansion $\rho = \bar{\rho} + \sigma$, treating $\bar{\rho}$ as a classical constant field (of order 1) and σ as small with respect to it.

At the lowest order, putting without loss of generality $U(\bar{\rho}) = 0$ and keeping only the quadratic terms in σ and W_μ, one has

$$\mathcal{L}^{(2)} = -\tfrac{1}{4}F_{\mu\nu}{}^2 + \tfrac{1}{2}e^2\bar{\rho}^2\,W_\mu^2 + \tfrac{1}{2}(\partial_\mu\sigma)^2 - \tfrac{1}{2}U''(\bar{\rho})\sigma^2. \tag{7.6.3}$$

This Lagrangian describes a massive vector boson and a massive scalar with (square) masses

$$M_W^2 = e^2\bar{\rho}^2, \quad m_\sigma^2 = U''(\bar{\rho}).$$

This result is taken as evidence that there are no massless particles in the theory described by the Lagrangian $\mathcal{L}$.

This argument, widely used in the literature,[57] is not without problems, because already at the classical level, for the full equivalence between the two forms of the Lagrangian, eqs. (7.6.1) and (7.6.2), one must add the constraint that ρ is positive, a property which seems difficult to combine with the time evolution given by the non-linear equation for ρ, obtained from the Lagrangian of eq. (7.6.2), treating ρ and W_μ as Lagrangian variables. Furthermore, the expansion which leads to the quadratic Lagrangian of eq. (7.6.3) requires that under the time evolution σ keeps being small with respect to $\bar{\rho}$—a condition which seems difficult to guarantee, even at the classical level.

Thus, the constrained system is rather singular and its mathematical control is doubtful. The situation becomes obviously more critical for the quantum version, since the definition of $|\varphi(x)|$ is also very problematic, for distributional reasons. In conclusion, ρ is a singular field and one should not consider it as a genuine (Lagrangian) field variable.

A better alternative is to decompose the field $\varphi = \varphi_1 + i\,\varphi_2$ in terms of hermitian fields, φ_i, $i = 1, 2$, and to consider the semiclassical expansion $\varphi_1 = \bar{\varphi} + \chi_1$, $\varphi_2 = \chi_2$, treating χ_i, $i = 1, 2$ as small. The quadratic Lagrangian defined by keeping only the quadratic terms in χ_1, χ_2, A_μ, has the following form:

$$-\tfrac{1}{4}F_{\mu\nu}{}^2 + \tfrac{1}{2}(\partial_\mu\chi_1)^2 + \tfrac{1}{2}(\partial_\mu\chi_2)^2 + \tfrac{1}{2}(e\bar{\varphi})^2 A_\mu^2 - e\bar{\varphi}\,A^\mu\partial_\mu\chi_2 - \tfrac{1}{2}U''(\bar{\varphi})\chi_1^2.$$

By introducing the field $W_\mu \equiv A_\mu - (e\bar{\varphi})^{-1}\partial_\mu\chi_2$, one eliminates χ_2 from $\mathcal{L}^{(2)}$, which takes exactly the same form of eq. (7.6.3), with $\bar{\rho}$ replaced by $\bar{\varphi}$ and σ by χ_1.

If indeed the fields χ_i can be treated as small, by appealing to the perturbative (loop) expansion (with $\hbar$ resulting as the small expansion parameter), one has that $< \varphi > \sim \bar{\varphi} \neq 0$, i.e., the vacuum expectation of φ is not invariant under the $U(1)$ global group (*symmetry breaking*).

---

[57] See, e.g., S. Coleman, *Aspects of Symmetry: Selected Erice lectures*, Cambridge University Press 1985, Sect. 2.4.

Thus, the expansion can be seen as an expansion around a (symmetry breaking) mean field ansatz, and it is very important that a renormalized perturbation theory based on it exists and yields a non-vanishing symmetry breaking order parameter $< \varphi > \neq 0$ at all orders. This is the standard (perturbative) analysis of the Higgs mechanism.

The extraordinary success of the standard model motivates an examination of the Higgs mechanism from a general non-perturbative point of view. In this perspective, one of the problems is that mean field expansions may yield misleading results about the occurrence of symmetry breaking and the energy spectrum.[58]

As a matter of fact, the Euclidean functional integral approach defined by the Lagrangian of eq. (7.6.1) gives symmetric correlation functions and in particular $< \varphi >= 0$ (*Elithur–De Angelis–De Falco–Guerra (EDDG) theorem*).[59]

The crux of the argument is that gauge invariance decouples the transformations of the fields inside a volume V (in a Euclidean functional integral approach) from the transformations on the boundary, so that the boundary conditions are ineffective and cannot trigger non-symmetric correlation functions. This means that the mean field *ansatz* is incompatible with the non-perturbative quantum effects and the approximation leading to the quadratic Lagrangian is not correct.

The same negative conclusion would be reached if (as an alternative to the transformation which at the classical level leads to eq. (7.6.2)), by means of a gauge transformation one reduces $\varphi(x)$ to a real, not necessarily positive, field $\varphi_r(x)$. In this case, the local gauge invariance has not been completely eliminated, and the corresponding Lagrangian, of the same form of eq. (7.6.2) with ρ replaced by φ_r, is invariant under a residual Z_2 local gauge group ($\varphi_r(x) \to e^{is(x)} \varphi_r(x)$, $W_\mu(x) \to W_\mu(x) - e^{-1}\partial_\mu s(x)$, $e^{2is(x)} = 1$). Then, an adaptation of the proof of the EDDG theorem gives $< \varphi >= 0$ and no symmetry breaking.

In order to avoid the vanishing of a symmetry breaking order parameter, one must reconsider the problem by adding, to the Lagrangian of eq. (7.6.1), a gauge-fixing $\mathcal{L}_{GF}$ which breaks local gauge invariance, so that the EDDG theorem does not apply. Then, the discussion of the Higgs mechanism, and in particular of the way it is realized, necessarily becomes gauge-fixing dependent. This should not appear strange, since the vacuum expectation of φ is a gauge dependent quantity.

The above problem of consistency with the non-perturbative approach arises also for gauge fixings involving a mean field ansatz, as for the case of the unitary gauge and of the ξ gauges; in the formulation working with a real reducible representation of the symmetry algebra, the unitary gauge is defined by the condition that the scalar (Higgs) fields are orthogonal to the vacuum expectation $< \varphi >$.[60]

---

[58] E.g., the mean field ansatz on the Heisenberg spin model of ferromagnetism gives a wrong critical temperature and an energy gap. For the problems of the mean field *ansatz* and of the expansion based on it, see, e.g., F. Strocchi, *Symmetry Breaking*, 2nd. ed. Springer 2008, esp. Part II, Chaps. 10, 11.

[59] S. Elitzur, Phys. Rev. D **2**, 3978 (1975); G. F. De Angelis, D. De Falco, and F. Guerra, Phys. Rev. D **17**, 1624 (1978). For a simple account, see, e.g., F. Strocchi, *Elements of Quantum Mechanics of Infinite Systems*, World Scientific 1985, Part C, Sect. 2.5.

[60] For a detailed discussion, see S. Weinberg, *The Quantum theory of Fields*, Vol. II, Sect. 21.1.

In the following, we will present a *non-perturbative* proof of the absence of Goldstone bosons associated with the breaking of the $U(1)$ gauge symmetry, in the local gauges and in the Coulomb gauge. In the first case, the (unavoidable) Goldstone modes are forbidden to show up in the physical spectrum as a consequence of the (weak) Gauss law constraint satisfied by the physical states. In the second case we will show that the Goldstone spectrum, i.e., the energy–momentum spectrum appearing in (the support of) the Fourier transform of $< j_0(x)\, \varphi(y) >_0$, at low momenta, coincides with the energy–momentum spectrum of the two-point function of the vector boson field $F_{\mu\nu}$, and the latter cannot have a $\delta(k^2)$ singularity as a consequence of $U(1)$ breaking. In this way one obtains a proof of the full Higgs mechanism as formulated by Weinberg.[61]

6.1 Local gauges

The evasion of the Goldstone theorem by the Higgs mechanism can be understood by a non-perturbative argument in local (renormalizable) gauges, defined by a *gauge-fixing invariant under the global $U(1)$ group*.

For concreteness, we discuss the abelian Higgs–Kibble model in the so-called α gauges, obtained by the addition of the gauge-fixing $-\tfrac{1}{2}\alpha(\partial A)^2$ to the gauge invariant Lagrangian of eq. (7.6.1). Proceeding as before with a perturbative expansion based on the mean field ansatz $\varphi = \overline{\varphi} + \chi_1 + i\chi_2$, and performing the change of variables $W_\mu = A_\mu - (e\overline{\varphi})^{-1}\partial_\mu\chi_2$, one obtains a quadratic Lagrangian of the form of eq. (7.6.3) plus the gauge-fixing term $-\tfrac{1}{2}\alpha(\square\chi_2)^2$.

Thus, χ_2 does not disappear from the quadratic Lagrangian, and satisfies a "massless" field equation $\square^2\chi_2 = 0$; this means that there are massless modes. The problem is their physical interpretation; as an argument against their physical relevance one may note the α-dependence of the corresponding Lagrangian term. Moreover, the general solution of the equation $\square^2\chi_2 = 0$ is a massless field $(\sim \delta(k^2))$ plus a dipole field $(\sim \delta'(k^2))$ and a $\delta'(k^2)$ cannot appear in the physical spectrum, because the spacetime translations must be described by unitary operators in the space of physical vectors.[62]

Actually, one can find a general *non-perturbative* argument about the unphysical nature of the massless modes associated to the breaking of the $U(1)$ gauge group in the local (renormalizable) gauges, by exploiting locality and the weak Gauss law, i.e., the vanishing of $\mathcal{L}_\mu \equiv j_\mu - \partial^\nu F_{\mu\nu}$ in matrix elements of physical vector states.[63]

Proposition 6.1 *(Higgs theorem in local gauges) In local gauges, defined by a gauge-fixing invariant under the global $U(1)$ group, the spontaneous breaking of such a $U(1)$ group, with order parameter $< \delta A > \neq 0$, $A \in \mathcal{F}$, implies that the Fourier transform of the two-point function $< j_0(x)\, A >$ contains a $\delta(k^2)$* **Goldstone modes.** *However, such a singularity cannot be ascribed to the energy–momentum spectrum of the physical*

---

[61] S. Weinberg, Phys. Rev. Lett. **27**, 1688 (1971); Rev. Mod. Phys. **46**, 255 (1974).

[62] For the discussion of the physical interpretation of the fields of the quadratic Lagrangian, see T. W. Kibble, Phys. Rev. **155**, 1554 (1966).

[63] F. Strocchi, Comm. Math. Phys. **56**, 57 (1977). For the extension to the non-abelian case see: G. De Palma and F. Strocchi, Ann. Phys. **336**, 112–117 (2013)

vectors Ψ, *which satisfy the weak Gauss law* $< \Psi, (j_\mu - \partial^\nu F_{\mu\nu})\Psi >= 0$, *i.e., the Goldstone modes are not physical.*

Proof. As a first step, one remarks that in the local gauges the field algebra $\mathcal{F}$ satisfies locality, and therefore there is no problem for the existence of $\lim_{R\to\infty}[j_0(f_R,\alpha), F]$, $\forall F \in \mathcal{F}$.

Furthermore, by locality and $\partial^\mu j_\mu = 0$ (implied by $U(1)$ invariance of the gauge fixing), the limit is independent of the smearing function α, with the normalization condition $\tilde{\alpha}(0) = 1$; i.e., the commutators $[j_0(f_R, t), F]$ are independent of t, in the limit $R \to \infty$. Hence, the $U(1)$ global gauge group is locally generated by the conserved current j_μ (i.e., eq. (7.2.4) holds), the assumption of the Goldstone theorem are fulfilled and the existence of a $\delta(k^2)$ singularity, in the Fourier transform of the two-point function $< j_0(x)\, A >$, follows from a slight extension of the proof of the Goldstone theorem in the absence of positivity.[64] Then, one has to discuss its physical consequences.

By the locality of A one has

$$\lim_{R\to\infty} < [\partial^i F_{0i}(f_R\alpha), A] >= 0;$$

therefore $\partial^i F_{0i}(f_R\alpha)$ cannot contribute to the $\delta(k^2)$ singularity in the commutator $< [j_0(f_R\alpha), A] >=< [\partial^i F_{0i}(f_R\alpha) + \mathcal{L}_0(f_R\alpha), A] >$, for large R, and only $\mathcal{L}_R \equiv \mathcal{L}_0(f_R\alpha)$ is responsible for it.

For the implication of the $\delta(k^2)$ on the energy–momentum spectrum, as in the standard proof, one should analyze the intermediate vectors which contribute to $\mathrm{Im} < j_0(f_R\alpha)A >\sim \mathrm{Im} < \mathcal{L}_R A >$. The insertion of a complete set of vectors Φ_n requires to make reference to a Hilbert structure, and, since in local gauges positivity does not hold, one has to refer to the Hilbert–Krein space $\mathcal{K}$ (with Hilbert scalar product $(.,.)$, see Section 3.1) obtained by a Hilbert–Krein closure of $\mathcal{D}_0$; the conclusions are independent of the chosen Hilbert–Krein structure. It is convenient to choose the Φ_n according to the (orthogonal) decomposition of $\mathcal{K} = K_{phys} \oplus K_{phys}^\perp$, where K_{phys} denotes the subspace of physical vectors (i.e., those satisfying the subsidiary condition). Then, the generic insertion takes the form

$$(\Psi_0, \eta\mathcal{L}_R\Phi_n)(\Phi_n, A\Psi_0) =< \Psi_0, \mathcal{L}_R\Phi_n > (\Phi_n, A\Psi_0),$$

and by the weak Gauss law the physical vectors cannot contribute. Thus, the Goldstone modes associated with the $\delta(k^2)$ singularity cannot be physical.

6.2   Coulomb gauge; a theorem on the Higgs phenomenon

The Coulomb gauge can be obtained by adding the gauge fixing condition

$$\partial_i A^i(x) = 0, \tag{7.6.4}$$

equivalently by adding a Lagrangian multiplier $\mathcal{L} \to \mathcal{L} + \xi(\partial_i A^i) \equiv \mathcal{L}_C$; the variation with respect to ξ gives eq. (7.6.4). Proceeding as before, one obtains the following quadratic Lagrangian in the Coulomb gauge:

---

[64] See F. Strocchi, *Symmetry Breaking*, 2nd ed. Springer 2008, and references therein.

$$\mathcal{L}_C^{(2)} = -\tfrac{1}{4}F_{\mu\nu}^{\ 2} + \tfrac{1}{2}e^2\overline{\varphi}^2\,W_\mu^2 + \tfrac{1}{2}(\partial_\mu\chi_1)^2 - \tfrac{1}{2}U''(\overline{\varphi})\chi_1^2 +$$

$$+\xi(\partial_i W^i + (e\overline{\varphi})^{-1}\Delta\chi_2), \tag{7.6.5}$$

and eq. (7.6.4) becomes $e\overline{\varphi}\partial_i W^i + \Delta\chi_2 = 0$. This is a non-dynamical equation and is easily solved by

$$\chi_2(x) = e\,\overline{\varphi}\,[(-\Delta)^{-1}\partial_i W^i](x). \tag{7.6.6}$$

This implies that whereas χ_1 and W_μ are expected to be local fields, since they are necessarily so in the quadratic approximation given by eq. (7.6.5), χ_2 cannot be local, since it is a Coulomb delocalized functional of $\partial_i W^i$. Thus, $\varphi(x) = \overline{\varphi} + \chi_1 + i\chi_2$ is non-local with respect to $F_{\mu\nu}$; this reflects the general conflict between the Gauss law and locality for charged fields, discussed in Section 2.

Since the gauge-fixing breaks local gauge invariance, but not the invariance under the global gauge group, the EDDG theorem does not apply, and one may consider the possibility of a symmetry breaking order parameter $< \varphi > \neq 0$.

Now, another conceptual problem arises: the starting Lagrangian $\mathcal{L}$ is invariant under the $U(1)$ global group, and its breaking with a mass gap seems incompatible with the Goldstone theorem.[65] As an explanation of such an apparent conflict, one finds in the literature the statement that the Goldstone theorem does not apply if the two-point function $< j_0(x)\,\varphi(y) >$ is not Lorentz covariant as it happens in the physical gauges, like the Coulomb gauge.

As a matter of fact, the Goldstone–Salam–Weinberg proof of the Goldstone theorem crucially uses Lorentz covariance; however, the more general proof by Kastler, Robinson, and Swieca does not assume it, only locality being used, so that the quest of a better explanation remains.

Quite generally, the crucial issue is the relation between the generation of the continuous symmetry β^λ, $\lambda \in \mathbf{R}$, and the associated conserved Noether current j_μ, i.e., the validity of eq. (7.2.4), by which $< \delta A > \neq 0$ implies the existence of a $\delta(k^2)$ in the Fourier transform of the two-point function $< j_0(x)\,A >$.

As stressed in Section 2, for such a relation *both* the existence of the limit $\lim_{R\to\infty} < [\,j_0(f_R\alpha),\,A\,] >$ *and* its independence of the test function α, with $\tilde{\alpha}(0) = 1$, is required, the latter property being the proper way of stating that the commutator $\lim_{R\to\infty} < [\,j_0(f_R,t),\,A\,] >$ is independent of time. Both properties are automatically satisfied if the (symmetry breaking) order parameter A is *relatively local* with respect to the conserved current j_μ, but their validity is a delicate and crucial issue,[66] which, in our opinion, is not sufficiently emphasized in the literature.

---

[65] For a discussion of the Goldstone theorem, its proofs, and the mechanisms of its evasion, see F. Strocchi, *Symmetry Breaking*, 2nd ed., Springer 2008.

[66] J. A. Swieca, Comm. Math. Phys. **4**, 1 (1967); G. Morchio and F. Strocchi, Jour. Phys. A: Math. Theor. **40**, 3173 (2007); for a general discussion, see F. Strocchi, *Symmetry Breaking*, 2nd ed., Springer 2008.

Quite generally, for the proof of the Goldstone theorem one needs a fall-off of the commutator $[j_\mu(x + \mathbf{a}), A]$, when $|\mathbf{a}| \to \infty$, faster than $|\mathbf{a}|^{-2}$.[67] A failure of such a fall-off in the Higgs–Kibble model in the Coulomb gauge for $A = \varphi(f)$ can be guessed on the basis of the quadratic approximation given by eq. (7.6.5), since by eq. (7.6.6) χ_2, and therefore $\varphi(x) = \overline{\varphi} + \chi_1 + i\chi_2$, has a Coulomb delocalization..

Actually, the results of Section 5.3 allow a non-perturbative control of such a behavior and from this one may obtain a non-perturbative treatment of the Higgs phenomenon in the general case of a symmetry breaking order parameter, which, as in the Kastler–Robinson–Swieca general proof of the Goldstone theorem, need not to be an elementary field, like the Higgs field.

Theorem 6.2   *(Higgs phenomenon)*
A. *If the spectral measure $d\rho(m^2)$ of the vector boson field $F_{\mu\nu}$ has a $\delta(m^2)$ contribution, i.e., if there are corresponding* **massless vector bosons,** *then, in the Coulomb gauge,*
i) the $U(1)$ global gauge transformations are generated by the (suitably smeared) conserved current j_μ,

$$\delta F \equiv d\beta^\lambda(F)/d\lambda|_{\lambda=0} = i \lim_{\delta \to 0} \lim_{R \to \infty} [j_0(f_R \alpha_{R\delta}), F], \quad \forall F \in \mathcal{F}_C, \tag{7.6.7}$$

ii) the vacuum is invariant and therefore the $U(1)$ **global gauge symmetry is unbroken,**

$$strong - \lim_{R \to \infty} j_0(f_R \alpha_{R\delta}) \Psi_0 = 0, \tag{7.6.8}$$

iii) the $U(1)$ generator Q can be expressed in terms of j_0 not only in the commutators but also in the matrix elements of the Coulomb charged states,

$$(\Phi, Q\,\Psi) = \lim_{\delta \to 0} \lim_{R \to \infty} (\Phi, j_0(f_R\,\alpha_{\delta R})\,\Psi), \tag{7.6.9}$$

and Q is a **superselected charge.**

B. *If the global $U(1)$* **symmetry** *is* **broken,** *i.e., there is a field F of the Coulomb field algebra $\mathcal{F}_C$ such that*

$$< \delta F > \neq 0, \tag{7.6.10}$$

then
1) the spectral measure $d\rho(m^2)$ of the vector boson field cannot contain a $\delta(m^2)$, i.e., there are **no massless vector bosons** *associated with $F_{\mu\nu}$,*
2) the $U(1)$ global gauge transformations are **not** *generated by the current charge; actually, the charge corresponding to the (regularized) integral of j_0 commutes with the Coulomb field algebra*

$$\lim_{\delta \to 0} \lim_{R \to \infty} [j_0(f_R \alpha_{R\delta}), F] = 0, \quad \forall F \in \mathcal{F}_C, \tag{7.6.11}$$

---

[67] This was first pointed out by G. S. Guralnik, C. R. Hagen, and T. W. Kibble, Phys. Rev. Lett. **13**, 585 (1964); see also T. W. Kibble, Phys. Rev. **155**, 1554 (1966); G. S. Guralnik, C. R. Hagen, and T. W. Kibble, Broken symmetries and the Goldstone theorem, in *Advances in Particle Physics*, Vol. 2, R. L. Cool, and R. E. Marshak (eds.), Interscience 1968, p. 567.

and annihilates the vacuum, so that

$$\lim_{\delta \to 0} \lim_{R \to \infty} j_0(f_R \alpha_{\delta R}) \, \mathcal{F}_C \, \Psi_0 = 0, \tag{7.6.12}$$

i.e., one has the **current charge screening**.
*3) the two-point function $< j_\mu(x)F >$ does not vanish, and its Fourier spectrum, i.e.,
the Goldstone spectrum, coincides with the energy momentum spectrum of the vector
boson field, so that the absence of massless vector bosons coincides with the* **absence
of massless Goldstone bosons**.

Proof. The proof largely overlaps that of Proposition 5.4, but for the autonomous
interest of the theorem it may be useful to give the main lines of the argument, even
at the expense of repetitions.
A. As discussed in the proof of Proposition 5.3, the crucial ingredients are the explicit
delocalization of the Coulomb charged field φ_C, given by its construction in terms of
the (local) FGB field, and the ensuing large-distance behavior (in the matrix elements
of the Coulomb charged states) of the commutator

$$[F_{\mu\nu}(x), \varphi_C(y)] \sim \frac{ie}{4\pi} \int d^3z \, \partial_z^j \frac{1}{|\mathbf{y} - \mathbf{z}|} < [F_{\mu\nu}(x), A_j(\mathbf{z}, y_0)] > \varphi_C(y),$$

the corrections being at least $O(|\mathbf{x}|^{-4})$. This implies the following behavior of the
current charge commutator for $R \to \infty$:

$$[j_0(f_R, x_0), \varphi_C(y)] \sim -ie\partial_0 \int d^3x \, f_R(\mathbf{x}) F(x - y) \, \varphi_C(y), \tag{7.6.13}$$

where $F(x) \equiv \int d\rho(m^2) \, \Delta(x; m^2)$, and $d\rho(m^2)$ denotes the vector boson spectral measure. For details of the proof of such estimates, see Proposition 5.3, eqs. (7.5.4) and
(7.5.5). Then, one has ($Q_{RT} = j_0(f_R \alpha_{T(R)})$, $T(R) = \delta R$)

$$[j_0(f_R \alpha_{T(R)}), \varphi_C(y)] \sim$$

$$\sim e \int d\rho(m^2) \, d^3q \, \tilde{f}(\mathbf{q}) \, \mathrm{Re} \, [e^{-i\omega(\mathbf{q},m)y_0} \tilde{\alpha}(\delta \, R \, \omega_R(\mathbf{q}, m))] \, \varphi_C(y), \tag{7.6.14}$$

where $\omega_R(\mathbf{q}, m^2) \equiv \sqrt{\mathbf{q}^2 R^{-2} + m^2}$, and the limit for $R \to \infty$ gives $e\varphi_C(y)$ iff $d\rho(m^2)$
assignes measure one to the point $m^2 = 0$; if $d\rho$ does not contain a $\delta(m^2)$ contribution,
the limit vanishes, since $\tilde{\alpha}$ is of fast decrease.
Furthermore, it is easy to check that $\lim_{R \to \infty} [j_0(f_R, x_0), A_i^C(y)] = 0$, as
a consequence of the locality of the FGB fields and the decrease of
$[F_{0j}(\mathbf{x}, x_0), \partial_i(-\Delta^{-1}\partial^k A_k)(\mathbf{y}, y_0)]$ at least as $|\mathbf{x}|^{-3}$.
The Gauss law by alone provides the proof of eq. (7.6.8), since it implies
$< j_0(x) j_0(y) >_0 = \Delta \Box F^+(x - y)$ and therefore

$$||Q_{RT} \, \Psi_0||^2 = R \int m^2 d\rho(m^2) \int d^3q \, |\tilde{\alpha}(T(R) \, \omega(\mathbf{q}, m))|^2 \frac{|\mathbf{q}\tilde{f}(\mathbf{q})|^2}{2\omega(\mathbf{q}, m)}.$$

Since $d\rho(m^2)$ is tempered, it can be written as $(1+m^2)^M d\rho'(m^2)$, with $d\rho'$ a finite measure; then, the fast decrease of $\tilde{\alpha}$ gives

$$\|Q_{RT}\,\Psi_0\|^2 \leq Const\ (R/T) \int d\rho'(m^2)Tm/(1+T^2m^2)^2 \equiv (1/\delta)G(T)$$

and by the dominated convergence theorem $G(T) \to 0$ as $T \to \infty$, so that $\lim_{R\to\infty} Q_{RT}\Psi_0 = 0$.

The superselection of Q follows from Gauss law and locality (Theorem 5.5).

B. In the broken case, the spectral measure of $F_{\mu\nu}$ cannot have a $\delta(m^2)$ contribution, because, otherwise, by eqs. (7.6.7) and (7.6.8), the $U(1)$ symmetry would not be broken.

By the argument following eq. (7.6.14), the absence of $\delta(m^2)$ implies eq. (7.6.11) and then eq. (7.6.12) follows from eq. (7.6.8).

The Goldstone spectrum is given by the Fourier transform of the two-point function $< j_0(x)\,F >$. In the proof of the Goldstone theorem, the role of the local generation of the symmetry by the density of the corresponding Noether current is that of excluding the vanishing of $< j_\mu(x)\,F >$, as a consequence of $< \delta F > \neq 0$. In the present case, even if the $U(1)$ group is not generated by a suitable integral of the current charge density, nevertheless, due to the estimate of eq. (7.6.13), $< \varphi_C > \neq 0$ implies that the two-point function $< j_\mu(x)\,\varphi_C >$ cannot vanish, being proportional to the vector boson two-point function; such a conclusion applies also to elements F of the Coulomb field algebra. Thus, the Goldstone spectrum is not trivial, it is given by the spectral function of the vector boson field, and cannot contain a $\delta(m^2)$ contribution.

The above theorem emphasizes the need, underestimated in the literature, of not taking for granted the link between the generation of the $U(1)$ symmetry, defined by the l.h.s. of eq. (7.6.7), and the (Noether) current charge $Q_R = j_0(f_R, \alpha_{\delta R})$. As a consequence of the Gauss law, the latter *always* annihilates the vacuum, $\lim_{R\to\infty} Q_R\Psi_0 = 0$, and therefore it cannot generate a broken $U(1)$ symmetry.

6.3   Delocalization and gap in Coulomb systems

A natural question, following from the above discussion of symmetry breaking, is the general characterization of the dynamics which induces a delocalization leading to the failure of eq. (7.2.3), so that the symmetry is not locally generated by the corresponding Noether current and one may have symmetry breaking with energy gap.

We recall that an energy gap at low momenta, $\omega(\mathbf{k}) > \mu > 0, \mathbf{k} \to 0$, is incompatible with the time-independence of $\lim_{R\to\infty} < [Q_R(t), A] >$, which implies ($< \delta A > = \lim_{R\to\infty} i < [Q_R(0), A] >$)

$$-(2\pi)^2 \lim_{\mathbf{k}\to 0} Im(j_0\Psi_0, dE(\omega)dE(\mathbf{k})\,A\Psi_0) =$$

$$\lim_{\mathbf{k}\to 0} < [\tilde{j}_0(\mathbf{k},\omega), A] > = -i < \delta A > \delta(\omega), \tag{7.6.15}$$

where $dE(\omega), dE(\mathbf{k})$ are the energy and momentum spectral measures.

On the other hand, as a consequence of the continuity equation

$$[\dot{Q}_R(t),\, A] = \int d^3x\, \boldsymbol{\nabla} f_R(\mathbf{x})[\mathbf{j}(\mathbf{x},t),\, A],$$

and since supp $\partial_i f_R(\mathbf{x}) \subset \{R \leq |\mathbf{x}| \leq R(1+\varepsilon)\}$, the right-hand side vanishes in the limit $R \to \infty$ if

$$\lim_{|\mathbf{x}|\to\infty} |\mathbf{x}|^2\, [\mathbf{j}(\mathbf{x},t),\, A] = 0. \tag{7.6.16}$$

Clearly, if both j_μ and A are observables fields, Einstein causality requires that the above commutator vanishes as soon as $x + \mathbf{a}$ becomes spacelike with respect to the localization support of A, but in gauge theories and more generally in non-relativistic Coulomb systems, the field algebra involves non-observable fields, and therefore such a relative locality may fail because either A (as in the Higgs phenomenon) or the conserved current j_μ, which generates the symmetry at equal times (as in $U(1)$ problem in quantum chromodynamics), have a Coulomb delocalization induced by the dynamics.

The failure of locality, rather than the lack of manifest covariance, is the crucial structural property which explains the evasion of the Goldstone theorem in the Higgs mechanism, as well as in Coulomb systems and in the $U(1)$ problem.

In this perspective, whenever the field algebra is not manifestly covariant, *instantaneous* interactions are possible and there is no longer a deep distinction between relativistic and non-relativistic systems.

A *dynamics* will be called *of short range* if, for any two elements A, B of the field algebra, $A_{\mathbf{x}} \equiv U(\mathbf{x})AU(-\mathbf{x})$, $B_t \equiv U(t)BU(-t)$,

$$\lim_{|\mathbf{x}|\to\infty} |\mathbf{x}|^{d+\varepsilon}\, [A_{\mathbf{x}},\, B_t] = 0, \quad \varepsilon > 0,$$

with $d \geq s =$ space dimension, *of medium range* if $s - 1 < d < s$, and *of long range* otherwise.

It should be remarked that the presence of long range forces, which always accompanies the presence of massless particles, does not imply a long-range dynamics, as clearly shown by local field theories with massless particles. As discussed in Chapter 3, Section 2, the range of the forces is displayed by the decay rate of the cluster property, which may be that of the derivative of a Coulomb potential in a strictly local theory. Thus, at face value, the statement in the literature that, in analogy with Anderson discussion of superconductivity, the mechanism of evasion of the Goldstone theorem is the presence of long-range forces, is not completely convincing (massless Goldstone bosons and therefore long-range forces are required by the theorem itself), even if, as we shall see, there is some truth in it.

As discussed by Swieca, a relation between the range of the dynamics and the range of the forces can be argued if the latter are described by an instantaneous interaction given by a two-body potential, i.e., of the form

$$H_{int} = \tfrac{1}{2}e^2 \int d^3x\, d^3y\, j_0(\mathbf{x})V(\mathbf{x} - \mathbf{y})j_0(\mathbf{y}). \tag{7.6.17}$$

In this case, the (unequal time) field commutators fall off at large distances $|\mathbf{x} - \mathbf{y}|$ like $V(\mathbf{x} - \mathbf{y})$, and this suggests that the critical decay of the potential, for the failure of eq. (7.6.16), is $|\mathbf{x}|^{-2}$. Actually, since typically $\mathbf{j}$ involves space derivatives of the fields, the critical decay turns out to be $|\mathbf{x}|^{-1}$, i.e., that of the Coulomb potential.[68]

The occurrence of the interaction (7.6.17) characterizes both the Coulomb gauge of quantum electrodynamics and the theory of non-relativistic Coulomb systems, like the electron gas, superconductivity, etc. Such a common feature of long-range dynamics of inducing a Coulomb-like delocalization implies that the commutator $[\,j_\mu(x), F\,]$ of the Noether current associated with the symmetry, broken by $< \delta F > \neq 0$, does not fall off at spacelike distances in such a way to yield eq. (7.6.16). Therefore, by this analysis one obtains a unifying picture and a clarification of the analogies proposed by Anderson between the Higgs mechanism (in the Coulomb gauge) and the symmetry breaking with energy gap in many body Coulomb systems.[69]

Also the debated problem of $U(1)$ axial symmetry breaking in quantum chromodynamics without massless Goldstone bosons can be clarified by the realization of the non-locality of the associated axial current (see Chapter 8).

7   Gauss law and infraparticles

According to Wigner analysis of the representations of the Poincaré group, a stable elementary particle is described by an irreducible representation of such a group, and therefore has a definite mass and spin (or helicity). The electron and the proton seem to qualify as significant examples; however, as we shall see below, since they interact with the electromagnetic field, they cannot have a definite mass and cannot be considered as Wigner particles.

The first indication of this phenomenon for the electron comes from the perturbative approach to quantum electrodynamics. In fact, the perturbative recipe for the solution of the infrared problem, with the summation over the unobserved infrared

---

[68] For a more detailed discussion, see F. Strocchi, *Symmetry Breaking*, 2nd ed., Springer 2008, Part II, Sect. 15.2.

[69] P. W. Anderson, Phys. Rev. **130**, 439 (1963). For a discussion of the Goldstone theorem in non-relativistic systems, see R. Lange, Phys. Rev. Lett. **14**, 3 (1965); for a general discussion of the energy gap associated with symmetry breaking in the case of long-range dynamics, see G. Morchio and F. Strocchi, Infrared Problem, Higgs Phenomenon and Long Range Interactions, in *Fundamental Problems of Gauge Field Theory*, G. Velo and A. S. Wightman (eds.), Plenum 1986, p. 301; Removal of the infrared cutoff, seizing of the vacuum and symmetry breaking in Many Body and in Gauge theories, invited talk at the *IXth Int. Congress on Mathematical Physics*, Swansea 1988, B. Simon *et al.* (eds.), Adam Hilger 1989, p. 490 and references therein. For a review, see F. Strocchi, *Symmetry Breaking*, 2nd ed., Springer 2008. It is worth stressing that the failure of eq. (7.6.16) is a *necessary* condition for evading the Goldstone modes, or more generally the absence of an energy gap, but it is *not sufficient*. Other ingredients and/or arguments are needed for such a conclusion. The compatibility of an energy gap with symmetry breaking in Coulomb systems and in superconductivity has been pointed out by Lange in his discussion of the Goldstone theorem, where the energy gap is *assumed* on the basis of Anderson's work. A non-perturbative proof that in the case of the electron gas, the Goldstone spectrum, associated with the spontaneous breaking of the Galilei symmetry, has an energy gap (given by the plasma frequency), has been given by G. Morchio and F. Strocchi, Ann. Phys. **170**, 310 (1986); see also A. Cintio and G. Morchio, Jour. Math. Phys. **50**, 042102 (2009).

photons (see Chapter 6, Section 4) leads to an electron propagator with the following behavior near the mass shell:

$$S_F(p) \sim \frac{\not{p}+m}{\Gamma(1-\beta)} \frac{p_\mu}{2m^2} \frac{\partial}{\partial p_\mu} \frac{\theta(p_0)\theta(p^2-m^2)}{(p^2-m^2)^\beta},$$

where, in the Coulomb gauge,[70]

$$\beta = -(\alpha/\pi)\,[2-(1/v)\ln((1+v)/(1-v))], \quad v \equiv |\mathbf{p}|/p_0.$$

Thus, there is no longer a pole, and correspondingly the Fourier transform of the two-point function no longer contains a $\delta(p^2-m^2)$ contribution. Due to the interaction with the infrared photons, the point of the spectrum $p^2=m^2$ has dissolved in the continuum; such an infrared dissolution of the sharp mass transforms a Wigner particle into what is called an *infraparticle*.

This represents a serious problem for the construction of a non-trivial S-matrix. As discussed in Chapter 6, Section 2, the existence of non-trivial asymptotic limits, needed for the LSZ reduction formulas, requires mass-shell δ singularities, otherwise the corresponding asymptotic limits vanish and so do the S-matrix elements.

As a matter of fact, the asymptotic limits of charged fields in QED is still an open problem.[71]

The inevitable infraparticle structure of the electron has been brilliantly proved by Buchholz with a non-perturbative argument, which derives this result from the Gauss law.[72]

At the basis of Buchholz theorem are the following assumptions:

I. Any physical charged state Ψ gives a well-defined expectation of the (suitably smeared) electromagnetic field at spacelike infinity, i.e., the following limits exist: $\forall h \in \mathcal{D}(\mathbf{R}^4)$, with supp $h \subset \{x; x^2 < 0\}$,

$$\lim_{R\to\infty} (\Psi, F_{\mu\nu}(h_R)\,\Psi) = f_{\mu\nu}(\Psi, h), \quad h_R(x) = h(x/R)/R^2. \tag{7.7.1}$$

The physical interpretation of eq. (7.7.1) becomes clear if one uses test functions $h(x) = h_1(r,t)h_2(\theta,\varphi)$, with r, θ, φ the polar coordinates, supp $h_1 \subset \{r, t; r < t\}$. In this case,

$$(F_{\mu\nu})_{Rr,Rt} \equiv \int \sin\theta \, d\theta \, d\varphi \, h_2(\theta,\varphi) \, F_{\mu\nu}(Rr, \theta, \varphi, Rt)$$

[70] J. M. Jauch, *The Theory of Photons and Electrons*, Springer 1980; T. W. B. Kibble, Phys. Rev. **173**, 1527 (1968); Coherent States and Infrared Divergences, in Lectures in Theoretical Physics. Vol. XI-D. *Mathematical Methods in Theoretical Physics*, K. T. Mahanthappa and W. E. Brittin (eds.), Gordon and Breach 1969, p. 387, esp. Sect. IV.D.

[71] For a possible strategy see J. Fröhlich, G. Morchio, and F. Strocchi, Ann. Phys. **119**, 241 (1979); a review can be found in G. Morchio and F. Strocchi, Infrared problem, Higgs phenomenon and long-range interactions, in *Fundamental Problems of Gauge Field Theory*, Erice School 1985, G. Velo and A. S. Wightman (eds.), Plenum 1986.

[72] D. Buchholz, Phys. Lett. **B 174**, 331 (1986).

is the average of $F_{\mu\nu}$ on the sphere of radius Rr, at time Rt, with weight function $h_2(\theta, \varphi)$, and

$$F_{\mu\nu}(h_R) = \int dr \, dt \, h_1(r,t) \, (Rr)^2 \, (F_{\mu\nu})_{Rr\,Rt}$$

is a further averaging over a spacetime shell of thickness given by the (spacelike) support of the weight function $h_1(r,t)$.

The existence of the charge as a suitably smeared flux of the electric field at spacelike infinity (as proved for the charged states of the Coulomb gauge, see Proposition 5. 4) implies the existence of the above limits in the special cases of $F_{0i}(h_R^i)$ with $h^i(x,t) = \partial^i f(|\mathbf{x}|)\alpha(x_0)$, f, and α as in eq. (7.2.2). This adds further plausibility for assumption I.

II. The second assumption is that

$$\lim \sup_{R\to\infty} ||[F_{\mu\nu}(h_R) - f_{\mu\nu}(\Psi,h)]\Psi||^2 < \infty; \tag{7.7.2}$$

i.e., the (quantum) fluctuation of $F_{\mu\nu}(h_R)$ on the charged states are bounded in the limit $R \to \infty$.

As discussed by Buchholz, a support for this property comes from the bounded fluctuations on the states of the vacuum sector. In fact, by using the K-L representation given by eq. (7.5.10) one obtains

$$< F_{\mu\nu}(x) \, F_{\mu\nu}(y) >_0 = \int d\rho(m^2) \, d\Omega_m(k)(-k_\mu^2 g_{\nu\nu} - k_\nu^2 g_{\mu\mu})e^{ik(x-y)},$$

where $k_0 \equiv (\mathbf{k}^2 + m^2)^{1/2}$, $\mu \neq \nu$ and no index summation is understood. Then, by the same argument used in the proof of Prop.5.4, one obtains (with $k_0 = |\mathbf{k}|$)

$$\lim_{R\to\infty} ||F_{\mu\nu}(h_R)\Psi_0||^2 = \int d\Omega_0(k)(-k_\mu^2 g_{\nu\nu} - k_\nu^2 g_{\mu\mu}) \, |\tilde{h}(k)|^2 < \infty. \tag{7.7.3}$$

This implies that eq. (7.7.2) holds on the vacuum vector. Moreover, since by locality $F_{\mu\nu}(\varphi_R)$ commutes with the (local) observables in the limit $R \to \infty$, the above boundedness extends to the dense set of local states of the vacuum sector.

Finally, since the description of charge states can be done in terms of charged-particle configurations with compensating charge "behind the moon", i.e., in terms of local states of the vacuum sector, it is plausible that eq. (7.7.2) holds also on the charged states.

The important consequence of eqs. (7.7.1) and (7.7.2) is the existence of the following weak limit:

$$\text{w} - \lim_{R\to\infty} F_{\mu\nu}(h_R)\Psi = f_{\mu\nu}(h)\Psi, \tag{7.7.4}$$

where the function $f_{\mu\nu}$ is no longer dependent on the state Ψ. In fact, by standard arguments, eqs. (7.7.1) and (7.7.2) imply the existence of weakly convergent subsequences; by locality, the corresponding limit operators $f_{\mu\nu}^{R_n}(h)$ commute with the (local) algebra of observables which is irreducibly represented in a charged sector $\mathcal{H}_q$. Hence, any such a limit operator must be a multiple of the identity there, and therefore eq. (7.7.1) implies eq. (7.7.4).

Furthermore, one has

$$\int d^4x\, f_{\mu\nu}(\lambda x)\, h(x) = \lambda^{-4}\int d^4y\, f_{\mu\nu}(y)h(y/\lambda) =$$

$$\lambda^{-2}\text{w} - \lim_{\lambda R\to\infty}\int d^4y F_{\mu\nu}(y)h_{\lambda R}(y) = \lambda^{-2}\int d^4x\, f_{\mu\nu}(x)\, h(x),$$

i.e., the distribution $f_{\mu\nu}(x)$ is homogeneous of degree -2, $f_{\mu\nu}(\lambda x) = \lambda^{-2} f_{\mu\nu}(x)$.

Quite generally, the validity of eq. (7.7.1), with the right-hand side independent of the state, can be proved to hold for the Coulomb charged states $\Psi \in \mathcal{F}_C \Psi_0$, by exploiting the construction of the Coulomb fields in terms of the Feynman–Gupta–Bleuler (FGB) local fields (see Section 5.2, Proposition 5.4). A *Coulomb charged state* Ψ, say of charge e, obtained through the regularized Dirac–Symanzik–Steinmann (DSS) construction, is of the form

$$\Psi = A\, V(f)e^{ieA(g)}\Psi_0,$$

where A is a unitary element of the observable algebra, $V(f)$ is the local unitary operator defined through the polar decomposition of the local FGB field $\psi(x)$, and $e^{ieA(g)}$ is a regularized version of the Dirac factor $e^{-ie[(-\Delta)^{-1}\partial_i A_i](x)}$; e.g.,

$$A(g) \equiv A_i(g_i), \quad g_i(y) = -(4\pi)^{-1}y_i\, |\mathbf{y}|^{-3}\, l(\mathbf{y})\, g(y_0),$$

where $l(\mathbf{y})$ is a smooth function with $l(\mathbf{y}) = 0$ for $|\mathbf{y}| < 3S$, $l(\mathbf{y}) = 1$ for $|\mathbf{y}| > 4S$, $g \in \mathcal{D}(\mathbf{R})$, supp $g \subseteq [-T, T]$, $\int dy_0\, g(y_0) = 1$.[73]

Proposition 7.1 *Let Ψ be a Coulomb charged state, then*

$$\lim_{R\to\infty} (\Psi, F_{\mu\nu}(h_R)\Psi) = f_{\mu\nu}(h)(\Psi,\, \Psi). \tag{7.7.5}$$

Proof. By locality, for R large enough, $F_{\mu\nu}(h_R)$ commutes with A and with $V(f)$, and therefore, in computing the left-hand side of eq. (7.7.5), we are left with the non-local Dirac factor. As proved in Buchholz *et al.*, one has the following large $|\mathbf{x}|$ behavior:

$$(\Psi, F_{0i}(\mathbf{x}, x_0)\, \Psi) \sim -(e/4\pi)x_i|\mathbf{x}|^{-3}\int d^4z\, g(z_0 + x_0)\, \partial_0 F(z), \tag{7.7.6}$$

where F is the invariant commutator function of $F_{\mu\nu}$ (see eq. (7.5.10)).
Then, for large R, one has

$$-\frac{4\pi}{e}(\Psi, F_{0\,i}(h_R)\Psi) \sim \int d^4y d^4z d^3k\, \frac{y_i}{|\mathbf{y}|^3}\, h(y)\, g(z_0 + Ry_0)d\rho(m^2)e^{-ikz}$$

$$= (2\pi)^{3/2}\int d^4y dz_0\, d\rho(m^2)\, (y_i/|\mathbf{y}|^3)\, h(y)g(z_0 + Ry_0)2\cos(mz_0) =$$

$$= (2\pi)^3\int d^4y\, d\rho(m^2)\, (y_i/|\mathbf{y}|^3)\, h(y)\, 2\text{Re}\,(\tilde{g}(m)e^{imRy_0}).$$

---

[73] For a detailed discussion of such a construction, see D. Buchholz, S. Doplicher, G. Morchio, J. E. Roberts, and F. Strocchi, Ann. Phys. **290**, 53 (2001).

Thus, in the limit $R \to \infty$, only a $\delta(m^2)$ contribution in $d\rho(m^2)$ contributes, and the finite limit is independent of the state, since $\tilde{g}(0) = 1$.

For F_{ij}, it has been proved[74] that, for large $|\mathbf{x}|$, its expectation on a Coulomb charged state Ψ decays like $|\mathbf{x}|^{-6}$; therefore

$$\lim_{R \to \infty} (\Psi, F_{ij}(h_R)\,\Psi) = 0.$$

The validity of eq. (7.7.4) in matrix elements between a charged state Ψ and its translated one $\Psi(y) \equiv e^{iPy}\Psi$ is the only ingredient needed for the proof that Ψ cannot be a state with definite mass, $P^\sigma P_\sigma \Psi = m^2\,\Psi$.

Theorem 7.2[75] *(Infraparticle structure of charged particles) Under the above assumptions, the charged states cannot have definite mass.*

Proof. Let Ψ be a charged state, $\Psi(y) \equiv \exp(iPy)\Psi$ and P_μ the four-momentum operator; then the property of definite mass reads $P^\sigma P_\sigma \Psi = m^2\Psi$ and one has ($[P_\sigma, F] = i\partial_\sigma F$),

$$0 = (\Psi(y), [P^\sigma P_\sigma, F_{\mu\nu}]\Psi) = i(\Psi(y), [P^\sigma\,\partial_\sigma F_{\mu\nu} + \partial^\sigma F_{\mu\nu}\,P_\sigma]\Psi) =$$

$$= 2i(P^\sigma\Psi(y), \partial_\sigma F_{\mu\nu}\Psi) + (\Psi(y), \Box F_{\mu\nu}\Psi). \tag{7.7.7}$$

Then, by choosing h_R with the notations of eq. (7.7.1), so that $\partial_\sigma h_R = R^{-1}(\partial_\sigma h)_R$, one has

$$2i(P^\sigma\Psi(y), F_{\mu\nu}((\partial_\sigma h)_R)\Psi) = -(1/R)(\Psi(y), F_{\mu\nu}((\Box h)_R)\Psi). \tag{7.7.8}$$

By eq. (7.7.4) the matrix element (in round brackets) on the r.h.s. has a finite limit when $R \to \infty$, so that the r.h.s. vanishes in that limit and, by eqs. (7.7.5) and (7.7.8), one has

$$(P^\sigma\Psi(y), \Psi)f_{\mu\nu}(\partial_\sigma h) = \lim_{R \to \infty} (P^\sigma\Psi(y), F_{\mu\nu}((\partial_\sigma h)_R)\Psi) = 0. \tag{7.7.9}$$

Hence, the four-momentum p in the spectral support of Ψ must satisfy the constraint $\{p : p^2 = m^2,\ p^\sigma f_{\mu\nu}(\partial_\sigma h) = 0\}$, i.e., it must be contained in a two-dimensional submanifold of $p^2 = m^2$. Such a spectrum is incompatible with the Lebesgue absolute continuity of the joint spectrum of P_i (apart from the point $p_i = 0$), unless $f_{\mu\nu}(\partial_\sigma h) = 0$. Since $f_{\mu\nu}$ is an homogenoeus distribution of degree -2, by the Euler theorem it satisfies $x^\sigma\partial_\sigma f_{\mu\nu}(x) = -2\,f_{\mu\nu}(x)$; hence, $\partial_\sigma f_{\mu\nu} = 0$ implies $f_{\mu\nu} = 0$. In particular, one has $f_{0i}(\partial^i f\alpha) = 0$, and by the Gauss law and eq. (7.5.12) this is incompatible with a non-zero charge.

Under the same assumptions, Buchholz provided a simple proof of the breaking of the Lorentz group in the charged sectors.[76]

[74] D. Buchholz, S. Doplicher, G. Morchio, J. Roberts, and F. Strocchi, Ann. Phys. **290**, 53 (2001).

[75] D. Buchholz, Phys. Lett. **B 174**, 331 (1986).

[76] The earlier proof by J. Fröhlich, G. Morchio, and F. Strocchi, Phys. Lett. **89**, 61 (1979), was based on the existence of the asymptotic fields $F_{\mu\nu}^{as}$, (see Chapter 6, Section 3), in the charged sectors, rather than on eqs. (7.7.1) and (7.7.2).

Proposition 7.3   *The Lorentz group is broken in the charged sectors of QED.*

Proof. If the Lorentz transformations Λ are implemented by unitary operators $U(\Lambda)$, one has $U(\Lambda)F_{\mu\,\nu}(x) = (\Lambda^{-1})^{\rho}_{\mu}\,(\Lambda^{-1})^{\sigma}_{\nu}\,F_{\rho\sigma}(\Lambda x)U(\Lambda)$, and eq. (7.7.4) gives

$$U(\Lambda)f_{\mu\nu}(h)\Psi = \mathrm{w} - \lim_{R\to\infty} U(\Lambda)F_{\mu\nu}(h_R)\Psi =$$

$$= (\Lambda^{-1})^{\rho}_{\mu}\,(\Lambda^{-1})^{\sigma}_{\nu}\,f_{\rho\sigma}(h^{\Lambda})U(\Lambda)\Psi,$$

$h^{\Lambda}(x) \equiv h(\Lambda^{-1}x)$, i.e., $f_{\mu\nu}(x) = (\Lambda^{-1})^{\rho}_{\mu}\Lambda^{-1})^{\sigma}_{\nu}\,f_{\rho\sigma}(\Lambda x)$. Thus, $f_{\mu\nu}(x)$ is a sum of covariant monomials of x_{μ} times invariant functions and, since no one is antisymmetric, $f_{\mu\nu} = 0$, incompatibly with a non-zero charge.

8   Appendix: Quantization of the electromagnetic potential

The problem of the quantization of the electromagnetic potential has a long story, and has traditionally been ascribed to the difficulty of imposing the Lorentz condition $\partial_{\mu}A^{\mu}(x) = 0$,[77] which is, on the other hand, required in order to eliminate the would be longitudinal photons. The somewhat *ad hoc* solutions proposed in the literature leave partially unclear the origin of the difficulty in terms of the general principles of quantum field theory.

From a classical point of view, the first Maxwell equation $dF \equiv \partial_{\mu}\varepsilon^{\mu\nu\rho\sigma}F_{\rho\sigma} = 0$ implies the existence of a four vector potential A_{μ}, undetermined by a gradient $\partial_{\mu}\varphi$, such that $F_{\mu\,\nu} = \partial_{\mu}A_{\nu} - \partial_{\nu}A_{\mu}$. The field φ can be chosen in such a way to ensure the Lorentz condition, so that the Maxwell equations $\partial^{\mu}F_{\mu\,\nu} = 0$ reduce to the free wave equation for A_{μ}: $\Box A_{\mu} = 0$.[78]

It is then natural to enquire about the obstructions in the so different quantum case. As we shall see below, the difficulty is reduced to the conflict between the validity of the Maxwell equations for $F_{\mu\,\nu}$ and either the Lorentz covariance of the electromagnetic potential A_{μ} *or* its locality (in the sense of weak local commutativity). Such a conflict at the roots of the problem exists independently of the canonical quantization, the Fock representation, the uniqueness of the vacuum, the positivity of the metric (defined by the Wightman functions), and/or the possibility of imposing the subsidiary condition. The only ingredients for the proof of such a conflict are the spectral condition for the spacetime translations $U(a)$ and the covariance of A_{μ} under them.

Theorem 8.1[79] *Let A_{μ} be an operator-valued distribution in a vector space $\mathcal{D}_0$, with a vacuum vector Ψ_0, such that the corresponding two-point function of A_{μ} is invariant under spacetime translations and satisfies the spectral condition*

---

[77] W. Heisenberg and W. Pauli, Z. Physik **56**, 1 (1929); **59**, 169 (1930); E. Fermi, Accad. Nazl. Lincei **2**, 881 (1920); Rev. Mod. Phys. **4**, 87 (1932); P. A. M. Dirac, V. A. Fock, and B. Podolsky, Z. Physik Sowjetunion **2** 468 (1932); P. A. M. Dirac, Proc. Roy. Soc. (London) **A114**, 243, 710 (1927).

[78] From the point of view of differential goemetry, the existence of the four vector potential A_{μ} follows from the Poincaré lemma applied to the two-form $F_{\mu\,\nu}$, since F is closed by the first Maxwell equation $dF = 0$ and the spacetime is homeomorphic to $\mathbf{R}^4$. For more details, see, e.g., Y. Choquet-Bruhat and C. DeWitt-Morette. *Analysis, Manifolds and Physics, Part I : Basic*, North-Holland 1991, pp. 222–4, 271.

[79] F. Strocchi, Phys. Rev. **162** 1429 (1967); Phys. Rev. **D 2**, 2334 (1970). For the following simplified version of the argument of the second paper I am indebted to Professor V. Glaser for useful comments.

$$< A_\mu(x)\, A_\nu(y) >_0 = W_{\mu\nu}(y - x), \qquad (7.8.1)$$

$$\tilde{W}_{\mu\nu}(q) = 0, \quad \text{if } q \notin \overline{V}_+.$$

If in addition $W_{\mu\nu}$ is required to satisfy either

i) **covariance** *under the restricted Lorentz group $L_+^\uparrow$,*

$$W_{\mu\nu}(x) = (\Lambda^{-1})^\rho_\mu\, (\Lambda^{-1})^\sigma_\nu\, W_{\rho\sigma}(\Lambda x), \quad \Lambda \in L_+^\uparrow, \qquad (7.8.2)$$

or

ii) temperedness and **locality** *with respect to $F_{\mu\nu}$, i.e.,*

$$D_{\mu\rho\sigma}(x - y) \equiv\, < [\, A_\mu(x),\, F_{\rho\sigma}(y)\,] >_0 = 0, \quad \text{if } (x - y)^2 < 0, \qquad (7.8.3)$$

Lorentz covariance being required only for the two-point function of the observable field $F_{\mu\nu}$, then the Maxwell equations

$$\partial^\mu F_{\mu\nu} = 0, \quad F_{\mu\nu} \equiv \partial_\mu A_\nu - \partial_\nu A_\mu, \qquad (7.8.4)$$

imply the vanishing of the two-point function of the electromagnetic field

$$< F_{\mu\nu}(x)\, F_{\rho\sigma}(y) >_0 = 0, \qquad (7.8.5)$$

i.e., a trivial theory.

Proof. We start by considering case i) i.e., Lorentz covariance of $W_{\mu\nu}$. The first step is to argue that the two-point function $W_{\mu\nu}(x)$ has the following form:

$$W_{\mu\nu}(x) = g_{\mu\nu} W_1(x) + \partial_\mu \partial_\nu W_2(x), \qquad (7.8.6)$$

where W_i, $i = 1, 2$ are Lorentz-invariant distributions.
A simple justification of eq. (7.8.6) is that in momentum space this is the most general (second-rank) tensor one can construct in terms of the four-vector variable p_μ.[80]

---

[80] By using the subgroup of rotations in eq. (7.8.2) and standard properties of its representations, one easily proves that the Fourier transform $\tilde{W}_{ij}, i, j = 1, 2, 3$ is of the form

$$\tilde{W}_{ij}(p_0, \mathbf{p}) = \delta_{ij} F_1(p_0, \mathbf{p}^2) + p_i p_j F_2(p_0, \mathbf{p}^2) + \varepsilon_{ijk} p_k F_3(p_0, \mathbf{p}^2).$$

Furthermore, by using the boosts in the directions $k \neq i$, one obtains $\tilde{W}_{ii}(p_0, \mathbf{p}) = \tilde{W}_{ii}((\Lambda p)_0, (\Lambda p)_k, p_j, p_i)$, and a similar relation from covariance under the boosts in the directions $j \neq i, k$. Thus, F_1, F_2 must depend on the variables p_0, p_k, p_j through the combination $p_0^2 - p_k^2 - p_j^2$; actually, by rotational invariance, through $p_0^2 - \mathbf{p}^2$. Hence, F_1, F_2 are Lorentz-invariant functions. By similarly exploiting the covariance of W_{12}, under boosts in the 3-direction one finds $\Lambda_3^0 p_0 F_3((\Lambda p)_0, 0, p_1^2 + p_2^2) = 0$; i.e., $F_3 = 0$. Moreover, $g^{\mu\nu} \tilde{W}_{\mu\nu}(p) \equiv G(p)$ is a Lorentz-invariant function, and therefore one has $g^{00} \tilde{W}_{00}(p_0, \mathbf{p}) = G(p) + (p_0^2 - \mathbf{p}^2) F_2 - (g^{\mu\nu} g_{\mu\nu} - g^{00} g_{00}) F_1$; i.e.,

$$\tilde{W}_{00}(p) = -g_{00} F_1(p) + p_0^2 F_2(p) + H(p), \quad H(p) \equiv G(p) - p^2 F_2 - 4 F_1.$$

The Lorentz-invariant function H must vanish in order to obtain the correct tensor properties of W_{00}. Finally, by Lorentz covariance one obtains $W_{0j} = p_0 p_j F_2$ and eq. (7.8.6).

A more general proof, which takes into account the distributional character (and possible singularities) of $W_{\mu\,\nu}$, exploits eq. (7.8.2) and the spectral condition, which imply that the analytic continuation of $W_{\mu\,\nu}$ to the extended tube $\mathcal{T}_1^{ext}$ (see Chapter 3, Section 4.2), satisfies

$$W_{\mu\nu}(\zeta) = (\Lambda^{-1})^\rho_\mu \, (\Lambda^{-1})^\sigma_\nu \, W_{\rho\sigma}(\Lambda\zeta), \quad \Lambda \in L_+(C).$$

By a general result,[81] analytic functions transforming as representations of the complex Lorentz group can be written as combinations of Lorentz-covariant polynomials of ζ times Lorentz-invariant analytic functions G_i, which are therefore functions of ζ^2. Then, by using the trivial identity $\zeta_\mu \zeta_\nu H(\zeta^2) = -4\partial_\mu \partial_\nu H(\zeta^2) + 2g_{\mu\nu} H'(\zeta^2)$, one can write

$$W_{\mu\,\nu}(\zeta) = g_{\mu\,\nu} W_1(\zeta^2) + \partial_\mu \partial_\nu W_2(\zeta^2), \tag{7.8.7}$$

and clearly this form remains valid also for the boundary value distributions.
Then, the Maxwell equations (7.8.4) imply

$$(\Box g_{\mu\,\nu} - \partial_\mu \partial_\nu)W_1 = \partial^\lambda < A_\mu\, F_{\lambda\nu} >_0 = 0. \tag{7.8.8}$$

By contracting eq. (7.8.8) by $g^{\mu\nu}$, one obtains $\Box W_1 = 0$ and, by eq. (7.8.8), $\partial_\mu \partial_\nu W_1 = 0$. Hence, eq. (7.8.5) follows.
For the proof of case ii), we first remark that the expectation value of the commutator of $F_{\mu\,\nu}$ has the following form:

$$D_{\mu\nu\rho\sigma}(x - y) \equiv < [F_{\mu\,\nu}(x), F_{\rho\,\sigma}(y)] >_0 = id_{\mu\nu\rho\sigma}F(x - y),$$

$$d_{\mu\nu\rho\sigma} \equiv g_{\nu\rho}\partial_\mu\partial_\sigma + g_{\mu\sigma}\partial_\nu\partial_\rho - g_{\nu\sigma}\partial_\mu\partial_\rho - g_{\mu\rho}\partial_\nu\partial_\sigma, \tag{7.8.9}$$

with F a Lorentz-invariant distribution.[82] The Maxwell eqs. (7.8.4) imply $\Box F_{\mu\,\nu} = 0$ and $\Box F = c_1 + bx^2$. Since, by locality $F(x) = 0$ for $x^2 < 0$, one obtains $c_1 = 0 = b$,

---

[81] K. Hepp, Helv. Phys. Acta **36**, 355 (1963); the simple case of the two-point function was known to H. Araki (communication to K. Hepp).

[82] The above form is the standard one; it follows in general from Lorentz covariance, the Araki–Hepp theorem, the antisymmetry properties of $F_{\mu\,\nu}$, the Maxwell equation $\partial_\lambda \varepsilon^{\lambda\tau\mu\nu} D_{\mu\nu\rho\sigma} = 0$, and locality. In fact, by the Araki–Hepp theorem the two-point function, and therefore the commutator function $D_{\mu\nu\rho\sigma}$, has the following form:

$$D_{\mu\nu\rho\sigma} = d_{\mu\nu\rho\sigma}C_1 + (\varepsilon_{\mu\nu\rho\tau}\partial^\tau\partial_\sigma + \varepsilon_{\rho\sigma\nu\tau}\partial^\tau\partial_\mu - \varepsilon_{\mu\nu\sigma\tau}\partial^\tau\partial_\rho + \varepsilon_{\rho\sigma\mu\tau}\partial^\tau\partial_\nu)C_2 +$$

$$+(g_{\mu\rho}g_{\nu\sigma} - g_{\mu\sigma}g_{\nu\rho})C_3 + \varepsilon_{\mu\nu\rho\sigma}C_4,$$

where $C_i, i = 1,\ldots 4$ are Lorentz-invariant functions, which vanish at spacelike points by locality. Then, the Maxwell equation $\partial_\lambda \varepsilon^{\lambda\tau\mu\nu} D_{\mu\nu\rho\sigma} = 0$ implies

$$(g_{\tau\sigma}\partial_\rho - g_{\tau\rho}\partial_\sigma)(\Box C_2 - C_4) + \partial^\lambda \varepsilon_{\lambda\tau\rho\sigma}C_3 = 0.$$

Then, $\sigma \neq \tau \neq \rho$ gives $C_3 = c_3$ and then $\Box C_2 - C_4 = c_2$; by locality the constants $c_i, i = 2, 3$, vanish. Finally, the trivial identity $(\varepsilon_{\mu\nu\rho\tau}\partial^\tau\partial_\sigma + \varepsilon_{\rho\sigma\nu\tau}\partial^\tau\partial_\mu - \varepsilon_{\mu\nu\sigma\tau}\partial^\tau\partial_\rho + \varepsilon_{\rho\sigma\mu\tau}\partial^\tau\partial_\nu)C_2 + \varepsilon_{\mu\nu\rho\sigma}\Box C_2 = 0$, gives eq. (7.8.9).

i.e., $\Box F = 0$. Moreover, locality, covariance, and spectral condition uniquely fix F up to a multiplicative constant:

$$F(x) = \lambda D(x),$$

where D is the massless commutator function.

Now, eq. (7.8.9) requires that the two-point function

$$C_{\nu\rho\sigma}(x-y) \equiv -i < [A_\nu(x), F_{\rho\sigma}(y)] >_0 - (g_{\nu\sigma}\partial_\rho - g_{\nu\rho}\partial_\sigma)\lambda D(x-y) \qquad (7.8.10)$$

satisfies $\partial_\mu C_{\nu\rho\sigma} - \partial_\nu C_{\mu\rho\sigma} = 0$ and, therefore there exists a distribution $C_{\rho\sigma}$ such that

$$C_{\nu\rho\sigma} = \partial_\nu C_{\rho\sigma}.$$

Clearly, $C_{\rho\sigma}$ is determined up to a constant $c_{\rho\sigma} = -c_{\sigma\rho}$, and this ambiguity may be resolved by requiring $C_{\rho\sigma}(x) = 0$, for $x^2 < 0$.

Furthermore, eq. (7.8.10) and $\Box F_{\rho\sigma} = 0$ give $\Box C_{\mu\rho\sigma} = 0$, so that $\Box C_{\rho\sigma} = a_{\rho\sigma}$; again, by locality the constants $a_{\rho\sigma}$ must vanish since $C_{\rho\sigma}(x) = 0$ for $x^2 < 0$. Furthermore, by locality and the wave equation one has

$$C_{\rho\sigma}(x) = \mathcal{P}_{\rho\sigma}(\partial)D(x), \qquad (7.8.11)$$

with $\mathcal{P}_{\rho\sigma}$ polynomials. In fact, by locality supp $C_{\rho\sigma} \subseteq \overline{V_+} \cup \overline{V_-}$ and as a (tempered) solution of the wave equation, $C_{\rho\sigma}$ is uniquely determined by the restrictions $C_{\rho\sigma}(\mathbf{x}, 0)$, $\partial_0 C_{\rho\sigma}(\mathbf{x}, 0)$, which exist as tempered distributions in $\mathbf{x}$, by general results. Since by locality such initial data vanish outside the origin $\mathbf{x} = 0$, they are polynomials in the partial derivatives applied to $\delta(\mathbf{x})$, and eq. (7.8.11) is obtained by the formula for the solution of the initial value problem

$$C_{\rho\sigma}(x) = \int_{y_0=x_0} d^3y\, [D(x-y)\partial_0 C_{\rho\sigma}(\mathbf{y}, 0) - \partial_0 D(x-y)C_{\rho\sigma}(\mathbf{y}, 0)].$$

The Maxwell equations (7.8.4) and eq. (7.8.10) give $\partial_\nu(\partial^\rho C_{\rho\sigma} - \lambda\partial_\sigma D) = 0$; i.e.,

$$\partial^\rho C_{\rho\sigma} - \lambda\partial_\sigma D = c_\rho \qquad (7.8.12)$$

and $c_\rho = 0$ by locality. Finally, as shown below, such an equation has no solution unless $\lambda = 0$; this implies that the two-point commutator function $D_{\mu\nu\rho\sigma}$ vanishes, and therefore so do its positive- and negative-energy parts. Hence the two-point function of $F_{\mu\nu}$ vanishes.

It remains to see why $\lambda = 0$. For this purpose, we note that the Fourier transform of eq. (7.8.11) gives

$$\tilde{C}_{\rho\sigma}(p) = r_{\rho\sigma}(p)\, \mathrm{sign} p_0\, \delta(p^2)$$

where $r_{\rho\sigma}$ are antisymmetric polynomials, and for $p \neq 0$ eq. (7.8.12) reads

$$p^\rho r_{\rho\sigma}(p)\, \mathrm{sign} p_0\, \delta(p^2) = \lambda p_\sigma\, \mathrm{sign} p_0\, \delta(p^2). \qquad (7.8.13)$$

Hence, the polynomial $p^\rho[r_{\rho\sigma}(p) - \lambda g_{\rho\sigma}]$ must vanish on the light cone $p^2 = 0$, which implies

$$p^\rho[r_{\rho\sigma}(0) - \lambda g_{\rho\sigma}] = \lim_{\varepsilon \to 0} \varepsilon^{-1}[\varepsilon p^\rho(r_{\rho\sigma}(\varepsilon p) - \lambda g_{\rho\sigma})] = 0, \qquad (7.8.14)$$

for all p in the light cone. This requires that the constant matrix $r_{\rho\sigma}(0) - \lambda g_{\rho\sigma}$ must vanish, and since $r_{\rho\sigma}$ is antisymmetric and $g_{\rho\sigma}$ is symmetric, both terms must vanish, i.e., $\lambda = 0$.

It is worth remarking that the above proof does not require any assumption about the positivity condition of the Wightman functions of A_μ; furthermore, as it clear from the proof of case ii), the covariance condition, eq. (7.8.2), could be replaced by the weaker condition of covariance of the two-point function $< A_\mu(x)F_{\rho\sigma}(y) >_0$, without affecting the conclusions. Also, the uniqueness of the vacuum is not assumed. Actually, even the spectral condition can be dispensed with, due to the characterization of the form of the two-point functions which transform as finite-dimensional representations of $SL(2, C)$, due to Oksak and Todorov.[83]

The conflict between the Maxwell equations and the covariance or locality of the vector potential can be regarded as a conflict between gauge invariance, expressed by the Maxwell equations, and covariance or locality of the field algebra generated by A_μ, similarly to the implications of the Gauss law discussed in Section 2.

The above no-go theorem can also be seen as the failure of the Poincaré lemma within the framework of covariant or local (quantum) differential forms, namely, the closedness condition $dF = 0$ does not admit a solution $F = dA$, if the one-form A is required to satisfy covariance or locality.

The above theorem shows that a local or covariant quantization of the vector potential must necessarily give up Maxwell equations. The latter may be required to hold only on a subspace $\mathcal{D}'_0$ of the vector space $\mathcal{D}_0$ (obtained by applying polynomials of the vector potential to the vacuum), and only the vectors of $\mathcal{D}'_0$ have a physical interpretation. Thus, a local or covariant quantization requires the introduction of unphysical vectors (actually, as we shall see, those describing "longitudinal" photons).

8.1 Coulomb gauge

The Coulomb gauge quantization of the vector potential is obtained by treating A_0 as a dependent variable (its conjugate momentum defined by the gauge-invariant Lagrangian vanishes) and by imposing the transversality condition $\mathrm{div}\mathbf{A} = 0$. By Maxwell's eqs. (7.8.4), $\Box A_i = 0$, $\Delta A_0 = 0$, and without loss of generality one may take $A_0 = 0$ (a non-covariant condition). By rotational covariance and parity (as in the proof of Theorem 8.1) and by the transversality condition, the two-point function of A_i is of the form

$$< A_i(x)A_j(y) >_0 = (\Delta\delta_{ij} - \partial_i\partial_j)G(x - y), \qquad (7.8.15)$$

where G is a rotational invariant function $G(\mathbf{x}, x_0) = G(|\mathbf{x}|, x_0)$. The derived two-point function of $F_{\mu\nu}$ must agree with eq. (7.8.9)), so that $\Delta G(x) = i\, D^+(x)$ and the solution is $G(x) = i\,(\Delta^{-1}D^+)(x) + G_1$, with $\mathbf{k}^2\tilde{G}_1(|\mathbf{k}|) = 0$, i.e., $\tilde{G}_1(|\mathbf{k}|) = a\delta(\mathbf{k})$. Then, G_1 does not contribute to the two-point function of A_i, eq. (7.8.15), and one may take

$$G(x) = i(\Delta^{-1} D^+)(x). \qquad (7.8.16)$$

---

[83] A. I. Oksak and I. T. Todorov, Comm. Math. Phys. **14**, 271 (1969), Sect. 3.

The vanishing of A_0 and all its two-point functions imply the lack of Lorentz covariance for the electromagnetic potential in the Coulomb gauge. It is instructive to see that also *locality is lost*. In fact, the vacuum expectation of the commutator has the same form of eq. (7.8.15), with D^+ replaced by $D(x) \equiv D^+(x) - D^+(-x)$,

$$D(x) = -\varepsilon(x_0)\, \delta(x^2)/(2\pi) = (4\pi|\mathbf{x}|)^{-1}[\delta(|\mathbf{x}| + x_0) - \delta(|\mathbf{x}| - x_0)]$$

and, putting $r \equiv |\mathbf{x}|$, $r' \equiv |\mathbf{x}'|$, one has[84]

$$(\Delta^{-1}D)(x) = \frac{1}{(4\pi)^2} \int \frac{d^3x'}{|\mathbf{x} - \mathbf{x}'|} \frac{1}{r'} \left[\delta(r' + x_0) - \delta(r' - x_0)\right] =$$

$$= [\varepsilon(x_0)\, \theta(|x_0| - r) + (t/r)\, \theta(r - |x_0|)]/4\pi.$$

Thus, the term $\partial_i\partial_j(\Delta^{-1}D)(x)$ vanishes for $x^2 > 0$, but it is non-zero for $x^2 < 0$.

By an adaptation of the argument of Chapter 4, Section 1, one can show that the commutator is a c-number:[85]

$$[A_i(x), A_j(y)] = i(\delta_{ij}D(x - y) - \partial_i\partial_j(\Delta^{-1}D)(x - y)), \tag{7.8.17}$$

and one has the following equal-time commutators:

$$[A_i(\mathbf{x}, t), \dot{A}_j(\mathbf{y}, t)] = i\,\delta_{ij}\delta(\mathbf{x} - \mathbf{y}) + i\,\partial_i\partial_j(4\pi|\mathbf{x} - \mathbf{y}|)^{-1}. \tag{7.8.18}$$

The representation of the field algebra generated by the vector potential is uniquely fixed by the two-point function, apart from isometries. A convenient choice is to expand A_i in terms of plane-wave solutions of the field equations $\Box A_i(x) = 0$, $\partial_i A_i(x) = 0$:

$$A_i(x) = \frac{1}{(2\pi)^{3/2}} \int \frac{d^3k}{\sqrt{2k_0}} \sum_{\sigma=\pm} [\varepsilon_i(\mathbf{k}, \sigma)a(\mathbf{k}, \sigma)e^{-ikx} + \varepsilon_i^*(\mathbf{k}, \sigma)a^*(\mathbf{k}, \sigma)e^{ikx}],$$

where $k_0 = |\mathbf{k}|$, $k_i\varepsilon_i(\mathbf{k}, \sigma) = 0$ and

$$\sum_{\sigma=\pm} \varepsilon_i(\mathbf{k}, \sigma)\, \varepsilon^*(\mathbf{k}, \sigma) = \delta_{ij} - k_ik_j/|\mathbf{k}|^2.$$

The destruction and creation operators a, a^* obey the following commutation relations:

$$[a(\mathbf{k}, \sigma), a^*(\mathbf{k}', \sigma')] = \delta_{\sigma\sigma'}\, \delta(\mathbf{k} - \mathbf{k}'), \tag{7.8.19}$$

all other commutators vanishing. One can easily check that the two-point function (7.8.15) is reproduced. The vectors $\varepsilon_i(\mathbf{k}, \sigma)$ have the meaning of polarization vectors and can be chosen as $\varepsilon_i(\mathbf{k}, \pm) = R(\mathbf{k})\bar{\varepsilon}(\pm)$, where $\bar{\varepsilon}_i(\pm) = (1/\sqrt{2}, \pm i/\sqrt{2}, 0)$ and $R(\mathbf{k})$ is the rotation which carries the third axis into the direction of $\mathbf{k}$.

---

[84] It may be helpful to note that $\int_{r'=|t|} d^3x'\, |\mathbf{x} - \mathbf{x}'|^{-1}$ is the electrostatic potential at the point $\mathbf{x}$ due to a uniform distribution of charge, $\sigma = 1$, on the surface of the sphere of radius $r' = |t|$, centered at the origin.

[85] Eq. (7.8.15) implies that $< F_{ki}(x)\, A_j(y) >= i\,(\partial_k\delta_{ij} - \partial_i\delta_{kj})D^+(x - y)$ satisfies the free wave equation and locality, so that $[F_{ki}, A_j]$ is a c-number given by its vacuum expectation. Hence, for any vector Ψ the function $F_{ij}(x, y) \equiv (\Psi, [A_i^+(x), A_j^+(y)]\Psi_0)$ is the boundary value of an analytic function $F_{ij}(z_1, z_2)$, Im $z_i \in V^+$, $i = 1, 2$, which has vanishing curl and divergence and must therefore be a constant. By the uniqueness of the vacuum its boundary value must vanish.

In the interacting case, the non-locality of the vector potential gives rise to the non-locality of the charged fields coupled to it. In fact, in the presence of the interaction, the Maxwell equations imply the following equation for the dependent variable A_0: $\Delta A_0 = -j_0$, so that A_0 is a non-local function of j_0:

$$A_0(\mathbf{x}, x_0) = -(\Delta^{-1} j_0)(\mathbf{x}, x_0).$$

The electric field can be decomposed into its longitudinal and transverse part

$$E_i = \partial_i A_0 - \partial_0 A_i$$

and, e.g., for the Maxwell–Dirac theory, the Hamiltonian can be written in terms of the *kinematically independent variables* A_i, $\dot{A}_i$, ψ, ψ^*

$$H = H_0(\psi, \psi^*) + H_0(A_i) + \int d^3x[-j_i(x)A^i(x)+$$

$$+\tfrac{1}{2} \int d^3y\, j_0(\mathbf{x}, x_0)\, (4\pi\, |\mathbf{x} - \mathbf{y}|)^{-1}\, j_0(\mathbf{y}, x_0)]. \tag{7.8.20}$$

The equal-time commutators between A_i, $\partial_0 A_i$ and ψ, ψ^* vanish, since they describe independent degrees of freedom, but the fermions have the non-local Coulomb interaction, described by the last term. Thus, even if the equal-time (anti)commutators are local, the dynamics induces a Coulomb-like delocalization at unequal times and a violation of locality.

On the other hand, if the electromagnetic field is described by the variables A_i, E_i, they still obey canonical commutation relations, but E_i is not kinematically independent from ψ, since A_0 is a non-local function of j_0, and in fact one has a non-local commutation relation

$$[E_i(\mathbf{x}, t),\, \psi(\mathbf{y}, t)] = -e\, \partial_i (4\pi|\mathbf{x} - \mathbf{y}|)^{-1} \psi(\mathbf{y}, t).$$

In terms of these variables, the non-local Coulomb interaction term gets absorbed by the free electromagnetic Hamiltonian

$$H_0(\mathbf{E}, \mathbf{B}) = \tfrac{1}{2} \int d^3x(\mathbf{E}^2 + \mathbf{B}^2).$$

The Hamiltonian is a local function of the variables A_i, E_i, ψ, ψ^*, and the equal-time commutator $[E_i, H_0(\psi, \psi^*]$ vanishes; then, the time evolution of (the observable field) E_i is local, but the anticommutators of ψ, ψ^* becomes non-local at unequal times, because the equal-time commutator $[E_i, \psi]$ is non-local, and therefore so is $[H, \psi]$.

8.2 Feynman–Gupta–Bleuler quantization

The idea of the FGB quantization goes back to Fermi,[86] who proposed to treat the four components of the vector potential as four independent fields and to impose the initial-time subsidiary conditions,

---

[86] E. Fermi, Rend. Accad. Naz. Lincei, XII, 431 (1930) (reprinted in *Selected Papers on Quantum Electrodynamics*, J. Schwinger (ed.), Dover 1958, p. 24); Rev. Mod. Phys. **4**, 87 (1932).

$$\chi(\mathbf{x},0) \equiv \partial^\mu A_\mu(\mathbf{x},0) = 0, \quad \partial_0 \chi(\mathbf{x},0) = 0.$$

Since χ is a free field, the vanishing of its initial data imply its vanishing for all times. This is very simple and clear in the classical case, but the correct formulation of the subsidiary condition in the quantum case is a delicate issue.

The vanishing of $\partial^\mu A_\mu$ as an operator equation, together with $\Box A_\mu = 0$, is incompatible with Lorentz covariance and/or locality (Theorem 8.1).[87] Furthermore, the Fermi choice of imposing the subsidiary condition as an operator equation on the physical states, and in particular on the vacuum state ($\partial^\mu A_\mu \Psi_0 = 0$), is not acceptable; in a covariant theory it would imply a transverse two-point function of A_μ, $\partial^\mu W_{\mu\nu} = 0$, which is incompatible with $\Box W_{\mu\nu} = 0$ (see Theorem 8.1 above).[88]

The FGB quantization provides the correct version of the Fermi quantization. It is defined by the following two-point function:

$$< A_\mu(x)\, A_\nu(y) > = -g_{\mu\nu}\, iD^+(x-y), \tag{7.8.21}$$

which satisfies covariance, locality, and spectral condition, and corresponds to the standard Feynman propagator. This underlies Feynman perturbative calculations of the S-matrix elements.

By an easy adaptation of the Jost–Schroer theorem (see Chapter 4, Section 1), one obtains

$$[\, A_\mu(x),\, A_\nu(y)\,] = -g_{\mu\nu}\, iD(x-y) \tag{7.8.22}$$

and the factorization of the n-point (Wightman) functions in terms of the two-point function.

Such Wightman functions define a vector space $\mathcal{D}_0 = \mathcal{F}\Psi_0$, where $\mathcal{F}$ is the field algebra generated by A_μ, and (with the hermiticity of A_μ) an inner product $< \,.\,,\,.\, >$ on $\mathcal{D}_0$, $< F\Psi_0, G\Psi_0 > \equiv < F^*G >$. Not all the vectors of $\mathcal{D}_0$ have a physical interpretation; the physical vectors $\Psi \in \mathcal{D}_0$ are selected by the property of yielding vanishing expectations of $\partial^\mu A_\mu$, $< \Psi,\, \partial^\mu A_\mu \Psi > = 0$.

The corresponding quantum-mechanical interpretation has been clarified by Gupta and Bleuler, who suggested the *GB subsidiary condition* for the physical vectors Ψ:

$$\partial^\mu A_\mu^- \Psi = 0, \tag{7.8.23}$$

where $\partial^\mu A_\mu^-$ denotes the negative-energy part of the free field ∂A.

Such a condition is satisfied by the vacuum if the spectral condition holds, and, since by eq. (7.8.22) $[\partial A^-, F_{\mu\nu}] = 0$, by any state obtained from the vacuum by applying polynomials of $F_{\mu\nu}$. Furthermore, the hermiticity of A_μ with respect to the inner product defined by the vacuum expectations of A_μ (see Chapter 3, Section 3), gives

$$\partial^\mu A_\mu = \partial^\mu A_\mu^+ + \partial^\mu A_\mu^-, \quad (\partial^\mu A_\mu^-)^* = \partial^\mu A_\mu^+, \tag{7.8.24}$$

---

[87] The tricky way by which $\partial^\mu A_\mu = 0$ holds in the Landau gauge quantization is that $\Box A_\mu \neq 0$, so that the Maxwell equations do not hold.

[88] Another argument against the Fermi condition is that by the spectral condition, locality, and the Reeh–Schlieder theorem, $\partial A \Psi_0 = 0$ imply $\partial A = 0$, which is incompatible with $\Box A_\mu = 0$ (see Theorem 8.1).

and the Gupta–Bleuler choice is equivalent to the vanishing of the matrix elements of ∂A on the states selected by the GB subsidiary condition.

The vectors of $\mathcal{D}_0$ which satisfy the GB subsidiary condition are the polynomials of A_μ smeared with test functions f^μ satisfying the *transversality condition* $k^\mu \tilde{f}^\mu(k)|_{k^2=0} = 0$.[89] The inner product is non-negative on the subspace $\mathcal{D}_0'$ of physical vectors of $\mathcal{D}_0$, but is indefinite on $\mathcal{D}_0$, (e.g., is negative on $A_0(f)\Psi_0$). Moreover, if $g^\mu = \partial^\mu h$, one has $< A_\mu(f^\mu) A_\nu(g^\nu) > = 0$, for all transverse f_μ, and therefore $\mathcal{D}_0'$ contains a null subspace $\mathcal{D}_0''$ (*"longitudinal photons"*).

The physical Hilbert space is obtained by the standard procedure from the pre-Hilbert space $\mathcal{D}_0'/\mathcal{D}_0''$, equipped with the positive inner product $< ., . >$. It is not difficult to see that it is isomorphic to the Hilbert space of the Coulomb gauge; e.g., through the following choice of elements in the equivalence classes $[\tilde{f}^\mu(k)] \leftrightarrow \tilde{f}^\mu(k) - (k^\mu/k_0)\tilde{f}^0(k)$.

A (non-covariant) Hilbert product $(\Psi, \Phi) \equiv < \Psi, \eta\Phi >$ can be defined on $\mathcal{D}_0$ by introducing the metric operator $\eta = \eta^*$ defined by

$$\eta A_\mu(x)\eta = -g^{\mu\nu} A_\nu(x), \quad \eta\Psi_0 = \Psi_0; \tag{7.8.25}$$

clearly, $\eta^2 = 1$. The closure of $\mathcal{D}_0$ with respect to such a Hilbert topology is a *Hilbert–Krein space* $\mathcal{K}$, since it is equipped with a metric η, with $\eta^2 = 1$ and a corresponding indefinite inner product topology (briefly called *Krein topology*).[90] Thus, the framework of the FGB quantization is an indefinite inner product (Krein) space, containing as a subspace the Hilbert space of physical states.[91]

The above discussion clarifies the quantum-mechanical structure and interpretation of the quantization of A_μ defined by the Feynman propagator, and can be extended to the general case of a local and covariant two-point function.

The same representation can be obtained by canonical quantization, starting from the Fermi Lagrangian

$$\mathcal{L} = -\tfrac{1}{4}F_{\mu\nu}F^{\mu\nu} - \tfrac{1}{2}(\partial_\mu A^\mu)^2. \tag{7.8.26}$$

---

[89] In fact, one has (the wide hat denotes omission)

$$\partial A(g)^- A(f_1)\dots A(f_n)\Psi_0 = \sum_j [\partial A(g)^-, A(f_j)A(f_1)\dots \widehat{A(f_j)}\dots A(f_n)\Psi_0$$

and $[\partial A(g)^-, A(f_j)] = 0$, $\forall g \in \mathcal{S}(\mathbf{R}^4)$, iff $k^\mu \tilde{f}_j^\mu(k) = 0$ on the light cone $k^2 = 0$.

[90] An inner product space carrying a Hilbert structure defined by a metric $\eta = \eta^*$, $\eta^2 = 1$ is called a *Krein (or Hilbert–Krein) space*; see J. Bognar, *Indefinite Inner Product Spaces*, Springer 1974, and F. Strocchi, *Selected Topics on the General Properties of Quantum Field Theory*, World Scientific 1993, esp. Appendix 1, p. 105.

[91] It is worth remarking that according to the general argument of the reconstruction theorem, the Lorentz transformations (as well as the spacetime translations) are described by operators $U(a, \Lambda)$ which preserve the indefinite inner product (i.e., $U^{-1} = \eta U^* \eta$). However, $U(\Lambda)$ does not commute with η, and is not a unitary operator in $\mathcal{K}$; actually, it is not a bounded operator in $\mathcal{K}$, since in each "n-particle" subspace $K^{(n)}$ one has $\|U(\Lambda)\|_{K^{(n)}} = C^n$, $C > 1$. It becomes a (bounded) unitary operator when restricted to the "physical" subspace $\mathcal{D}_0'$.

Since the canonical conjugated momenta are $\pi_\mu = -\partial_0 A_\mu$, canonical covariant quantization gives

$$[A_\mu(x),\, \partial_0 A_\nu(y)]_{x_0 = y_0} = -i g_{\mu\nu}\, \delta(\mathbf{x} - \mathbf{y}), \qquad (7.8.27)$$

where the minus sign is required by the equal-time commutators of $F_{\mu\nu}$, all other equal-time commutators vanishing. The above canonical quantization gives eq. (7.8.22), and by the Fock condition, $A_\mu^- \Psi_0 = 0$ (equivalently, by the spectral condition), the two-point function (7.8.21).

By decomposing A_μ into its positive- and negative-energy parts, one finds that the corresponding annihilation and creation operators $a_\mu(\mathbf{k})$, $a_\mu^*(\mathbf{k})$ obey the following covariant canonical commutation relations:

$$[a_\mu(\mathbf{k}),\, a_\nu^*(\mathbf{q})] = -k_0\, g_{\mu\nu}\, \delta(\mathbf{k} - \mathbf{q}).$$

Thus, $[a_0, a_0^*]$ differs from the standard commutator by a minus sign, leading to the non-positivity of $< a_0^*(f)\Psi_0,\, a_0^*(f)\Psi_0 >$. The (Wick ordered) Hamiltonian

$$H = -\int d\Omega(k)\, k_0\, a_\mu(\mathbf{k})^*\, a^\mu(\mathbf{k}) \qquad (7.8.28)$$

has a positive action on the states of a Fock representation, since $H a_\mu^*(\mathbf{k})\Psi_0 = k_0 a_\mu^*(\mathbf{k})\Psi_0$. Actually, H is a positive definite operator in the Hilbert–Krein space $\mathcal{K}$, since, denoting by $\dagger$ the Hilbert space adjoint (i.e., with respect to the Hilbert product $(.,.)$), one has $a_\mu^* = \eta a_\mu^\dagger \eta = -g^{\mu\nu} a_\nu^\dagger$; thus, the Hamiltonian density takes the form $k_0 a_\mu^\dagger(\mathbf{k}) a_\mu(\mathbf{k})$, which displays its positive definiteness.[92]

As a possible alternative, it has been suggested to keep the standard interpretation of destruction and creation operators for a_i, a_i^*, but reverse it for the zero-th components, i.e., to interpret a_0 and a_0^* as creation and destruction operators, respectively. This choice apparently avoids the indefiniteness, but it leads to an Hamiltonian which is unbounded from below in the corresponding Fock representation.

The most general covariant quantization of the free vector potential is defined by a two-point function of the form

$$< A_\mu(x) A_\nu(y) > = -g_{\mu\nu}\, i D^+(x - y) + (1 - \alpha)\partial_\mu\, \partial_\nu G(x - y), \qquad (7.8.29)$$

---

[92] For detailed discussions of the FGB quantization, see N. N. Bogoliubov and D. V. Shirkov, *Introduction to the Theory of Quantized Fields*, Interscience 1959, Chap. II, Sect. 13: S. Schweber, *An Introduction to Relativistic Quantum Field Theory*, Harper and Row 1961, Chap. 9. A mathematical formulation of the Gupta–Bleuler quantization of free QED is discussed in A. S. Wightman and L. Gårding, Arkiv f. Fysik **28**, 129 (1964); however, in such a formulation of indefinite quantum field theories, the proposed modified axioms are on one side too restrictive, since the Poincaré transformations are assumed to be described by bounded operators (in contrast with the FGB quantization), and on the other side too weak because they do not include the Hilbert space structure condition necessary for the reconstruction of fields as Hilbert space operators. For a discussion of these points and a proposal of modified axioms for indefinite metric quantum field theories, see F. Strocchi, *Selected Topics on the General Properties of Quantum Field Theory*, World Scientific 1993, Sects. 6.2–6.4 and Appendix 1, and references therein.

where the first term on the r.h.s. is fixed by the two-point function of $F_{\mu\nu}$, α is a real (gauge) parameter, and the Lorentz-invariant (gauge) function G is constrained by the choice of the gauge and the corresponding field equations.

A class of gauges is selected by the gauge condition $\Box\partial_\mu A^\mu = 0$ (called the linear conformal invariant gauges), and for them the corresponding Hilbert–Krein realization can be obtained by a procedure similar to that adopted above.[93]

For simplicity, we follow the canonical approach starting from the Lagrangian

$$\mathcal{L} = -\tfrac{1}{4}F_{\mu\nu}F^{\mu\nu} + \tfrac{1}{2}\alpha B^2 + B\,\partial_\mu A^\mu, \tag{7.8.30}$$

where B is an auxiliary (hermitian) field, and α is a real (gauge) parameter.[94] The corresponding Euler–Lagrange equations are

$$\partial^\mu F_{\mu\nu} = \partial_\nu B, \qquad \partial A + \alpha B = 0 \tag{7.8.31}$$

and canonical quantization gives the following equal-time commutators (the canonical momentum conjugated to A_0 being $\pi_0 = B$)

$$[\,A_i(\mathbf{x}),\, \dot A_j(\mathbf{y})\,] = i\delta_{ij}\,\delta(\mathbf{x}-\mathbf{y}), \quad [\,A_0(\mathbf{x}),\, B(\mathbf{y})\,] = i\delta(\mathbf{x}-\mathbf{y}),$$

$$[\,A_0(x),\, \dot A_0(\mathbf{y})\,] = -i\alpha\delta(\mathbf{x}-\mathbf{y}), \quad [\,A_j(\mathbf{x}),\dot B(\mathbf{y})\,] = -(i/\alpha)\,\partial_j\delta(\mathbf{x}-\mathbf{y}), \tag{7.8.32}$$

all other equal-time commutators vanishing.

From eqs. (7.8.31) one obtains $\Box B = 0$ and $\Box A_\mu - (1-\alpha)\partial_\mu B = 0$, so that apart from the case $\alpha = 1$, A_μ does not satisfy the free wave equation, but rather the dipole field equation $\Box^2 A_\mu = 0$. Hence, the initial data for $\partial_i\partial_j H$, $H(x) \equiv G(x) - G(-x)$, given by the equal-time commutators, yield the following solution for the (invariant) function G appearing in eq. (7.8.29):

$$G(x) = -i(4\pi)^{-2}\log(-x^2 + i\varepsilon x_0).$$

Formally $G(x) = i\,\Box^{-1}D^+(x)$; this form displays the transversality of the two-point function of the Landau gauge, defined by $\alpha = 0$.

Furthermore, one has, from eq. (7.8.29) and $\Box B = 0$,

$$< B(x)\,B(y) >= 0, \quad < A_\mu(x)\,B(y) >= -i\,\partial_\mu D^+(x-y).$$

The subsidiary condition which selects the physical vectors now reads

$$B^-(x)\Psi = 0. \tag{7.8.33}$$

---

[93] See, e.g., G. Rideau, Covariant quantizations of the Maxwell field, in *field Theory, Quantization and Statistical Physics*, E. Tirapegui (ed.), D. Reidel 1981, p. 201, esp. Part II.

[94] The corresponding quantizations define the so-called α (or ξ) gauges; $\alpha = 1$ gives the FGB gauge, and $\alpha = 0$ the Landau gauge. The introduction of the B field has been advocated by Nakanishi and Lautrup (B is also called the Nakanishi–Lautrup field), and we refer to the very nice presentation by N. Nakanishi and I. Ojima, *Covariant Operator Formalism of Gauge Theories and Quantum Gravity*, World Scientific 1990, Sects. 2.3.1–2.3.5, for related references and for a more detailed discussion; see also K. Symanzik, Lectures on Lagrangian Quantum Field Theory, Desy Report T-71/1 1971, esp. Sect. 5.3.

In the B-field formulation of the α gauges the (c-number) gauge transformations $A_\mu(x) \to A_\mu(x) + \partial_\mu \Lambda(x)$, with $\Box\Lambda(x) = 0$, are generated by the B field:

$$\lim_{R\to\infty} [-i \int d^4x [f_R(\mathbf{x})\alpha(x_0)(\Lambda(x) \overset{\leftrightarrow}{\partial_0} B(x)), A_\mu(y)] = \partial_\mu\Lambda(y).$$

Again, as discussed in Section 3.4, a vector $\Psi \in \mathcal{D}_0$ satisfies the subsidiary condition (7.8.33) if it is of the form $\Psi = F_{gi}\Psi_0$, with F_{gi} a gauge-invariant element of the field algebra.

The Hilbert–Krein representation of fields with dipole (infrared) singularities leads to Fock states with a doubling of the components of the wave function (i.e., one must introduce both $\tilde{f}(k)|_{k_0=|\mathbf{k}|}$ and $(k_0\partial\tilde{f}(k)/\partial k_0)|_{k_0=|\mathbf{k}|})$.[95]

8.3 Temporal gauge

For the discussion of non-perturbative aspects of QED, typically by the functional integral approach, and in particular for the discussion of the existence of a symmetry breaking order parameter in the Higgs phenomenon, the temporal gauge has been widely used, but the textbook presentations of such a gauge have somewhat neglected its peculiar mathematical features, and in particular the mathematical consistency of the proposed realizations.[96]

The *temporal gauge* is defined by the gauge condition $A_0 = 0$, both in the free as well as in the interacting case, without requiring the transversality of A_i. Thus, in the free case it is similar to the Coulomb gauge (but longitudinal photons are allowed). Manifest covariance is obviously lost, but locality has been claimed to hold. The equations of motion read

$$\partial_0^2 A_i - \Delta A_i + \partial_i \mathrm{div} A = -j_i, \tag{7.8.34}$$

where j_μ is the conserved gauge-invariant electromagnetic current constructed in terms of the charged fields.

The canonical commutation relations for the vector potential

$$[A_i(\mathbf{x},0), \partial_0 A_j(\mathbf{y},0)] = i\delta_{ij}\delta(\mathbf{x}-\mathbf{y}), \quad [A_i(\mathbf{x},0), A_j(\mathbf{y},0)] = 0 \tag{7.8.35}$$

are clearly incompatible with the Gauss law $\mathrm{div} E = 0$, $E_i = \partial_0 A_i$ in the free case, and such an incompatibility persists in the interacting case.

In fact, the problem of the Gauss law can be argued quite generally, independently of the existence of equal-time restrictions of the fields and of a full canonical structure. Eqs. (7.8.34) and the current continuity equation imply that

$$G(\mathbf{x}, x_0) \equiv \mathrm{div} E(x) - j_0(x)$$

[95] See U. Moschella and F. Strocchi, Lett. Math. Phys. **19**, 143 (1990); a Hilbert–Krein realization of the Landau gauge ($\alpha = 0$) has been given by D. Pierotti, Jour. Math. Phys. **26**, 143 (1985) (see references therein for earlier attempts).

[96] See, e.g., the comprehensive book by A. Bassetto, G. Nardelli, and R. Soldati, *Yang–Mills Theories in Algebraic Non-covariant Gauges*, World Scientific 1991.

is time-independent, so that

$$G(f, x_0) \equiv G(f, h), \quad f \in \mathcal{S}(\mathbf{R}^3), \quad h \in \mathcal{S}(\mathbf{R}), \quad \int dt \, h(t) = 1,$$

is a well- (densely) defined operator, and its equal-time commutators with the fields are well-defined operator-valued distributions. Such commutators are fixed by the condition, which can be taken as part of the definition of the temporal gauge, that G generates the time-independent gauge transformations.[97] Thus, quite generally, one has

$$[\, A_i(x, t), G(\mathbf{y}, t)\,] = -i\, \partial_i \delta(\mathbf{x} - \mathbf{y}), \tag{7.8.36}$$

and by taking the vacuum expectation of the above equation, one obtains a conflict with the validity of the Gauss law on the vacuum:

$$G(f)\Psi_0 = 0.$$

The standard textbook solutions of such a conflict raise non-trivial mathematical problems, and some of them are mathematically inconsistent. As we shall prove, there are only two (mathematically consistent) alternative solutions[98] of such a conflict between eq. (7.8.36) and the Gauss law, with peculiar (widely unnoticed) mathematical properties.[99]

1) **Regular indefinite quantization**

The first alternative is characterized by the requirement of existence of the vector potential as a field operator, i.e., with well-defined vacuum correlation functions (*regular representation of the field algebra*). Then, as shown below, one has to give up the Gauss law and positivity (as in the FGB quantization); furthermore, one has the positivity of the energy spectrum, but not the relativistic spectral condition.

The gauge condition $A_0 = 0$ does not completely fix the gauge, since it still allows time-independent operator gauge transformations $A_i(x) \to A_i(x) + \partial_i \varphi(\mathbf{x})$ (and the corresponding one for fermions), so that one can exploit this residual gauge invariance to simplify the form of the two-point function of A_i.

Proposition 8.2 *The two-point function $W_{ij} \equiv\, < A_i(x)\, A_j(y) >$ of the vector potential, with $A_0 = 0$, is completely characterized by the following general requirements:*
i) invariance under spacetime translations and parity and covariance under rotations,

---

[97] Such a property follows from canonical quantization and (in the interacting case) a gauge-invariant point-splitting regularization of the fermion current involving additional terms linear in $A_i, \partial_0 A_i$.

[98] J. Löffelholz, G. Morchio, and F. Strocchi, Jour. Math. Phys. **44**, 5095 (2003), hereafter referred to as LMS.

[99] Non-regular positive quantizations of the temporal gauge in the free case has been discussed by D. Buchholz and K. Fredenhagen, Comm. Math. Phys. **84**, 1 (1982); H. Grundling and C. A. Hurst, Lett. Math. Phys. **15**, 205 (1988); F. Acerbi, G. Morchio, and F. Strocchi, J. Math. Phys. **34**, 899 (1993). The unique selection of the non-regular representation considered in such papers, by the condition of positivity of the energy spectrum, has been proved in LMS.

ii) the derived two-point function $W_{\mu\nu\,\lambda\sigma}(x)$ of $F_{\mu\nu}$ is the standard one, eq. (7.5.10),

$$W_{\mu\nu\,\lambda\sigma}(x) = (g_{\nu\lambda}\partial_\mu\partial_\sigma + g_{\mu\sigma}\partial_\nu\partial_\lambda - g_{\nu\sigma}\partial_\mu\partial_\lambda - g_{\mu\lambda}\partial_\nu\partial_\sigma)F(x-y), \qquad (7.8.37)$$

dictated by the Wightman axioms for observable fields,
iii) locality.

   By exploiting the residual gauge invariance, the two-point function can be written in the following form:

$$W_{ij}(x) = (\delta_{ij} - \partial_i\,\partial_j/\partial_0^2)F(x) - i\,x_0\,\partial_i\,\partial_j G(\mathbf{x}^2),$$

$$\tilde{G}(\mathbf{k}^2) = \mathcal{P}(\mathbf{k}^2) + \int d\rho(m^2)\,(\mathbf{k}^2 + m^2)^{-1}, \qquad (7.8.38)$$

where $d\rho(m^2)$ is the spectral measure of F, and $\mathcal{P}$ is a polynomial, which is actually a constant Z, if at equal times one has a canonical structure, apart from renormalization constants.

   Positivity of the energy spectrum holds, but the relativistic spectral condition does not. As displayed by the free field case, positivity is lost.

Lemma 8.3 *The equations*

$$\delta_{ij}G(\mathbf{x}^2) + \partial_i\partial_j H(\mathbf{x}^2) = 0 \qquad (7.8.39)$$

imply that G is independent of $\mathbf{x}$ and $2H = -G\mathbf{x}^2 + const.$
   In particular, the equation

$$(\delta_{ij}\Delta - \partial_i\partial_j)K(\mathbf{x}^2) = 0 \qquad (7.8.40)$$

implies that K is independent of $\mathbf{x}$.

Proof. In fact, eqs. (7.8.39) for $i \neq j$ imply that $\partial_j H$ can depend only on x_j and, for $i = j$, that $G = \partial_j^2 H$; hence, G can only depend on x_j, and since j is arbitrary, it is independent of $\mathbf{x}$, and $2H = -G\mathbf{x}^2 +$ const.
This implies, for eq. (7.8.40), $2K = \Delta K\mathbf{x}^2+$ const; on the other side, by contracting eq. (7.8.40) with δ_{ij}, one has $\Delta K = 0$; then, $K = $ const.

Proof. (Proof of Proposition 8.2) By condition i), the two-point function has the following form:

$$W_{ij}(x) = \delta_{ij}H(\mathbf{x}^2, x_0) + \partial_i\partial_j L(\mathbf{x}^2, x_0),$$

where H, L are defined up to the following redefinition:

$$H \to H + h(x_0), \quad L \to L - \tfrac{1}{2}h(x_0)\,\mathbf{x}^2.$$

A comparison with the two-point functions of the electromagnetic field

$$< E_i\,E_j >= (-\partial_0^2\delta_{ij} + \partial_i\partial_j)F, \quad < B_i\,B_j >= (\delta_{ij}\Delta - \partial_i\partial_j)F$$

gives

$$\delta_{ij}\partial_0^2(H - F) + \partial_i\partial_j(F + \partial_0^2 L) = 0, \quad (\delta_{ij}\Delta - \partial_i\partial_j)(H - F) = 0.$$

Then, by Lemma 8.3 and by the above allowed redefinition one may put

$$H = F, \quad \partial_i \partial_j \partial_0^2 L = -\partial_i \partial_j F.$$

The operator $\partial_i \partial_j / \partial_0^2$ is well defined in momentum space, where it corresponds to multiplication of the spectral measure $d\rho(m^2)$ by the bounded function $k_i k_j / (\mathbf{k}^2 + m^2)$, and therefore the general (rotationally covariant) solution of the last equation is

$$\partial_i \partial_j L = -(\partial_i \partial_j / \partial_0^2) \, F + a_{ij}(\mathbf{x}) + i x_0 b(\mathbf{x}),$$

$$a_{ij}(\mathbf{x}) = \delta_{ij} a_1(\mathbf{x}^2) + \partial_i \partial_j a_2(\mathbf{x}^2), \quad b_{ij}(\mathbf{x}) = \delta_{ij} b_1(\mathbf{x}^2) + \partial_i \partial_j b_2(\mathbf{x}^2).$$

Again the comparison with $< B_i B_j >$ gives $a_1 = 0 = b_1$, and the function $a_2(\mathbf{x}^2)$ can be eliminated by an operator gauge transformation.

The vanishing of the equal-time commutator $< [A_i, A_j] >$ is easily checked, and $< [A_i(\mathbf{x}, x_0), \partial_0 A_j(\mathbf{y}, x_0)] > = 0$, for all $\mathbf{x} - \mathbf{y} \neq 0$, gives the following condition:

$$2 k_i k_j \tilde{b}_2(\mathbf{k}^2) = k_i k_j \, \mathcal{P}(\mathbf{k}^2) + \int d\rho(m^2) k_i k_j / (\mathbf{k}^2 + m^2).$$

The Fourier transform of the term linear in time has support on the plane $k_0 = 0$, $\mathbf{k}$ arbitrary, so that the relativistic spectral condition does not hold.

In the free field case one has

$$< A_i(\tfrac{1}{2}x) \, A_j(-\tfrac{1}{2}x) > = (\delta_{ij} - \partial_i \partial_j / \Delta) D^+(x) + \tfrac{1}{2} i \, (x_0) \, (\partial_i \partial_j / \Delta) \delta(\mathbf{x})$$

and positivity cannot hold, since both $< \partial_i A^i(f, t) \, \partial_j A^j(f, t) >$ and $< \partial_i E^i \, \partial_j E^j >$ vanish, but one cannot have $\partial_i E^i \Psi_0 = 0$.

Thus, a realization of the temporal gauge in which the two-point function or the propagator of the vector potential exists and the vacuum is invariant under translations and rotations, inevitably violates positivity, the vector space $\mathcal{D}_0$ defined by the vacuum correlation functions contains unphysical vectors, and the Gauss law can be required to hold only as a weak constraint on the physical vectors:

$$< \Psi, G(f) \Phi > = 0,$$

for any pair of physical vectors Ψ, Φ, as in the FGB quantization.

2) Positive non-regular quantization

In order to discuss the positive (second) alternative, it is convenient to introduce the following field algebras:

- *the field algebra* $\mathcal{F}$, generated by the smeared fields $A(f) \equiv A_i(f_i)$, and $\psi(g), \overline{\psi}(g)$, $f_i, g \in \mathcal{S}(\mathbf{R}^4)$;
- *the observable field subalgebra* $\mathcal{F}_{obs}$ characterized by its invariance under time-independent gauge transformations: $[G(f), \mathcal{F}_{obs}] = 0$;
- *the exponential field algebra* $\mathcal{W}$ generated by $\exp i A(f), \psi(g), \overline{\psi}(g)$;
- *the exponential longitudinal algebra* $\mathcal{W}_L$ generated by $\exp i G(f), \exp i A(\partial h)$, $A(\partial h) \equiv A_i(\partial_i h)$.

A temporal gauge quantization will be called *regular* if the corresponding vacuum correlation functions of the field algebra $\mathcal{F}$ are well defined and *non-regular* if the vacuum correlation functions of the exponential field algebra $\mathcal{W}$ are well defined, but not those of $\mathcal{F}$.

In the free case the exponential longitudinal algebra is a canonical Weyl algebra, and its non-regular quantization defines a non-regular representation of such a Weyl algebra, in the technical mathematical sense that the unitary operators $\exp{(i\beta G(f))}$, $\exp{(i\alpha A(\partial h))}$, α, $\beta \in \mathbf{R}$ are not weakly continuous in α, β, so that the generators do not exist.

The inevitable violation of positivity for the regular quantizations and the general properties of the positive quantizations are stated by the following:

Proposition 8.4 *Let Ω be a positive vacuum state on the observable subalgebra $\mathcal{F}_{obs}$ (therefore satisfying covariance, locality, spectral condition, and cluster property), then one has*
i) any extension of Ω to the field algebra $\mathcal{F}$ violates positivity,
ii) all the positive extensions ω to the exponential field algebra $\mathcal{W}$ are invariant under time-independent gauge transformations, in particular, if Ψ_ω denotes the GNS cyclic vector which represents the state ω, one has

$$G(f)\Psi_\omega = 0, \tag{7.8.41}$$

iii) any positive extension ω of Ω to the exponential longitudinal algebra $\mathcal{W}_L$ is non-regular, since it must satisfy

$$\omega(e^{i\alpha A(\partial h)}) = 0, \quad \text{if } \alpha \neq 0; \quad = 1, \text{ if } \alpha = 0, \tag{7.8.42}$$

iv) in the GNS representation defined by the a positive extension ω to $\mathcal{W}$, the unitary operators $U(\mathbf{a})$ which implement the space translations are not strongly continuous, and therefore their generator, the momentum, cannot be defined in the whole representation space $\mathcal{H}_\omega$.

Proof. i) Since $G(f)$ commutes with $\mathcal{F}_{obs}$, by Theorem 4.5 in Streater and Wightman, *PCT, Spin and Statistics, and All That*, if Ψ_Ω denotes the cyclic vector which represents the state Ω on $\mathcal{F}_{obs}$,

$$G(f)\Psi_\Omega = c(f)\Psi_\Omega, \quad c(f) \in \mathbf{C},$$

and by Lorentz invariance of Ω one has $c(f) = 0$. Hence, by Schwarz's inequality, a positive extension ω to $\mathcal{F}$ should satisfy

$$\omega(G(f)F) = 0, \quad \forall F \in \mathcal{F}, \tag{7.8.43}$$

which is incompatible with the ω-expectation of eq. (7.8.36), so that there is no positive extension to $\mathcal{F}$.

ii) Any positive extension to $\mathcal{W}$ satisfies eq. (7.8.43), with F replaced by any $B \in \mathcal{W}$, so that eq. (7.8.41) follows by the cyclicity of Ψ_ω.

iii) Putting $V(f) \equiv \exp iG(f)$, one has

$$\omega(e^{i\alpha A(\partial h)}) = \omega(V(-f)\,e^{i\alpha A(\partial h)}\,V(f)) = e^{i\alpha \int d^4x\, h\Delta f}\,\omega(e^{i\alpha A(\partial h)})$$

and eq. (7.8.42) follows.

iv) Putting $h_{\mathbf{a}}(x) \equiv h(x + \mathbf{a})$ one has

$$\omega(e^{iA(\partial h)}U(-\mathbf{a})e^{-iA(\partial h)}) = \omega(e^{iA(\partial h - \partial h_{\mathbf{a}})}) = 0, \ \text{if } \mathbf{a} \neq 0; \ = 1, \ \text{if } \mathbf{a} = 0.$$

The invoked escape of a non-normalizable vacuum vector is mathematically unacceptable for the following reasons. Non-normalizable state vectors are often used in quantum mechanics, such as the Dirac δ function, but their mathematical oddness is harmless, provided they are considered (as they should) only as extrapolated limits of normalizable vectors of a well defined Hilbert space. It is mathematically inconsistent to consider quantizations built up on a *cyclic* non-normalizable vector Ψ_0, because then *all* the so-obtained vectors are non-normalizable, and all transition amplitudes are divergent. In particular, if the vacuum is not normalizable, the divergent expectations are not confined to those of the gauge-dependent fields, but also affect those of the observable fields. A mathematically acceptable solution is to keep the normalizability of the vacuum and give up the definiteness of the gauge-dependent fields, only their exponentials having well-defined vacuum correlation functions. This strategy has a well-founded mathematical status, and it is under complete control for the quantization of canonical systems; in this case it corresponds to consider *non-regular representations* of the canonical Weyl algebra. The Weyl quantization differs from Heisenberg quantization because it involves only the exponentials of the canonical variables, e.g., $U(\alpha)$, $V(\beta)$, formally corresponding to the exponentials $\exp i\alpha q$, $\exp i\beta p$ of the canonical variables (q, p), and does not require the existence of the latter; the corresponding algebra, called the Weyl C^*-algebra, admits representations which are not weakly continuous in the parameters α, β (called *non-regular representations*), in which the canonical variables cannot be defined (as self-adjoint operators), only their exponentials being well defined.[100]

In the free field case, the positive non-regular representation of the exponential field algebra $\mathcal{W}$ is uniquely determined solely by the condition of positivity of the energy spectrum.[101]

---

[100] See, e.g., F. Strocchi, *Introduction to the Mathematical Structure of Quantum Mechanics*, 2nd ed., World Scientific 2008; F. Strocchi, Gauge invariance and Weyl quantization, SNS lectures 2007–2008. For a detailed discussion with applications to relevant physical systems see R. Beaume, J. Manuceau, A. Pellet, and M. Sirugue, Comm. Math. Phys. **38**, 29 (1974); C. Hurst, Dirac's theory of constraints, in *Recent Developments in Mathematical Physics*, H. Mitter *et al.* (eds.), Springer 1987, p. 18: F. Nill, Int. J. Mod. Phys. **B 6**, 2159 (1992); F. Acerbi. G. Morchio, and F. Strocchi, Lett. Math. Phys. **27**, 1 (1993): Jour. Math. Phys, **34**, 899 (1993).

[101] LMS, Proposition 3.3.

8

Chiral symmetry breaking and vacuum structure in QCD

1 The $U(1)$ problem

The extraordinary success of the standard model is its best justification, but since some problems remain open (the flavor problem, the hierarchy problem, the $U(1)$ problem, the strong CP problem) it is perhaps useful to recall the main motivations for its birth and the ideas that led to its present form.

The unification of the electromagnetic and weak interactions requires a spontaneously broken symmetry, because the effective strength of the two interactions differs by two orders of magnitude—a feature which does not fit in a (broken) Wigner symmetry. Moreover, the absence of the corresponding Goldstone bosons requires a Coulomb-delocalized time evolution and therefore a gauge symmetry[1]; the gauge symmetry $SU(2) \times U(1)$ is the minimal realization of such requirements.

Less direct and simple are the motivations at the origin of color $SU(3)$. The Gell-Mann and Zweig quark model very successfully accounts for the hadron classification, as given by the Eightfold way: it requires i) three quarks (u, d, s) as basic constituents of hadrons, transforming as the fundamental representation of the flavor $SU(3)$, ii) they must have fractional charges ($q_u = 2/3$, $q_d = q_s = 1/3$), iii) they must have spin $1/2$.[2]

The first problem in the developments leading to quantum chromodynamics (QCD) is that the baryon spectrum requires totally symmetric wave functions, with respect to quark permutations; to avoid violation of the spin–statistics theorem, Greenberg[3] suggested that quarks have an additional quantum number, called *color*, so that the baryons' and mesons' wave functions are respectively of the form $\varepsilon_{\alpha\beta\gamma} q_{f_1}^\alpha q_{f_2}^\beta q_{f_3}^\gamma$, $\bar{q}_{f_i}^\alpha q_{f_i}^\alpha$, where f_k denotes the flavor $SU(3)$ indices and $\alpha, \beta, \gamma = 1, 2, 3$ are color indices.

The second problem arose due to failure of detecting quarks, only the color singlets (baryon and mesons) being observed. This led to postulate that only color singlets exist in nature (*weak confinement*). A dynamical explanation of such a phenomenon requires a rather peculiar color interaction, with properties very different from standard quantum field theories. The only known possible mechanism to give weak confinement is that the color interaction be described by a Yang–Mills theory (*quantum*

---

[1] F. Strocchi, *Symmetry Breaking*, 2nd ed., Springer 2008.

[2] For a very comprehensive account of the developments which led to QCD see S. Narison, *QCD as a Theory of Hadrons: From Partons to Confinement*, Cambridge University Press 2004.

[3] O.W. Greenberg, Phys. Rev. Lett. **13**, 598 (1964).

chromodynamics)[4]. The absence of massless particles in the hadron spectrum led to postulate a *strong* form of *confinement*, embodying the existence of a mass gap; unfortunately, there is no proof that QCD has this property.[5]

The role of quarks as hadron constituents received an impressive confirmation by the deep inelastic scattering and the related parton picture. Furthermore, as the $SU(2) \times U(1)$ electroweak gauge theory, also QCD is asymptotically free, and therefore, from a quantum-field-theory point of view, qualifies as a candidate of a non-trivial consistent theory, without the triviality problems of φ^4 in four spacetime dimensions.

With the discovery of the J/ψ particle and also for explaining the absence of flavor-changing neutral currents by the Glashow–Iliopoulos–Maiani (GIM) mechanism, it was necessary to introduce other quarks and the present picture consists of three families of quarks doublets $((u,d),(c,s),(t,b))$, each transforming as an (isospin) $SU(2)$ doublet. The quark masses are believed to be generated by the electroweak interaction, through the vacuum expectations of the Higgs field, and therefore, in the absence of such an interaction, the pure QCD Lagrangian reads

$$\mathcal{L}_{QCD} = -\tfrac{1}{4} F_{a\,\mu\,\nu}\, F_a^{\mu\,\nu} + \tfrac{1}{2}\bar\psi i\gamma^\mu D_\mu\,\psi, \tag{8.1.1}$$

where a is a color gauge index, D_μ the (color) covariant derivative, and ψ denotes the quark field with components labeled by flavor ($SU(2)$ and family) and color indices.

Such a Lagrangian is invariant under the (unbroken) color group and the flavor group $U(N)$, where N is the number of flavors. Since there is no quark mass, $\mathcal{L}_{QCD}$ is also invariant under axial $U_A(N)$. An interesting puzzle is the consequence of such symmetries on the hadron spectrum. With respect to the QCD scale, the Higgs induced masses of the u,d,s quarks are small, and therefore in the massless limit one has a chiral $U(3)_V \times U(3)_A = SU(3)_V \times SU(3)_A \times U(1)_V \times U(1)_A$ symmetry (the vector/axial nature of the current which generates the flavor/chiral transformations has been spelled out by the index V/A).

In the QCD framework (without the electroweak interaction) the $SU(3)_V \times SU(3)_A$ symmetry is believed to be spontaneously broken down to $SU(3) \equiv SU(3)_V$ (the group at the basis of the Gell-Mann–Ne'eman Eightfold way), with the octet of pseudoscalar mesons ($\pi^\pm$, π_0, $K^\pm$, K_0, $\bar K_0$, η) being the corresponding Goldstone bosons; this explains their (relatively) small masses (due to the electroweak interaction). The experimental and theoretical evidence for spontaneous symmetry breaking is much stronger for the $SU(2)_V \times SU(2)_A$ subgroup, with the pions playing the role of the Goldstone bosons; indeed, their masses are very small ($m_\pi^2 = 0,01885$ Gev^2) compared with the hadron spectrum, and their interpretation as Goldstone bosons is

[4] H. Fritzsch, M. Gell-Mann, and H. Leutwyler, Phys. Lett. **B 47**, 365 (1973); A. Casher, J. Kogut and L. Susskind, Phys. Rev Lett. **31**, 792 (1973); K. G. Wilson, Quarks: from paradox to myth: Quarks and strings on a lattice, in *New Phenomena in Subnuclear Physics*, Proceedings of the 1975 Int. School of Subnuclear Physics, Erice, A. Zichichi (ed.), Plenum 1977, pp. 13–32, 69–125.

[5] For a deep account of the theoretical situation and problems see A. Jaffe and E. Witten, Quantum Yang–Mills Theory, in *The Millenium Prize Problems*, J. Carlson *et al.* (eds.), Amer. Math. Soc. 2006.

at the basis of the very successful soft-pion theorems.[6] The $U(1)_V$ is unbroken and corresponds to the baryon number transformation.

It remains to understand the status of the axial $U(1)_A$ symmetry:

$$\psi \to e^{\alpha\gamma_5}\,\psi, \quad \bar{\psi} \to \bar{\psi}e^{\alpha\gamma_5}, \quad \gamma_5^* = -\gamma_5, \quad \alpha \in \mathbf{R}. \tag{8.1.2}$$

An unbroken $U(1)_A$ at the QCD level would imply that for each particle there would be a mirror particle with opposite parity (*parity doublets*); such a pattern is not present in the hadron spectrum. The alternative is that the symmetry is spontaneously broken.

According to the standard version of the Goldstone theorem, the spontaneous breaking of the global $U(1)_A$ symmetry should be accompanied by a Goldstone pseudoscalar boson, transforming as an isospin singlet. By using chiral perturbation theory, its mass has been estimated to be less than $\sqrt{3}\,m_\pi \sim 240$ Mev; neither the pseudoscalar $\eta(549$ Mev), which rather accounts for the isospin singlet of the pseudoscalar octet of Goldstone bosons associated with the breaking of chiral $SU(3) \times SU(3)$, nor the isospin singlet $\eta'(958$ Mev), satisfy this bound.[7] This is the $U(1)$ *problem*.

The standard solution of the problem is ascribed to the chiral anomaly and to the effects of instanton configurations in the functional integration.[8]

According to the classical Noether theorem, the current associated with the chiral transformation (8.1.2) is the (gauge-invariant) chiral current $j_\mu^5 = i\,\bar{\psi}\gamma^5\gamma_\mu\,\psi$. As discussed in the abelian case, Chapter 4, Section 6.2, the quantum version requires a point-splitting regularization, and as in the abelian case a gauge-invariant procedure yields a regularized chiral current with anomaly

$$\partial^\mu j_\mu^5(x) = N\frac{g^2}{8\pi^2}\varepsilon^{\mu\nu\rho\sigma}\,\mathrm{Tr}\,(F_{\mu\nu}\,F_{\rho\sigma}(x)), \tag{8.1.3}$$

where N is the number of quarks involved in the chiral limit, i.e., $N = 2$ in the case of chiral $SU(2) \times SU(2)$, $N = 3$ in the case of chiral $SU(3) \times SU(3)$.

A non-perturbative proof of the chiral anomaly (8.1.3) has been given by using the functional integral approach,[9] and eq. (8.1.3) is a solid established feature of QCD.

The non-conservation of j_μ^5 has been interpreted as a sign that the action is not invariant under chiral transformations, and that such a symmetry of the classical action does not survive quantization. However, in our opinion, such a conclusion requires further argument, since the right-hand side of eq. (8.1.3) may be written as a four-divergence, and the action changes by a term proportional to such a four-divergence, which, as such, does not affect the equations of motion.

Furthermore, for the symmetry breaking Ward identities at the basis of the Goldstone theorem, all that is needed is that j_μ^5 generate the time independent $U(1)_A$

---

[6] See S. Weinberg, *The Quantum Theory of Fields: Modern Applications*, Cambridge University Press 1995, Chapter 19.

[7] For an excellent discussion, see S. Weinberg, *The Quantum Theory of Fields. Modern Applications*, Cambridge University Press 1995, Chapter 19.

[8] G. 't Hooft, Phys. Rev. Lett. **37**, 8 (1976); Phys. Rev. D **14**, 3432 (1976); S. Coleman, *Aspects of Symmetry*, Cambridge University Press 1985, Chapter 7, Sect. 3.5.

[9] K. Fujikawa, Phys. Rev. **D21**, 2848 (1980); see also S. Weinberg, *The Quantum Theory of Fields*, Vol.II, Cambridge University Press, Chap. 22.

transformations, and this would still hold if the (suitably regularized) integral of the right-hand side of eq. (8.1.3) commutes with the order parameter, at least in the corresponding vacuum expectation.

The non-trivial effects of the above four-divergence in both cases has been argued by 't Hooft on the basis of the non-trivial contributions of the instanton configurations in a semiclassical approximation of the functional integral. This has been taken as evidence that, due to quantum effects, the transformation of eq. (8.1.2) does not define an algebraic symmetry (technically an automorphism) of the field algebra $\mathcal{F}$, and not even of its observable subalgebra $\mathcal{F}_{obs}$, which contains the relevant symmetry breaking order parameter $(\bar{\psi}\,\psi)$.

Thus, according to this point of view there is no symmetry to be broken, the absence of parity doublet is not related to a spontaneous symmetry breaking, and for this reason the basic assumption of the Goldstone theorem fails.

As a matter of fact, as shown by Bardeen on the basis of perturbative renormalization in local gauges,[10] for the definition of the chiral transformations of the fields, an important role is played by the *conserved*, gauge-dependent, chiral current

$$J_\mu^5 = j_\mu^5 - (2\pi)^{-2}\varepsilon_{\mu\nu\rho\sigma}\,\mathrm{Tr}\,[A^\nu\partial^\rho A^\sigma + (2/3)A^\nu A^\rho A^\sigma] \equiv j_\mu^5 + K_\mu^5.$$

In local gauges, i.e., defined in terms of a local field algebra $\mathcal{F}$, by the standard argument (see Chapter 7, Section 2), since J_μ^5 is conserved, $\lim_{R\to\infty}[\,J_0^5(f_R,t),\,F\,]$ exists $\forall F \in \mathcal{F}$ and is independent of time; hence, the computation of such commutators may be reduced to equal-time commutators, which are governed by the canonical commutation relations. The result is that the above limit of the commutator yields the *infinitesimal axial transformations of the (local) fields, and, in particular, of the gauge-invariant observable fields*. Such general properties have been checked by Bardeen to be stable under renormalization.

The claim that has sometimes appeared in the literature, that the generation of a symmetry by a gauge-variant current is meaningless, is disputable. In our opinion, there is no logical reason for *a priori* rejecting the use of the gauge-dependent current J_μ^5 and of its associated Ward identities. Also, the invariance under a non-abelian gauge group gives rise to gauge-dependent conserved currents and the corresponding symmetry breaking Ward identities characterize the non-abelian Higgs phenomenon!

The existence of axial $U(1)$ transformations as automorphisms of the observable subalgebra $\mathcal{F}_{obs}$ implies that *the absence of parity doublets is a problem of spontaneous symmetry breaking*; then, the crucial issues for *the absence of massless Goldstone bosons* are the local generation of the symmetry and the existence of symmetry breaking Ward identities in the irreducible vacuum representations of the observable algebra.

Since the symmetry breaking Ward identity involves the gauge-dependent field J_μ^5, just as for the Higgs phenomenon, the discussion is likely to involve theoretical mechanisms which have different realizations in different gauges.

---

[10] W. A. Bardeen, Nucl. Phys. **B 75**, 246 (1974).

The distinctive property of the non-abelian anomaly, which plays a crucial role in the solution of the $U(1)$ problem, is the θ vacuum structure, as discovered by using the functional integral methods and the instanton semiclassical approximation.[11]

In this way one obtains that i) the irreducible vacuum representations of the observable algebra $\mathcal{F}_{obs}$ are labeled by an angle θ (θ vacua), ii) the chiral transformations do not leave the sectors invariant, i.e., in each θ sector the symmetry is broken, and iii) the corresponding symmetry breaking Ward identities at the basis of the Goldstone theorem become substantially modified by the occurrence of topological effects.

We briefly sketch the standard argument pointing out the assumptions and the approximations involved. The basic starting point is to rely on a semiclassical approximation, namely the assumption that the Euclidean functional integral is dominated by field configurations with finite action (which minimize the classical Euclidean action). Furthermore, such configurations are assumed to be continuous and, when the Euclidean radial variable $r \to \infty$, behave like

$$A_\mu \sim g\partial_\mu g^{-1} + O(1/r^2),$$

where g is a pure gauge configuration depending only on the angular variables, $g = g(\Omega)$.

Such configurations are therefore classified by the winding number n defined by (with the normalization Tr $(T^a\, T^b) = \delta_{a\,b}$)

$$-24\pi^2 n = \int d\theta_1\, d\theta_2\, d\theta_3\, \varepsilon^{i\,j\,k}\, \mathrm{Tr}\,[g\partial_i g^{-1}\, g\partial_j g^{-1}\, g\partial_k g^{-1}], \qquad (8.1.4)$$

where θ_i, $i = 1, 2, 3$ are three angles which parametrize S^3 and ∂_i the corresponding partial derivatives. The winding number is a topological invariant.

The winding number of a field configuration A_μ may also be written in terms of a volume integral. To this purpose, one introduces the so-called topological current

$$G_\mu = \varepsilon_{\mu\nu\lambda\sigma}\mathrm{Tr}\,[A_\nu\, F_{\lambda\sigma} - \tfrac{2}{3}A_\nu\, A_\lambda\, A_\sigma\,], \quad \partial_\mu G_\mu = \tfrac{1}{2}\varepsilon_{\mu\nu\lambda\sigma}\mathrm{Tr}\,[F_{\mu\nu}\, F_{\lambda\sigma}]; \qquad (8.1.5)$$

then the integral of $\partial_\mu G_\mu$ over the whole space is given by the flux at infinity of G_μ, which is governed by the asymptotic behavior of A_μ. For finite-action configurations, the first term in G_μ is $O(1/r^4)$ and gives a vanishing contribution to the flux at infinity; the other term gives $64\pi^2\, n$. Hence, one has (in euclidean space)

$$64\,\pi^2 n = \int d^4x\, \varepsilon_{\mu\nu\lambda\sigma}\mathrm{Tr}\,[F_{\mu\nu}\, F_{\lambda\sigma}]. \qquad (8.1.6)$$

In the semiclassical approximation of the functional integral in terms of finite-action configurations, one may first integrate over the class of configurations with given winding number n, and then sum over n. By semiclassical arguments, which exploit the fact that n is the integral of a local density or the cluster decomposition property, one concludes that the weight for each class labeled by n is of the form $e^{in\theta}$, where θ

[11] G. 't Hooft. Phys. Rev. Lett. **37**, 8 (1976); C. G. Callan, R. F. Dashen, and D. Gross, Phys. Lett. B **63**, 34 (1976); R. Jackiw and C. Rebbi, Phys. Rev. Lett. **37**, 172 (1976).

is a free parameter with the properties of an angle (θ *angle*), and the corresponding correlation functions define the so-called θ vacua.[12]

Hence, by eq. (8.1.6) the non-trivial topology of the finite-action configurations leads to a modification of the original Lagrangian by the addition of the so-called *topological term*

$$\mathcal{L}_{top} = i\theta(64\pi^2)^{-1}\int d^4x\, \varepsilon_{\mu\nu\lambda\sigma}\mathrm{Tr}\,[F_{\mu\nu}\,F_{\lambda\sigma}], \qquad (8.1.7)$$

which selects the corresponding θ vacuum.

A similar topological classification may be determined in the temporal gauge (defined by $A_0 = 0$) under the corresponding assumptions for the semiclassical approximation: i) the Euclidean field configurations $A_i(\mathbf{x}, \tau)$ are assumed to approach gauge configurations $g_\pm(\mathbf{x})\partial_i g_\pm^{-1}(\mathbf{x})$ in the limit $\tau \to \pm\infty$, ii) $A_i(\mathbf{x}, \tau)$ are assumed to vanish for $|\mathbf{x}| \to \infty$, so that $g_\pm(\mathbf{x})$ must approach a constant in that limit, and the three-dimensional space at $\tau \to \infty$ may be compactified, becoming isomorphic to a three-sphere S^3.

Such gauge-field configurations are then classified by the winding numbers $n_\pm$ defined by

$$-24\pi^2 n_\pm = \varepsilon_{ijk}\int d^3x\,\mathrm{Tr}\,[g_\pm\partial_i g_\pm^{-1}\,g_\pm\partial_j g_\pm^{-1}\,g_\pm\partial_k g_\pm^{-1}]$$

and, by eq. (8.1.6), one has $n = n_+ - n_-$.

In the presence of fermions, the Fujikawa argument on the transformation of the fermion functional measure under the chiral transformation (8.1.2) shows that the topological terms changes by a shift $\theta \to \theta + 2\alpha$. Hence, in each irreducible representation of the observable algebra, defined by a θ vacuum, chiral symmetry is broken, and at the same time the local integral $\int d^4x\,\varepsilon_{\mu\nu\lambda\sigma}\mathrm{Tr}\,[F_{\mu\nu}\,F_{\lambda\sigma}]$ is not irrelevant, since it acquires non-vanishing contributions by the instanton configurations with non-zero winding number.

A natural question is whether one may improve the derivation of the θ vacuum structure and the solution of the $U(1)$ problem without relying on the instanton calculus. In fact, as discussed in Chapter 5, Section 8, and also recognized by Coleman, even in the simple free field case, the set of configurations with finite classical (Euclidean) action has zero functional measure; the relevant configurations are not continuous, and a topological classification of them is precluded.

A treatment which does not use the instanton semiclassical approximation has been proposed by Jackiw, in terms of the topology of the gauge group, without involving the topological classification of the finite action configurations.[13]

Jackiw's analysis is carried out in the temporal gauge, and the very peculiar features of such a gauge require special care concerning mathematical problems of consistency, which, as we shall see, have a non-trivial impact on the general argument.

---

[12] See S. Coleman, *Aspects of Symmetry*, Cambridge University Press 1985, Chap. 7, Sect. 3; S. Weinberg, *The Quantum Theory of Fields*, Vol. II, Sect. 23.

[13] R. Jackiw, Topological investigations of quantized gauge theories, in S. B. Treiman, R. Jackiw, B. Zumino, and E. Witten, *Current Algebra and Anomalies*, World Scientific 1985, pp. 211–359.

Furthermore, the discussion is not free of assumptions about the behavior at space infinity of the (relevant) field configurations—a property which may possibly have a meaning only in a semiclassical approximation.

In the following, we will revisit the strategy and the arguments with attention to the mathematical fine points involved. As we shall see, a general non-perturbative analysis free of mathematical problems may indeed be obtained by exploiting the topology of the gauge group directly, rather than through the classification of the behavior at infinity of the (actually very rough) field configurations.

As explained below, one of the relevant ingredients, with respect to the Jackiw approach, is the realization that in the temporal gauge the time-independent gauge transformations, including the so-called large ones, can be localized in space, and therefore may be taken to be implemented by local operators. This allows a local analysis of the topology of the gauge group.

In this way one obtains a non-perturbative derivation of the θ vacua structure and a proof of the evasion of the Goldstone theorem, i.e., a solution of the $U(1)$. The result makes clear that the existence of the chiral anomaly is by itself not sufficient, and that further ingredients, basically the non-trivial topology of the gauge group, are needed for the solution.

From a phenomenological point of view, the next problem related to chiral symmetry is the estimation of the mass gap which accompanies the breaking; this amounts to a derivation of the η' mass from the chiral Ward identities. A possible approach has been proposed[14] which considers the T-ordered products of the gauge-invariant chiral currents and the quark bilinears arising in the current divergences due to the non-vanishing quark masses. The numerical results are in very good agreement with experiments, but non-trivial problems affect the Ward identities derived by such T-products.

In such an approach a crucial role is played by the equal-time commutators involving the gauge-invariant chiral current j_μ^5, but, as discussed by Bardeen, a consequence of the chiral anomaly is that such equal-time commutators are no longer protected by the non-renormalization theorem for conserved currents (see Chapter 7, Section 2) and are actually divergent.

Furthermore, the two-point function of K_μ^5 (not to speak of the space integral of K_0^5) defined by a gauge-invariant θ vacuum does not exist by general arguments of gauge-invariance (see Propositions 2.3 and 2.4 below), and the assumption of a pole $1/q^2$ in its Fourier transform is problematic.

2 Topology and chiral symmetry breaking in QCD

For the discussion of the non-perturbative aspects of QCD, in particular the θ vacua structure, its relation with the topology of the gauge group, and its role

---

[14] G. Veneziano, Nucl. Phys. **B159**, 213 (1979); D. I. D'yakonov and M. I. Eides, Sov. Phys. JEPT **54**, 232 (1982).

in chiral symmetry breaking, the *temporal gauge* has proved to be particularly convenient.[15]

One of the advantages of the temporal gauge is that the gauge fixing leaves a residual gauge group $\mathcal{G}$ consisting of the time-independent gauge transformations, and the non-trivial topology of the corresponding gauge functions emerges in a simple way. Since the gauge functions which parametrize the gauge transformations can (actually should) be taken infinitely differentiable, their topological classification is mathematically well founded.

On the other side, as noted before for the abelian (QED) temporal gauge, the conflict between the Gauss-law constraint and canonical quantization raises problems of mathematical consistency, which also have a non-trivial impact on the derivation of the general structures and have significant physical implications, like the evasion of the Goldstone theorem (see Section 2.4).

As a result of our analysis, several statements that have appeared in the literature must be revised, since they are not mathematically correct. The following analysis is organized according to the following pattern.[16]

1) The crucial point of our analysis is the realization that the *non-trivial topology of the gauge group $\mathcal{G}$* may be characterized in terms of group-valued *gauge functions $\mathcal{U}(\mathbf{x})$ with $\mathcal{U}(\mathbf{x}) - 1$ of compact support* (Section 2.2).

This greatly simplifies the analysis with respect to the standard discussion, where the role of the topology was historically associated to the existence of instanton solutions, which minimize the classical action and for this reason cannot have compact support. As a consequence, for the classification of (non-localized) gauge functions $\mathcal{U}(\mathbf{x})$, considered in the literature, and in particular in Jackiw analysis, one has to require that they have a limit for $|\mathbf{x}| \to \infty$, independent of the direction. In order to support such an assumption it has been assumed that the gauge potential $A_i^a(\mathbf{x})$ falls off faster than $1/r$, a condition which can possibly have a meaning only within a semiclassical approximation of the functional integral.

2) The group $\mathcal{G}(\mathcal{O})$ of gauge transformations localized in the spacetime region $\mathcal{O}$, modulo its subgroup $\mathcal{G}_0(\mathcal{O})$ with zero winding number, defines an abelian group $\mathcal{T}(\mathcal{O})$, whose local implementers $T_n(\mathcal{O})$, labeled by the winding number n, belong to the center $\mathcal{Z}$ of the local observable algebra $\mathcal{A}$. The θ angle appears as the spectrum of $\mathcal{Z}$, and as such it is a candidate for labeling the irreducible representations of $\mathcal{A}$ (Section 2.5). For the physical implications, the ensuing question is the non-triviality of $\mathcal{Z}$.

It has been claimed that such a non-triviality is displayed by the non-invariance of the topological charge under the large transformations, but the argument is not mathematically correct. In fact, the invariance of the vacuum under the small gauge transformations (implemented by the formal exponential of the Gauss operator) implies the *non-regularity of the representation of the field algebra* defined by the

---

[15] See the excellent lectures by R. Jackiw, Topological investigations of quantized gauge theories, in S. B. Treiman, R. Jackiw, B. Zumino, and E. Witten, *Current Algebra and Anomalies*, World Scientific 1985, pp. 211–359.

[16] We follow G. Morchio and F. Strocchi, Ann. Phys. **324**, 2236 (2009).

vacuum correlation functions. This implies that the (exponential of the) topological charge has vanishing matrix elements between states satisfying the Gauss law constraint (Proposition 2.3); hence, other ingredients are needed for arguing a non-trivial effect of the large gauge transformations on the structure of the physical states.

3) As we shall see, a *crucial role is played by the fermions and the associated chiral symmetry*. Following Bardeen, contrary to statements appearing in the literature, the chiral anomaly does not prevent the *chiral symmetry* β^λ, $\lambda \in \mathbf{R}$, from being *a well-defined time-independent automorphism of the field algebra* and of its observable subalgebra.

Actually, β^λ is locally implemented by unitary operators $V_R^5(\lambda)$, $\lambda \in \mathbf{R}$, formally the exponentials $e^{i\lambda J_0^5(f_R \alpha_R)}$ of the (regularized space integral of the) gauge-dependent conserved chiral current J_μ^5 (Section 2.3). The crucial effect of the chiral anomaly is that the local implementers of the large gauge transformations and the elements of $\mathcal{Z}$ defined by them transform non-trivially under the chiral symmetry, so that $\mathcal{Z}$ is not trivial.

Thus, the *non-trivial topology of $\mathcal{G}$* reflects in a *non-trivial center of the local observable algebras*, which is not left pointwise invariant by the chiral transformations; hence, *chiral symmetry is always broken in any irreducible (or even factorial) representation of the observable algebra.*

Such an analysis displays the role of the gauge group $\mathcal{G}$ in the classification of the irreducible (or factorial) representations of the observable algebra, not only through the invariants of the irreducible representations of $\mathcal{G}$ (as discussed in Chapter 7, Sections 4 and 5.2), but also through the topological invariants of $\mathcal{G}$ (Section 2.5).

4) A mechanism of evasion of the Goldstone theorem (*solution of the $U(1)$ problem*) is provided by the impossibility of writing the symmetry breaking Ward identities, since *the vacuum expectations $< J_0^5 \, \bar\psi \psi >_0$ do not exist*, as a consequence of the non-regularity of the $V_R^5(\lambda)$, which prevents the existence of their generators (Proposition 2.4).

5) As discussed in the abelian case, one may consider a realization of the temporal gauge obtained by a *vacuum functional* which is only required to be weakly invariant under Gauss transformations. This allows well-defined vacuum correlations of the fields A_i^a, ψ, not only of their exponentials, at the price of an indefinite metric (*regular temporal gauge*). In particular, the gauge-dependent conserved axial current J_μ^5 may be a well-defined local field operator, and Bardeen analysis applies directly. In this case, the vacuum always defines a reducible representation $\pi^{(0)}$ of the observable algebra, and chiral symmetry is broken in each irreducible subrepresentation of the observable algebra. The reduction of $\pi^{(0)}$ into θ sectors does not extend to the chiral current J_μ^5, which, as before, does not exist in a θ sector (Section 2.6).

In conclusion, the derivation of the physical consequences of the topology of $\mathcal{G}$ crucially relies on the presence of fermions and their chiral transformations. The essential point, which distinguishes the abelian case from the non-abelian one, beyond the existence of the chiral anomaly, is the existence of a center of the local observable algebra, which is a direct consequence of the topology of $\mathcal{G}$, and is not pointwise-invariant under chiral transformations. One may say that *the physical relevance of the topology of $\mathcal{G}$ is disclosed by the chiral symmetry.*

2.1   Temporal gauge and Gauss law

For simplicity, we start by considering the case with only gauge vector fields (no fermion or scalar field being present). Then, the QCD Lagrangian density reduces to the Yang–Mills form (see Section 1)

$$\mathcal{L} = -\tfrac{1}{4} \sum_a F^a_{\mu\nu} F^{\mu\nu\,a} = \tfrac{1}{2} \sum_a (\mathbf{E}^2_a - \mathbf{B}^2_a), \tag{8.2.1}$$

where, in the *temporal gauge* defined by $A^0_a = 0$,

$$\mathbf{E}_a = -\dot{\mathbf{A}}_a, \quad \mathbf{B}_a = \boldsymbol{\nabla} \times \mathbf{A}_a - \tfrac{1}{2} g f_{abc} \, \mathbf{A}_b \times \mathbf{A}_c. \tag{8.2.2}$$

The corresponding equations of motion,

$$\partial_t \mathbf{E}_a = \boldsymbol{\nabla} \times \mathbf{B}_a + g f_{abc} \mathbf{A}_b \times \mathbf{B}_c \equiv (\mathbf{D} \times \mathbf{B})_a, \tag{8.2.3}$$

imply

$$\partial_t G_a = 0, \quad G_a \equiv \boldsymbol{\nabla} \cdot \mathbf{E}_a + g f_{abc} \, \mathbf{A}_b \cdot \mathbf{E}_c \equiv (\mathbf{D} \cdot \mathbf{E})_a. \tag{8.2.4}$$

The operators G_a are called *Gauss (law) operators*. Since they are time-independent, they allow a sharp time restriction:

$$G_a(g, x_0) \equiv G_a(g, h), \quad g \in \mathcal{S}(\mathbf{R}^3), \ h \in \ S(\mathbf{R}), \ \int dt\, h(t) = 1.$$

Thus, even if the fields do not allow a sharp time restriction and canonical commutation relations are spoiled by the appearance of divergent renormalization constants, the equal-time commutators of G_a and the fields $\mathbf{A}$, $\mathbf{E}$ are well-defined operator valued distributions.

Such commutators are fixed by the condition, which can be taken as part of the definition of the temporal gauge (and follows from canonical quantization and gauge-invariant point-splitting regularization), that G_a generates the time-independent infinitesimal gauge transformations in the following sense:

$$-i\left[\mathbf{D} \cdot \mathbf{E}_a(\mathbf{x}, t), \, \mathbf{A}_b(\mathbf{y}, t)\right] = \delta_{ab} \, \boldsymbol{\nabla}\delta(\mathbf{x} - \mathbf{y}) + g f_{abc} \mathbf{A}_c(\mathbf{x}, t)\, \delta(\mathbf{x} - \mathbf{y}). \tag{8.2.5}$$

$$-i\left[\mathbf{D} \cdot \mathbf{E}_a(\mathbf{x}, t), \, \mathbf{E}_b(\mathbf{y}, t)\right] = g f_{abc} \, \mathbf{E}_c(\mathbf{x}, t)\, \delta(\mathbf{x} - \mathbf{y}). \tag{8.2.6}$$

In fact, if $\delta^{\Lambda, a}$ denotes the infinitesimal time-independent gauge transformation in the a-direction, with gauge function $\Lambda(\mathbf{x}) \in \mathcal{S}(\mathbf{R}^3)$, by the time-independence of $G_a(\Lambda, t)$, the above equations imply for the smeared fields $\mathbf{A}_b(f)$, $f \in \mathcal{S}(\mathbf{R}^4)$,

$$\delta^{\Lambda, a} \mathbf{A}_b(f) = \delta_{ab} \, (\boldsymbol{\nabla}\Lambda, f) + g f_{abc} \mathbf{A}^c(\Lambda f) = -i\left[G_a(\Lambda, t), \, \mathbf{A}_b(f)\right]. \tag{8.2.7}$$

Since the variables A^0_a are missing in the Lagrangian, one cannot exploit the stationarity of the action with respect to them, and therefore one does not obtain the Gauss law $G_a = 0$. Actually, the Gauss law is incompatible with eqs. (8.2.5) and (8.2.6) and therefore with canonical quantization, and more crucially with the Gauss operator being the generator of the time-independent gauge transformations, eq. (8.2.7).

A proposed solution of this conflict, widely adopted in the literature and in textbook discussions of the temporal gauge, is to require the Gauss-law constraint as an operator equation on the (subspace of) physical states and, in particular, on the vacuum state. However, such a solution is not free of paradoxes and mathematical inconsistency. In fact, the vacuum expectation of eq. (8.2.5) gives zero on the left-hand side and non-zero on the right-hand side.

It has been suggested[17] to cope with this paradox by admitting that the vacuum vector is not normalizable. In our opinion, such a solution is not acceptable, because it does not allow a representation of a field algebra containing both the gauge-dependent and the gauge-independent fields, since the non-normalizability of the vacuum vector would also produce divergent expectations of the observable algebra (which in particular contains the identity). As we shall see, a mathematically acceptable solution for the Gauss-law constraint is to adopt Weyl quantization and use "non-regular representations" (see Chapter 7, Section 8.3).

For this purpose, we start by considering the field algebra generated by the polynomials of the smeared fields $A_a^i(f)$, $f \in \mathcal{S}(\mathbf{R}^4)$. We assume that powers of $A_a^i(x)$ and of their derivatives can be defined through a suitable point-splitting procedure (in particular G_a, $\mathbf{B}_a$ should be expressible in this way as "functions" of $\mathbf{A}_a$).

In order to simplify the book-keeping of the indices, we introduce the following notations and at the same time specify the general mathematical setting which defines the temporal gauge.

- T^a denotes the hermitean representation matrices of the Lie algebra of the gauge group, normalized so that $\text{Tr}\, T^a T^b = \delta_{ab}$.
- The *local fields* $A^i(x) \equiv \sum_a A_i^a(x) T^a$ are Lie algebra-valued distributions on Lie algebra-valued test functions $f^i(x) \equiv \sum_a f_a^i(x) T^a$, $f_a^i \in \mathcal{S}(\mathbf{R}^4)$ (the sum over repeated indices is understood), and transform covariantly under spacetime translations α_y, $y \in \mathbf{R}^4$,

$$\mathbf{A}(\mathbf{f}) \equiv \int d^4x \text{Tr}[\,\mathbf{A}(x)\,\mathbf{f}(x)] = \int d^4x \sum_{i,a} A_a^i(x)\, f_a^i(x),$$

$$\alpha_y(A_a^i(f)) = A_a^i(f_y), \quad f_y \equiv f(x - y)$$

- $\mathcal{F}$ denotes the *local field algebra* (as Borchers field algebra) generated by the fields $\mathbf{A}(\mathbf{f})$ and by their functions $G_a, \mathbf{B}_a$.
- The (residual) *time-independent gauge transformations* $\alpha_\mathcal{U}$ are labeled by **gauge functions** $\mathcal{U}(\mathbf{x})$, which are group-valued unitary C^∞ functions and may be taken to differ from the identity only on a compact set $\mathcal{K}_\mathcal{U}$ ($\alpha_\mathcal{U}$ **localized**); the spacetime support of $\mathcal{U} - \mathbf{1}$ is given by the cylinder $C_\mathcal{U} \equiv \mathcal{K}_\mathcal{U} \times \mathbf{R}$ and

$$\alpha_\mathcal{U}(\mathbf{A}(\mathbf{f})) = \mathbf{A}(\mathcal{U}\mathbf{f}\mathcal{U}^{-1}) + \mathcal{U}\partial\mathcal{U}^{-1}(\mathbf{f}), \tag{8.2.8}$$

$$\mathcal{U}\partial\mathcal{U}^{-1}(\mathbf{f}) \equiv i \int d^4x \,\, \text{Tr}[\sum_i \mathcal{U}(\mathbf{x})\partial^i\mathcal{U}^{-1}(\mathbf{x})\, f^i(x)],$$

[17] R. Jackiw, Topological investigations..., 1985; A. Bassetto, G. Nardelli, and R. Soldati, *Yang–Mills Theories in Algebraic Non-covariant Gauges*, World Scientific 1991.

– $\mathcal{U}^\lambda$, $\lambda \in \mathbf{R}$ denote the gauge functions which correspond to one-parameter subgroups of the *residual gauge group* $\mathcal{G}$. They are of the form $\mathcal{U}_g^\lambda(\mathbf{x}) = e^{i\lambda\, g(\mathbf{x})}$, with $g(\mathbf{x}) = \sum_a g_a(\mathbf{x})\, T^a$ a Lie algebra-valued function, infinitely differentiable and of compact support $(g_a \in \mathcal{D}(\mathbf{R}^3))$; in the sequel, the index g will in general be omitted. All gauge transformations of compact support in a neighborhood of the identity, in the C^∞ topology, are of this form and generate the **Gauss subgroup** $\mathcal{G}_0$.

– $\mathcal{F}_W$ denotes the *exponential field algebra* generated by the unitary operators $W(\mathbf{f})$, $f^i(x) = \sum_a f_a^i(x)\, T^a$, $f_a^i \in \mathcal{S}(\mathbf{R}^4)$, formally the exponentials $e^{i\mathbf{A}(\mathbf{f})}$, and by the unitary operators $V(\mathcal{U}^\lambda)$, representing $\mathcal{G}_0$, formally the exponentials of the Gauss operators

$$V(\mathcal{U}^\lambda) = e^{i\,\lambda\, G(g)}, \quad G(g) = \sum_a G_a(g_a), \quad g_a \in \mathcal{D}(\mathbf{R}^3), \qquad (8.2.9)$$

transforming covariantly under space translations. The generation of the dynamics α_t by local gauge-invariant Hamiltonians H_R implies the time-independence of the $V(\mathcal{U}^\lambda)$

$$(d/dt)\alpha_t(V(\mathcal{U}^\lambda)) = i \lim_{R\to\infty} [H_R, \alpha_t(V(\mathcal{U}^\lambda))] = 0.$$

– $\mathcal{F}_W(\mathcal{O}) \subset \mathcal{F}_W$ denotes the localized (strongly closed) subalgebra generated by fields smeared with test functions supported in the bounded region $\mathcal{O}$.

A representation of $\mathcal{F}_W$ also defines a representation of $\mathcal{F}$ only if it is *regular*, i.e., if (the representatives of) the field exponentials $W(\lambda \mathbf{f})$, $\lambda \in \mathbf{R}$, define weakly continuous one-parameter groups.

A vacuum state is invariant under spacetime translations, which are therefore implemented by unitary operators $U(a)$, $a \in \mathbf{R}^4$ in the corresponding representation. A state ω on the exponential field algebra $\mathcal{F}_W$, in particular a vacuum state, satisfies the **Gauss law** in exponential form if $\omega(V(\mathcal{U}^\lambda)) = 1$, equivalently if its representative vector Ψ_ω in the (GNS) Hilbert space $\mathcal{H}_\omega$ defined by the expectations $\omega(\mathcal{F}_W)$ satisfies $V(\mathcal{U}^\lambda)\, \Psi_\omega = \Psi_\omega$, $\forall \mathcal{U}^\lambda$. Briefly, a state vector Ψ satisfying $V(\mathcal{U}^\lambda)\, \Psi_\omega = \Psi_\omega$, $\forall \mathcal{U}^\lambda$, is said to be **Gauss-invariant**. An operator in $\mathcal{H}$ is *Gauss-invariant* if it commutes with all the $V(\mathcal{U}^\lambda)$.

Proposition 2.1 *A vacuum state ω on the exponential field algebra $\mathcal{F}_W$, satisfying the Gauss law, defines a* **non-regular representation** *of $\mathcal{F}_W$, since*

$$\omega(W(\mathbf{f})) = 0, \quad \textit{if } \mathbf{f} \neq 0, \qquad (8.2.10)$$

$$= 1, \quad \textit{if } \mathbf{f} = 0.$$

The fields $\mathbf{A}(\mathbf{f})$, formally the generators of the $W(\mathbf{f})$, cannot be defined in the (GNS) Hilbert space $\mathcal{H}_\omega$ defined by ω, and in particular the two-point function of $\mathbf{A}$ does not exist, only (the vacuum expectations of) the exponential functions (and of course the gauge-invariant functions) of $\mathbf{A}$ can be defined.

In the free case, i.e., for $g = 0$, the exponential field algebra becomes a Weyl field algebra, generated by the exponentials of $\mathbf{A}_a$ and of its conjugate momenta

$\mathbf{E}_a$, *and eqs. (8.2.10) uniquely determine its representation as a* **non-regular Weyl quantization**.

Proof. For each $f^i = \sum_a f^a_i T^a$ there is a one-parameter subgroup $\mathcal{U}^\lambda$ such that $\mathcal{U}^\lambda \mathbf{f}\, \mathcal{U}^{-\lambda} = \mathbf{f}$, and $\exp i D_{\mathcal{U}^\lambda}(\mathbf{f}) \equiv \exp\left(\mathcal{U}^\lambda \partial\, \mathcal{U}^{-\lambda}(\mathbf{f})\right) \neq 1$; therefore, by eq. (8.2.8), one has

$$\omega(W(\mathbf{f})) = \omega(V(\mathcal{U}^\lambda)\, W(\mathbf{f})\, V(\mathcal{U}^\lambda)^*) = e^{i D_{\mathcal{U}^\lambda}(\mathbf{f})}\, \omega(W(\mathbf{f})),$$

and eqs. (8.2.10) follow.

Clearly, the one-parameter groups defined by $W(\mathbf{f})$ cannot be weakly continuous, and therefore the corresponding generators do not exist as operators in $\mathcal{H}_\omega$. In particular the "gluon" propagator does not exist. The free case can be worked out as for the abelian case.[18]

Remark 1. The one-parameter groups $V(\mathcal{U}^\lambda)$ are not assumed to be weakly continuous in λ; actually, continuity cannot hold if the global gauge group is simple and has rank at least two (as in the case of color $SU(3)$), since then one obtains the vanishing of $\omega(W(\mathbf{f})V(\mathcal{U}^\lambda)W(-\mathbf{f}))$, for $\lambda \neq 0$, $f^i_c(x) = \delta_{ca}f^i(x)$, $\mathcal{U}^\lambda f^i \mathcal{U}^{-\lambda} = f^i(x)T^b$, $[T^a, T^b] = 0$, from the invariance of ω under the subgroup generated by T^a, as in the proof of Proposition 2.1. Thus, in this case the Gauss law constraint can only be imposed in the exponential form.

Remark 2. In the following we will consider the realization of the temporal gauge defined by a vacuum state which is Gauss-invariant. The non-regularity of the exponential field algebra is shared by interesting quantum-mechanical models, like the electron in a periodic potential (Bloch electron), the quantum particle on a circle, the Quantum Hall electron, and so on, similarly implied by the invariance of the ground state under a group of gauge transformations.

The Hilbert space $\mathcal{H}$ of the representation of $\mathcal{F}_W$ defined by a Gauss-invariant vacuum state ω contains a subspace $\mathcal{H}'$ of Gauss-invariant vectors

$$V(\mathcal{U}^\lambda)\,\Psi = \Psi, \quad \forall \Psi \in \mathcal{H}', \quad \forall \mathcal{U}^\lambda. \tag{8.2.11}$$

It is worth remarking that the local operator $v(\mathcal{U}^\lambda) \equiv V(\mathcal{U}^\lambda) - 1$ is non-zero in $\mathcal{H}$, since $V(\mathcal{U}^\lambda)$ implements the time-independent gauge transformations, corresponding to the one-parameter subgroups, which are non-trivial on $\mathcal{F}_W$. Thus, the assumptions of the Reeh–Schlieder theorem, according to which a local operator which annihilates the vacuum must vanish, cannot be satisfied. One can easily check that the crucial point in the proof of the theorem fails, namely $\mathcal{F}_W(\mathcal{O})\,\Psi_0$, where $\mathcal{F}_W(\mathcal{O})$ is the exponential field algebra localized in the region $\mathcal{O}$, is not dense in $\mathcal{H}$. In fact, for any $\mathcal{O}$ disjoint from $C_{\mathcal{U}^\lambda}$, by locality one has

$$(v(\mathcal{U}^\lambda)\,\mathcal{F}_W\Psi_0,\, \mathcal{F}_W(\mathcal{O})\Psi_0) = (\mathcal{F}_W\Psi_0,\, \mathcal{F}_W(\mathcal{O})v(\mathcal{U}^\lambda)^*\,\Psi_0) = 0.$$

---

[18] J. Löffelholz, G. Morchio, and F. Strocchi, Jour. Math. Phys. **44**, 5095 (2003).

The fact that $\mathcal{F}_W(\mathcal{O})\Psi_0$ is not dense, even if locality holds, is due to the failure of the relativistic spectral condition.[19]

2.2   Topology of the gauge group

By definition, eq. (8.2.8), the gauge functions considered are "localized", and therefore they obviously extend to the one-point compactification of $\mathbf{R}^3$, $\dot{\mathbf{R}}^3$, which is isomorphic to the three-sphere S^3; i.e.,

$$\mathcal{U}(\mathbf{x}) : \dot{\mathbf{R}}^3 \sim S^3 \to \mathcal{G}.$$

Such maps fall into disjoint homotopy classes labeled by "winding" numbers n:

$$n(\mathcal{U}) = (24\pi^2)^{-1} \int d^3x\, \varepsilon^{ijk}\, \mathrm{Tr}[\mathcal{U}_i(\mathbf{x})\,\mathcal{U}_j(\mathbf{x})\,\mathcal{U}_k(\mathbf{x})] \equiv \int d^3x\, n_\mathcal{U}(\mathbf{x}), \qquad (8.2.12)$$

where $\mathcal{U}_i(\mathbf{x}) \equiv \mathcal{U}(\mathbf{x})^{-1}\, \partial_i\, \mathcal{U}(\mathbf{x})$. Hence, the gauge transformations may be labeled by their winding number n.

Those with $n \neq 0$ are called *large gauge transformations*. Those with zero winding number $\mathcal{U}_0 \equiv \mathcal{U}_{n=0}$ are called *small*; since they are contractible to the identity, they are products of $\mathcal{U}(\mathbf{x})$ which are close to the identity (in the C^∞ topology) and therefore are expressible as products of $\mathcal{U}^\lambda$.[20]

The so-obtained topological classification and all the following results do not change if one considers gauge functions $\mathcal{U}(\mathbf{x})$, which are only required to have a limit for $|\mathbf{x}| \to \infty$ independent of the direction; in fact, the corresponding transformations differ from those of compact support by transformations $\mathcal{U}_0$ (i.e., with $n = 0$) with the above behavior at infinity. The equivalence is then guaranteed by the existence of the corresponding (extended) Gauss operators $V(\mathcal{U}_0)$.

By definition of the temporal gauge, the small gauge transformations $\alpha_{\mathcal{U}_0}$ are implemented by the operators $V(\mathcal{U}^\lambda)$, and the next question is the implementability of the large gauge transformations and their distinction from the small ones on the Gauss-invariant states. Actually, the non-triviality of the large gauge transformations on the physical space turns out to be a rather subtle question (by the following Proposition).

---

[19] In fact, the implementers $U(\mathbf{a})$ of the space translations are not weakly continuous, since, as in the proof of Proposition 2.1,

$$\omega(W(\mathbf{f})U(\mathbf{a})W(-\mathbf{f})) = \omega(W(\mathbf{f})\,W(-\mathbf{f_a})) =$$

$$= e^{iD_{\mathcal{U}^\lambda}(\mathbf{f}-\mathbf{f_a})}\, \omega(W(\mathbf{f})\,W(-\mathbf{f_a})),$$

so that the right-hand side vanishes if $\mathbf{a} \neq 0$, and is $= 1$ otherwise. The spectral condition for the Fourier transforms of the matrix elements of $U(\mathbf{a}, t) = U(\mathbf{a})\, U(t)$ is violated if there is strong continuity in t, since it would imply that after smearing in time, the Fourier transform in $\mathbf{a}$ is a finite measure, and therefore one has continuity in $\mathbf{a}$ of the above matrix elements.

[20] For the geometry of the gauge transformations, see S. Coleman, *Aspects of Symmetry*, Cambridge University Press 1985, Chapter 7, Sect. 3; R. Jackiw, Topological investigations..., 1985; S. Weinberg, *The Quantum Theory of Fields. Vol II*, Cambridge University Press 1996, Sect. 23.4; T. Frankel, *The Geometry of Physics. An Introduction*, 2nd ed., Cambridge University Press 2004.

Proposition 2.2   *A local field operator invariant under Gauss gauge transformations is also invariant under large gauge transformations.*

A vector $\Psi \in \mathcal{H}$, *satisfying the Gauss law in exponential form, in particular the vacuum vector* Ψ_0, *defines a state, i.e., expectations, on* $\mathcal{F}_W$ *invariant also under the large transformations.*

Proof.   In fact, if $\mathcal{U}_n(\mathbf{x})$ and $\mathcal{U}_n^a(\mathbf{x}) = \mathcal{U}_n(\mathbf{x} - \mathbf{a})$ denote a gauge function and its space translated one by $\mathbf{a}$, the combined gauge transformations $\alpha_{\mathcal{U}_n^a}^{-1}\,\alpha_{\mathcal{U}_n}$ and $\alpha_{\mathcal{U}_n}\,\alpha_{\mathcal{U}_n^a}^{-1}$ have zero winding number and therefore are small gauge transformations, say $\alpha_{\mathcal{U}_0}$ and $\alpha_{\mathcal{U}_0'}$ respectively. Then, for any local field operator F invariant under small gauge transformations, one has

$$\alpha_{\mathcal{U}_n}(F) = \alpha_{\mathcal{U}_n^a}\,\alpha_{\mathcal{U}_0}(F) = \alpha_{\mathcal{U}_n^a}(F),$$

and for $|\mathbf{a}|$ sufficiently large, by locality $\alpha_{\mathcal{U}_n^a}(F) = F$.

Quite generally, for any (local) operator $F \in \mathcal{F}_W$ one has, for $|\mathbf{a}|$ sufficiently large, $\alpha_{\mathcal{U}_n}(F) = \alpha_{\mathcal{U}_0'}\,\alpha_{\mathcal{U}_n^a}(F) = \alpha_{\mathcal{U}_0'}(F)$; then, if a state ω is invariant under Gauss transformations, one has $\omega(\alpha_{\mathcal{U}_n}(F)) = \omega(F)$.

By a standard argument, the invariance of the vacuum under the large gauge transformations implies that they are implemented by unitary operators; they can be chosen to yield a representation of the gauge group $\mathcal{G}$, to coincide with the $V(\mathcal{U}^\lambda)$, for all Gauss transformations, and to transform covariantly under space translations. Furthermore, the implementers commute with the time translations as a consequence of the gauge invariance of the local Hamiltonians H_R.

The implementers are unique, up to multiples of the identity, if the field algebra is irreducible in $\mathcal{H}$; in this case, one can show that the implementers of the large gauge transformations are multiples of the identity in $\mathcal{H}'$ (see Section 2.5 below).

An important question is the implication, if any, of the non-trivial topology of the gauge group on the physical states—more generally, on the representations of the observable algebra.

One of the standard (and historically the first) arguments in favor of a non-trivial role of the topology of the gauge group relies on semiclassical approximations, in terms of boundary conditions of the field configurations and their effect on the (Euclidean) functional integral.

This approach has been excellently reviewed by Coleman,[21] who also pointed out the limitation of such arguments, since smooth field configurations, in particular those of finite action, have zero functional measure. The relevance of the classification of the (smooth) field configurations in terms of their pure gauge behavior at space infinity, and therefore of the corresponding winding number, is therefore a non-trivial mathematical problem, especially in the infinite volume limit.[22]

---

[21] S. Coleman, *Aspects of Symmetry*, Cambridge University Press 1985, Chapter 7, Sect. 3. See also S. Weinberg, *The Quantum Theory of Fields. Vol. II. Modern Applications*, Cambridge University Press 1996, Sect. 23.6.

[22] The topological classification is done in finite volume, but in the infinite volume limit a non-trivial instanton density implies that in the dilute gas approximation, the functional measure is concentrated on configurations with divergent topological number. The association of the non-trivial topology with the existence of instanton solutions, which minimize the classical action and as such

An important step in the direction of a non-perturbative argument was taken by R. Jackiw,[23] by directly exploiting the topology of the gauge group, without reference to the classification of the field configurations entering in the functional integral.

The basic ingredients in Jackiw analysis are the use of the temporal gauge and the transformation properties of the topological charge under large gauge transformations. We have already commented on the fine mathematical points, which have to be correctly dealt with in order to obtain a mathematically consistent argument in the temporal gauge. We shall now discuss the mathematical problems of the topological charge.

The so-called *topological current* is formally defined by

$$C^\mu(x) = -(16\pi^2)^{-1}\varepsilon^{\mu\nu\rho\sigma}\mathrm{Tr}[F_{\nu\rho}(x)\,A_\sigma(x) - \tfrac{2}{3}A_\nu(x)\,A_\rho(x)\,A_\sigma(x)]. \tag{8.2.13}$$

$$\partial_\mu C^\mu(x) = -(16\pi^2)^{-1}\,\mathrm{Tr}[{}^*F_{\mu\nu}(x)\,F^{\mu\nu}(x)] \equiv \mathcal{P}(x),$$

where $A_\mu = (0, A_i)$, ${}^*F_{\mu\nu} \equiv \tfrac{1}{2}\varepsilon_{\mu\nu\rho\sigma}\,F^{\rho\sigma}$. In the mathematical literature for classical fields, $\mathcal{P}$ is called the "Pontryagin density" and C_μ the "Chern–Simons secondary characteristic class".

At the classical level one can easily prove the following transformation law of $C_0(x)$ under gauge transformations $\alpha_{\mathcal{U}}$, defined by eq. (8.2.8):

$$\alpha_{\mathcal{U}}(C^0(x)) = C^0(x) - (8\pi^2)^{-1}\partial_i\,\varepsilon^{ijk}\mathrm{Tr}[\partial_j\mathcal{U}(\mathbf{x})\,\mathcal{U}(\mathbf{x})^{-1}A_k(x)] + n_{\mathcal{U}}(\mathbf{x}), \tag{8.2.14}$$

($n_{\mathcal{U}}(\mathbf{x})$ defined by eq. (8.2.12)). Therefore, at the classical level the space integral of $C_0(\mathbf{x}, x_0)$ is invariant under small transformations, but it is shifted by n under gauge transformations with winding number n.

In the quantum case one meets non-trivial mathematical problems. First of all, the formal expression in the right-hand side of eq. (8.2.13) requires a point-splitting regularization. It is very reasonable to assume that this can be achieved by keeping the transformation properties of the formal expression under large gauge transformations, eq. (8.2.14).

The next problem is the space integral of $C_0(x)$; namely, the existence of the topological charge. The space integrals of charge densities, even for conserved currents, are known to diverge, and suitable regularizations are needed, including a time smearing (as discussed in Chapter 7, Sections 2 and 5.3).

In the case of conserved currents, under some general conditions one may obtain the convergence of a suitably regularized integral of the charge density, in matrix elements on states with suitable localization[24]; but in the general case of non-conserved currents the problem is seriously open.

Actually, in the quantum case an even more serious problem arises as a consequence of the gauge-dependence of $C_\mu(x)$. By Proposition 2.1, the regularized space integral of $C_0(x)$ (as in Chapter 7, Section 2)

do not have compact support, is probably the reason why localized large gauge transformations have not been considered in the literature.

[23] R. Jackiw, Topological Investigations of Quantized Gauge Theories, 1985.

[24] B. Schroer and P. Stichel, Comm. Math. Phys. **3**, 258 (1966); H. D. Maison, Nuovo Cim. **11A**, 389 (1972); M. Requardt, Comm. Math. Phys. **50**, 259 (1976); G. Morchio and F. Strocchi, Jour. Math. Phys. **44**, 5569 (2003).

$$C^0(f_R \alpha_R) \equiv \int d^4x \, f_R(\mathbf{x}) \, \alpha_R(x_0) \, C^0(x),$$

where $f_R(\mathbf{x}) = f(|\mathbf{x}|/R)$, $f(x) = 1$, for $|x| \leq 1$, $= 0$, for $|x| \geq 1 + \varepsilon$, $\alpha_R(x_0) = \alpha(x_0/R)/R$, $\int dx_0 \, \alpha(x_0) = \tilde{\alpha}(0) = 1$, cannot exist as an operator in $\mathcal{H}$; only its formal exponentials $V^C(f_R \alpha_R) \sim \exp iC^0(f_R \, \alpha_R)$ may be defined. The transformation properties of such unitary operators under gauge transformations reflect those of $\exp iC^0(f_R \, \alpha_R)$; in particular, for any f_R with $f_R = 1$ on $K_{\mathcal{U}}$, so that $f_R \, \partial_j \mathcal{U} = \partial_j \mathcal{U}$, $\partial_i f_R \partial_j \mathcal{U} = 0$, one has

$$\alpha_{\mathcal{U}}(V^C(f_R \alpha_R)) = e^{in\mathcal{U}} \, V^C(f_R \, \alpha_R). \tag{8.2.15}$$

Furthermore, as shown by the following Proposition 2.3, such exponentials have vanishing expectation on Gauss-invariant states, i.e., their restriction to the physical states vanishes.

Thus, the non-invariance of the formal space integral of $C^0(x)$ does not by itself allow the construction of an operator on $\mathcal{H}'$ invariant under the Gauss gauge transformations, but not invariant under the large gauge transformations. Then, a non-trivial action of the implementers of the large gauge transformations in $\mathcal{H}'$ cannot be derived merely by the non-invariance of the (non-existing) topological charge.

As we shall see in the following sections, one cannot ignore the non-regularity of $\exp i \, \lambda C_0(f_R \alpha_R)$, $\lambda \in \mathbf{R}$, its non-observability and the non-existence of the limit $R \to \infty$. As discussed in Section 2.5 below, a crucial additional ingredient is provided by the chiral symmetry and the associated conserved gauge-dependent current.

Proposition 2.3   *The operators $V^C(f_R \alpha_R)$, formally the exponentials of the regularized space integrals $C_0(f_R \alpha_R)$ of $C_0(x)$, and therefore assumed to transform under gauge transformations as such exponentials, are not weakly continuous in λ, and therefore the field $C_0(f_R \alpha_R)$ cannot be defined.*

Furthermore, for all Gauss-invariant vectors Ψ, Φ, one has

$$(\Psi, \, V^C(f_R \alpha_R) \, \Phi) = 0. \tag{8.2.16}$$

Proof.   In fact, if $C_0(f)$, $f \in \mathcal{D}(\mathbf{R}^4)$, exists, by using the Gauss gauge invariance of the vacuum state ω, the vanishing of $\omega(A_k)$ by rotational invariance and eq. (8.2.14), one has

$$\omega(C_0(f)) = \omega(V(\mathcal{U}^\lambda) \, C_0(f) V(\mathcal{U}^\lambda)^{-1}) =$$

$$= \omega(C_0(f)) + \int d^4x f(\mathbf{x}, t) \, n_{\mathcal{U}^\lambda}(\mathbf{x}).$$

Since for any f there is at least one $\mathcal{U}^\lambda(\mathbf{x})$ such that the last term on the right-hand side does not vanish, one obtains a contradiction. Thus, only the exponential of $C_0(f)$ can be defined.

Moreover, given f_R one can find a small gauge transformation $\mathcal{U}_0(\mathbf{x})$ and localized $\mathcal{U}_n, \mathcal{U}_{-n}$, such that

$$\mathcal{U}_0(\mathbf{x}) = \mathcal{U}_n(\mathbf{x})\,\mathcal{U}_{-n}(\mathbf{x}), \quad n \neq 0,$$

$$f_R\left(\mathcal{U}_{-n} - 1\right) = 0, \quad f_R\left(\mathcal{U}_n - 1\right) = \mathcal{U}_n - 1.$$

Then, $\partial_i f_R \partial_j \mathcal{U}_{-n} = 0$, $\partial_i f_R\,\partial_j \mathcal{U}_n = 0$ and the second term on the right-hand side of eq. (8.2.14) vanishes; furthermore, $\int d^3x\, n_{\mathcal{U}_0}(\mathbf{x}) f_R(\mathbf{x}) = n$. Hence, one has

$$\left(\Psi,\, V^C(f_R \alpha_R)\,\Phi\right) = \left(\Psi,\, V(\mathcal{U}_0)\, V^C(f_R \alpha_R)\, V(\mathcal{U}_0)^{-1}\,\Phi\right) =$$

$$= e^{in}\left(\Psi,\, V^C(f_R \alpha_R)\,\Phi\right),$$

and eq. (8.2.16) follows.

As we shall see in the following sections, the presence of fermions and the related chiral symmetry allows us to display the physical relevance of the topology of the gauge group, and in particular its role in classifying the irreducible (or factorial) representations of the gauge-invariant algebra.

2.3 Fermions and chiral symmetry

The situation changes substantially in the presence of massless fermions, since the role of the non-conserved topological current is taken by a *conserved* current; hence, there is a symmetry associated with it, and the relevant point is its relations with the large gauge transformations.

In this case, the Lagrangian of eq. (8.2.1) becomes modified by the addition of the (gauge-invariant) fermion Lagrangian, eq. (7.1.17) with $m = 0$, and the Gauss operators become

$$G_a = (\mathbf{D}\cdot\mathbf{E})_a - j_a^0, \quad j_\mu^a = ig\bar{\psi}\gamma_\mu t^a \psi.$$

The time-independent gauge transformations of the fermion fields in the fundamental representation of the gauge group are

$$\alpha_{\mathcal{U}}(\psi(x)) = \mathcal{U}(\mathbf{x})\psi(x). \tag{8.2.17}$$

At the classical level the Lagrangian is invariant under the one-parameter group of *chiral transformations* β^λ, $\lambda \in \mathbf{R}$,

$$\beta^\lambda(\psi) = e^{\lambda\gamma_5}\,\psi, \quad \beta^\lambda(\bar{\psi}) = \bar{\psi}\,e^{\lambda\gamma_5}, \quad \gamma_5^* = -\gamma_5, \quad \beta^\lambda(\mathbf{A}) = \mathbf{A}. \tag{8.2.18}$$

Correspondingly, there is a conserved current $j_\mu^5 = ig\bar{\psi}\gamma^5\gamma_\mu\psi$—the gauge-invariant fermion axial current.

In the quantum case, as we have seen in the abelian case (Chapter 4, Section 6.2), a gauge-invariant point-splitting regularization for the definition of j_μ^5 inevitably leads to an *anomaly*,[25]

$$\partial^\mu j_\mu^5 = -2\partial^\mu C_\mu = -2\mathcal{P}.$$

---

[25] For a general review of this phenomenon, see R. Jackiw, Field Theoretical Investigations in Current Algebra, in S. B. Treiman, R. Jackiw, B. Zumino, and E. Witten, *Current Algebra and Anomalies*, World Scientific 1985, pp. 81–210.

The conserved axial current is now the gauge-dependent current

$$J_\mu^5(x) = j_\mu^5(x) + 2C_\mu,$$

its conservation being equivalent to the anomaly equation for j_μ^5.

For the discussion of the Weyl quantization, we take as local exponential field algebra $\mathcal{F}_W$ the algebra generated by the operators $W(\mathbf{f})$, by the Gauss operators $V(\mathcal{U}^\lambda)$ (see eq. (8.2.9)), by the gauge-invariant bilinear functions of the fermion fields, and by the unitary operators $V^5(f)$, formally given by $\exp i\, J_0^5(f)$. As before, and as is standard, the local field algebras $\mathcal{F}_W(\mathcal{O})$ are taken strongly closed.

As shown by Bardeen[26] on the basis of perturbative renormalization in local gauges, the chiral transformations are generated by the conserved gauge-dependent current J_μ^5 (not by j_μ^5); the continuity equation of J_μ^5 plays a crucial role in Bardeen analysis.

Bardeen analysis justifies our assumption that the exponential field algebra contains the unitary operators $V^5(f)$ and their role in the local generation of the (time-independent) chiral symmetry β^λ.

By the proof of Proposition 2.3, the one-parameter group of unitary operators $V_R^5(\lambda) \equiv V^5(\lambda\, f_R \alpha_R)$ are not weakly continuous in λ, since this would imply the existence of its generator

$$J_0^5(f_R \alpha_R) = j_0^5(f_R \alpha_R) + 2C_0(f_R \alpha_R),$$

which is excluded by the non-existence of $C_0(f_R \alpha_R)$.

However, $V_R^5(\lambda)$ act as local implementers of β^λ on the local field algebra $\mathcal{F}_W$. In fact, since J_μ^5 is conserved, most of the standard wisdom is available; in particular, the commutation relations of $V^5(\lambda)$ with the local fields are governed by the canonical (anti)commutation relations[27] and, as a result of Bardeen analysis, one may write

$$\lim_{R\to\infty} V_R^5(\lambda)\, F\, V_R^5(-\lambda) = \beta^\lambda(F), \quad \forall F \in \mathcal{F}_W. \tag{8.2.19}$$

It is important to stress that due to locality the above limit is reached for finite values of R, and that it preserves locality and gauge invariance. Thus, contrary to statements that appeared in the literature, the presence of the chiral anomaly does not prevent the chiral symmetry from being a well-defined time-independent automorphism of the field algebra $\mathcal{F}_W$ and of its observable subalgebra $\mathcal{A}$.

The loss of chiral symmetry is therefore a genuine phenomenon of spontaneous symmetry breaking, and the confrontation with the Goldstone theorem becomes a crucial issue: the so-called $U(1)$ problem.

For the analysis of the interplay between large gauge transformations and chiral symmetry, we note that the gauge transformations of $\mathcal{F}_W$ are given by eqs. (8.2.8), together with the following relation formally reflecting the transformation properties of $C_0(f)$ and the gauge-invariance of j_0^5: for R large enough so that $f_R(\mathbf{x}) = 1$ on the localization region of $\mathcal{U}_n(\mathbf{x})$, eq. (8.2.15) gives

[26] W. A. Bardeen, Nucl. Phys. **B 75**, 246 (1974).
[27] G. Morchio and F. Strocchi, Jour. Phys. A: Math. Theor. **40**, 3173 (2007).

$$\alpha_{\mathcal{U}_n}(V_R^5(\lambda)) = e^{i\lambda\, 2n}\, V_R^5(\lambda). \tag{8.2.20}$$

It is worth stressing that eq. (8.2.20) is merely a consequence of the localizability of the large gauge transformations, and the assumption that $V_R^5(\lambda)$ has the same transformation properties under them as the formal exponential $\exp i\,\lambda\, J_0^5(f_R\alpha_R)$. It codifies the *crucial consequence of the axial anomaly*, at the local level; namely, that the implementers of the large gauge transformations (which exist by Proposition 2.2) do not commute with the local implementers of the chiral transformations.

2.4   Solution of the $U(1)$ problem

As mentioned in Section 1, the absence of parity doublets requires that the chiral symmetry be broken, and the $U(1)$ problem amounts to explaining the absence of the corresponding Goldstone massless bosons.

Now, as discussed above, eq. (8.2.19) implies the existence of a one-parameter group of automorphisms of the algebra of observables, which commute with space and time translations, so that one of the basic assumptions of the Goldstone theorem is satisfied.

The second crucial property, needed for the proof of the theorem,[28] is the local generation of the symmetry by a conserved current, at least in expectations on the vacuum state; i.e., for any local field $A \in \mathcal{F}_W$,

$$<\delta^5 A> \equiv \frac{d}{d\lambda}<\beta^\lambda(A)>|_{\lambda=0} = i\lim_{R\to\infty}<[\,J_0^5(f_R\alpha_R),\, A\,]>. \tag{8.2.21}$$

Since the chiral automorphism β^λ is C^∞ in the group parameter λ, $\delta^5 A$ is a well-defined derivation on $\mathcal{F}_W$, and the problem is its relation with the (formal) generator of the one-parameter group of unitary operators $V_R^5(\lambda)$, which locally implement the symmetry on $\mathcal{F}_W$. The non-regularity of $V_R^5(\lambda)$, implied by Proposition 2.3, prevents the existence of correlation functions of $J_0^5(f_R\alpha_R)$ on a Gauss-invariant state; then, one cannot write the symmetry breaking Ward identities (8.2.21) and obtain the Goldstone energy–momentum spectrum of $<J_0^5(x)\,A>$. The eq. (8.2.20) provides a direct proof, which applies to any gauge, provided that eq. (8.2.20) holds and the expectations are defined by a gauge-invariant vacuum, e.g., a θ vacuum (see below).

Proposition 2.4   *If ω is a vacuum state invariant under Gauss gauge transformations, and A is a gauge-invariant symmetry breaking order parameter, i.e., $\omega(\beta^\lambda(A)) \neq \omega(A) \neq 0$ (the standard candidate being $\bar{\psi}\psi$), then the expectations $\omega(J_0^5(x)\,A)$ cannot be defined, and eq. (8.2.21) does not hold.*

Proof.   By Proposition 2.2, ω is also invariant under the large gauge transformations, and if the expectations $\omega(J_0^5(f_R\alpha_R)\,A)$ were defined, by eq. (8.2.20), for R sufficiently large, one would produce a contradiction

---

[28] For a discussion of the role of this property, see, e.g., F. Strocchi, *Symmetry Breaking*, 2nd ed., Springer 2008, esp. Part II, Chapter 15.

$$\omega(J_0^5(f_R\alpha_R)\,A) = \omega(\alpha_{\mathcal{U}_n}(J_0^5(f_R\alpha_R)\,A)) = \omega(J_0^5(f_R\alpha_R)\,A) + 2n\,\omega(A). \qquad (8.2.22)$$

Clearly, by Proposition 2.2, the above proposition applies to the Gauss-invariant state Ψ_0. The impossibility of writing expectations involving J_μ^5 on a gauge-invariant vacuum state, like a θ vacuum, solves the problems raised by R. J. Crewther in his analysis of chiral Ward identities.[29]

It is worthwhile to remark that for the evasion of the Goldstone theorem discussed above, the occurrence of the so-called chiral anomaly (which is present also in the abelian case) is not enough; the crucial ingredient is eq. (8.2.20), which directly implies the non-regularity of the unitary operators $V_R^5(\lambda)$ and the non-existence of the local charges $J_0^5(f_R\alpha_R)$ generating the chiral symmetry, in expectations on a gauge-invariant vacuum state.

2.5 Topology and vacuum structure

The solution of the $U(1)$ problem discussed above does not involve the θ structure of the vacuum, the Gauss invariance of the vacuum state being enough. Further information on the breaking of chiral symmetry can be added by further exploiting eq. (8.2.20).

The invariance of the vacuum under large gauge transformations (Proposition 2.2) implies their implementation by unitary operators $V(\mathcal{U}_n)$. The first step is to characterize the implementers of the abelian group $\mathcal{T}$ defined by $\mathcal{G}$ modulo $\mathcal{G}_0$.

Proposition 2.5 *The implementers $V(\mathcal{U})$ of the gauge transformations*
i) have the following form:

$$V(\mathcal{U})\,F\Psi_0 = T(\mathcal{U})\,\alpha_{\mathcal{U}}(F)\Psi_0, \quad [T(\mathcal{U}),\,F]=0, \quad \forall F \in \mathcal{F}_W, \qquad (8.2.23)$$

with $T(\mathcal{U})^* = T(\mathcal{U})^{-1}$, and

$$V(\mathcal{U})\mathcal{H}' \subset \mathcal{H}', \quad V(\mathcal{U})\Psi = T(\mathcal{U})\Psi, \quad \forall\Psi \in \mathcal{H}'; \qquad (8.2.24)$$

ii) the operators $T(\mathcal{U})$ depend only on the equivalence class of $\mathcal{U}$, are gauge-invariant, commute with the spacetime translations, and define an abelian group,

$$T(\mathcal{U}_n) = T(\mathcal{U}_n') \equiv T_n, \quad T_n\,T_m = T_{n+m}; \qquad (8.2.25)$$

iii) if the representation of the field algebra is irreducible, the T_n are multiples of the identity, and $T_n\Psi_0 = e^{i2n\theta}\,\Psi_0$, $\theta \in [0,\pi)$.

Proof. i) Given $\alpha_{\mathcal{U}_n}$, the transformation $\alpha_{\mathcal{U}_n}\alpha_{\mathcal{U}_n^a}^{-1}$, $\mathcal{U}_n^a(\mathbf{x}) \equiv \mathcal{U}(\mathbf{x} - \mathbf{a})$, has zero winding number and therefore is implementable by an element $V^{(a)}$ of the Gauss group, which leaves the vacuum invariant. Then, $\forall F \in \mathcal{F}_W$, by locality

[29] R. J. Crewther, Chiral properties on quantum chromodynamics, in *Field Theoretical Methods in Particle Physics*, W. Rühl (ed.), Reidel 1980, pp. 529–90. The impossibility of writing the Ward identities for chiral symmetry breaking on a gauge-invariant state has been explicitly checked in the Schwinger model; see F. Strocchi, *Selected Topics on the General Properties of Quantum Field Theory*, World Scientific 1993, Chapter 7, Section 7.4, (iv).

$$\alpha_{\mathcal{U}_n}(F) = \lim_{|\mathbf{a}| \to \infty} V^a F (V^a)^{-1} \text{ and}$$

$$V^{(a)} F \Psi_0 = V^{(a)} F (V^{(a)})^{-1} \Psi_0 = \alpha_{\mathcal{U}_n}\, \alpha_{\mathcal{U}_n^a}^{-1}(F)\Psi_0$$

converges strongly, as $|\mathbf{a}| \to \infty$, to $\alpha_{\mathcal{U}_n}(F)\Psi_0$. Hence,

$$S(\mathcal{U}_n) \equiv \mathrm{s} - \lim_{|\mathbf{a}| \to \infty} V^{(a)}$$

satisfies

$$S(\mathcal{U}_n)\Psi_0 = \Psi_0, \quad S(\mathcal{U}_n) F S(\mathcal{U}_n)^{-1} = \alpha_{\mathcal{U}_n}(F).$$

Clearly, $T(\mathcal{U}_n) \equiv V(\mathcal{U}_n)\, S(\mathcal{U}_n)^{-1}$ commutes with $\mathcal{F}_W$,

$$T(\mathcal{U}_n) F T(\mathcal{U}_n)^{-1} = \alpha_{\mathcal{U}_n}(S(\mathcal{U}_n)^{-1} F\, S(\mathcal{U}_n)) = F,$$

and eqs. (8.2.23) hold.

Since $V(\mathcal{U}^\lambda)V(\mathcal{U}_n) = V(\mathcal{U}_n)V((\mathcal{U}')^\lambda)$, one has, $\forall \Psi \in \mathcal{H}'$,

$$V(\mathcal{U}^\lambda)V(\mathcal{U}_n)\Psi = V(\mathcal{U}_n)V((\mathcal{U}')^\lambda)\Psi = V(\mathcal{U}_n)\Psi,$$

and eqs. (8.2.24) hold.

ii) Since $T(\mathcal{U}_n)$ commutes with $\mathcal{F}_W$, its action (on $\mathcal{F}_W \Psi_0$) is completely determined by its action on Ψ_0. The products $V(\mathcal{U}_n')^{-1}V(\mathcal{U}_n)$ and $V(\mathcal{U}_n^a)^{-1}V(\mathcal{U}_n)$ define transformations implemented by elements of the Gauss group, which leave Ψ_0 invariant; then, $V(\mathcal{U}_n')^{-1}V(\mathcal{U}_n)\Psi_0 = \Psi_0$ and $V(\mathcal{U}_n^a)^{-1}V(\mathcal{U}_n)\Psi_0 = \Psi_0$ imply

$$V(\mathcal{U}_n')\Psi_0 = V(\mathcal{U}_n)\Psi_0 = T_n \Psi_0, \quad V(\mathcal{U}_n^a)\Psi_0 = U(\mathbf{a})T_n \Psi_0 = T_n \Psi_0.$$

The time-independence on the $V(\mathcal{U}_n)$ and of the T_n follows by the same argument used in Section 2.1 for the $V(\mathcal{U}^\lambda)$. Furthermore, since the product $T_m\, T_n$ is completely determined by its action on the vacuum and $V(\mathcal{U}_n)\Psi_0 = T_n \Psi_0 \in \mathcal{H}'$, by the second of eqs. (8.2.24) one has

$$T_m\, T_n \Psi_0 = V(\mathcal{U}_m)V(\mathcal{U}_n)\Psi_0 = V(\mathcal{U}_{n+m}')\Psi_0 = T_{n+m}\,\Psi_0.$$

Similarly, one has

$$\alpha_{\mathcal{U}_m}(T_n)\Psi_0 = V(\mathcal{U}_m)V(\mathcal{U}_n)(V(\mathcal{U}_m))^{-1}\,\Psi_0 = T_m\, T_n\, T_{-m}\,\Psi_0 = T_n \Psi_0.$$

iii) Since the T_n commute with $\mathcal{F}_W$, irreducibility implies that they are multiples of the identity; by eq. (8.2.25) and unitarity, $T_n \Psi_0 = e^{i2n\theta}\,\Psi_0$.

The relevant question is whether the angle θ has some physical meaning or whether, without loss of generality, one may put $\theta = 0$. More generally, the question is whether the angle θ labels inequivalent representations of the algebra of observables $\mathcal{A}$; the fact that θ belongs to the spectrum of the commutant of $\mathcal{F}_W$, and therefore of commutant of $\mathcal{A} \subset \mathcal{F}_W$, is not enough for the proof of inequivalence.

For this result a crucial role is played by chiral symmetry—in particular, by the chiral transformations of the $V(\mathcal{U}_n)$. One might think that such a relation is plainly given by eq. (8.2.20), but the problem is more subtle. Chiral transformations, defined on the field algebra $\mathcal{F}_W$ by eq. (8.2.18), are locally generated on $\mathcal{F}_W$ by the unitary operators $V_R^5(\lambda)$; however, the possibility of obtaining the chiral transformations of an operator by using such a local generation does not extend to generic operators in $\mathcal{H}$, even if they belong to the strong closure of $\mathcal{F}_W$ (unless the symmetry is unbroken). This is a general fact in local quantum field theory. Hence, the equation

$$V_R^5(\lambda)\, V(\mathcal{U}_n) V_R^5(-\lambda) = e^{-i2n\lambda}\, V(\mathcal{U}_n),$$

which follows from eq. (8.2.20), for R large enough, does not yield

$$\beta^\lambda(V(\mathcal{U}_n)) = e^{-i2n\lambda}\, V(\mathcal{U}_n). \tag{8.2.26}$$

Such an equation, which is the correct counterpart of the equation $[\,V(\mathcal{U}_n),\, Q^5\,] = 2n$, at the basis of most of the standard discussions of chiral symmetry breaking in QCD, is therefore missing, without further ingredients.

To this purpose one may appeal to the localization of the large gauge transformations and to the possibility of implementing them by *local* operators $V(\mathcal{U}_n)$, called *local implementers.*

This allows us to exploit, at the local level, the relation between chiral symmetry and large gauge transformations in a mathematically consistent way. In fact, due to the locality of the $V(\mathcal{U}_n)$, eq. (8.2.19) applies and gives

$$\beta^\lambda(V(\mathcal{U}_n)) = V_R^5(\lambda)\, V(\mathcal{U}_n)\, V_R^5(-\lambda) = e^{-i2n\lambda}\, V(\mathcal{U}_n), \tag{8.2.27}$$

for R large enough with respect to the localization region of $V(\mathcal{U}_n)$.

The non-trivial interplay between chiral symmetry and the topology of the gauge group, as displayed by eq. (8.2.27), codifies the effect of the chiral anomaly, namely, the generation of the chiral transformations by a gauge dependent current.

Eq. (8.2.26) has exactly the same form of the Weyl relation arising in the description of a quantum particle on a circle, with $V(\mathcal{U}_n)$ playing the role of $e^{i2n\pi p_\varphi}$ and $U^5(\lambda)$ playing the role of the formal exponential $e^{i\lambda\varphi}$.[30]

The next step for the derivation of the physical meaning of the angle θ is the proof that it belongs to the spectrum of the center of the observables. In fact, we will prove that the implementers $V(\mathcal{U}_n)$ localized in $\mathcal{O}$ give rise not only to the gauge-invariant operators T_n, but also to local operators $T_n(\mathcal{O})$ which belong to the local observable algebra $\mathcal{A}(\mathcal{O})$, and therefore to the center $\mathcal{Z}$ of $\mathcal{A}$, and transform as the $V(\mathcal{U}_n)$ under β^λ, eq. (8.2.26). This implies that $\mathcal{Z}$ is not pointwise-invariant under chiral transformations, so that the latter intertwine between representations labeled by different θ; this implies the inevitable breaking of the chiral symmetry in any irreducible representation of the observable algebra, and that θ plays the role of a symmetry breaking order parameter. This provides a non-perturbative derivation of the θ vacuum structure.

In order to justify the local implementation of the gauge transformations, we recall that if the gauge function $\mathcal{U}$ is $\mathbf{1}$ outside the cylinder $C_{\mathcal{U}} = K_{\mathcal{U}} \times \mathbf{R}$, the corresponding implementer $V(\mathcal{U})$ commutes with the fields localized in regions disjoint

---

[30] For a discussion of this model and its relation with QCD structures, see F. Strocchi, *An Introduction to the Mathematical Structure of Quantum Mechanics*, 2nd ed., World Scientific 2008, Section 6.8.

from $C_{\mathcal{U}}$; thus, $V(\mathcal{U})$ is local with respect to the field algebra, and any bounded cylinder $\mathcal{O}$ with base $K_{\mathcal{U}}$ may be taken as its localization region.

It is therefore reasonable that one may construct implementers $V(\mathcal{U})$ as "functions" of the fields localized in $\mathcal{O}$; more precisely, that $V(\mathcal{U})$ can be chosen to belong to the (strongly closed) algebra $\mathcal{F}_W(\mathcal{O})$, generated by the polynomials of the fields $W(\mathbf{f})$, $V^5(f)$, and of the gauge-invariant bilinears of the fermion fields, all smeared with test functions with support in $\mathcal{O}$.

As is standard, the center of $\mathcal{F}_W(\mathcal{O})$ is assumed to be trivial; therefore the local implementers are uniquely determined and, clearly, for Gauss transformations $\mathcal{U}^\lambda$ they coincide with the previously introduced Gauss operators $V(\mathcal{U}^\lambda)$ (see eq. (8.2.9)).[31]

i) Topology of the gauge group and center of the observables

According to the strategy outlined above, we start by deriving the existence of (local) central elements of the observables defined in terms of the local implementers of the large gauge transformations.

For this purpose, let $\mathcal{O} = \mathcal{O}_0 \times (t_1, t_2)$, with $\mathcal{O}_0$ a compact set in space, $\mathcal{H}'(\mathcal{O})$ the subspace of $\mathcal{H}$ consisting of vectors invariant under all the Gauss operators $V(\mathcal{U}^\lambda)$ with supp $(\mathcal{U}^\lambda - 1) \subset \mathcal{O}_0$ (briefly belonging to $\mathcal{G}_0(\mathcal{O}_0)$), and $P(\mathcal{O})$ the corresponding projection, which belongs to (the strongly closed) $\mathcal{F}_W(\mathcal{O})$.

Lemma 2.6 *If an operator A and its adjoint A^* satisfy*

$$V(\mathcal{U}^\lambda)\, A = A\, V(\mathcal{U}_1^\lambda), \quad V(\mathcal{U}^\lambda)\, A^* = A^*\, V(\mathcal{U}_2^\lambda), \tag{8.2.28}$$

$\forall \mathcal{U}^\lambda$ *with supp* $(\mathcal{U}^\lambda - 1) \subset \mathcal{O}_0$, *and supp* $(\mathcal{U}_i^\lambda - 1) \subset \mathcal{O}_0$, $i = 1, 2$, *then*

$$[A, P(\mathcal{O})] = 0. \tag{8.2.29}$$

Proof. In fact, $\forall \Psi \in \mathcal{H}$ one has

$$V(\mathcal{U}^\lambda)\, A\, P(\mathcal{O})\, \Psi = A\, V(\mathcal{U}_1^\lambda)\, P(\mathcal{O})\, \Psi = A\, P(\mathcal{O})\, \Psi,$$

i.e., $A\, P(\mathcal{O})\, \Psi \in P(\mathcal{O})\, \mathcal{H}$ and, therefore, $A\, P(\mathcal{O}) = P(\mathcal{O})\, A\, P(\mathcal{O})$. Similarly, one has $A^*\, P(\mathcal{O}) = P(\mathcal{O})\, A^*\, P(\mathcal{O})$, which implies $P(\mathcal{O})\, A = P(\mathcal{O})\, A\, P(\mathcal{O})$, and eq. (8.2.29) follows.

Proposition 2.7 *The operators* $T_{\mathcal{U}_n}(\mathcal{O}) \equiv V(\mathcal{U}_n) P(\mathcal{O})$, *with* $V(\mathcal{U}_n)$ *local implementers localized in* $\mathcal{O}$, *have the following properties:*
i) they depend only on n: $T_{\mathcal{U}_n}(\mathcal{O}) = T_{\mathcal{U}'_n}(\mathcal{O}) \equiv T_n(\mathcal{O})$, *and satisfy*

$$T_n(\mathcal{O}) T_m(\mathcal{O}) = T_{n+m}(\mathcal{O}); \tag{8.2.30}$$

ii) they belong to the center $\mathcal{Z}(\mathcal{O})$ of the local gauge-invariant algebra $\mathcal{A}(\mathcal{O}) \subset \mathcal{F}_W(\mathcal{O})$, as well as to the center $\mathcal{Z}$ of the algebra of observables $\mathcal{A}$;

---

[31] The existence of implementers of a gauge symmetry belonging to the local field algebras has been proved in general for local covariant field algebras, by exploiting the Reeh–Schlieder property; see D. Buchholz, S. Doplicher, R. Longo, and J. E. Roberts, Rev. Math. Phys. **4**, 49 (1992).

iii) under chiral transformations they transform as

$$\beta^\lambda(T_n(\mathcal{O})) = e^{-i2n\lambda} T_n(\mathcal{O}).\qquad(8.2.31)$$

This implies that the $T_n(\mathcal{O})$ are not trivial.

Proof. i) $\forall \mathcal{U}_n, \mathcal{U}'_n$ both $= 1$ outside $\mathcal{O}_0$, the product $V(\mathcal{U}'_n)^{-1} V(\mathcal{U}_n) = V(\mathcal{U}_0) \in \mathcal{G}_0(\mathcal{O}_0)$, and therefore

$$V(\mathcal{U}'_n)^{-1} V(\mathcal{U}_n) P(\mathcal{O})\Psi = V(\mathcal{U}_0) P(\mathcal{O})\Psi = P(\mathcal{O})\Psi, \quad \forall \Psi \in \mathcal{H};$$

hence

$$V(\mathcal{U}_n) P(\mathcal{O}) = V(\mathcal{U}'_n) P(\mathcal{O}) \equiv T_n(\mathcal{O})$$

depends only on n. By construction, $T_n(\mathcal{O}) \in \mathcal{F}_W(\mathcal{O})$ since so does $P(\mathcal{O})$ as well $V(\mathcal{U}_n)$, supp $(\mathcal{U}_n - 1) \subset \mathcal{O}_0$.
Moreover, $\forall \mathcal{U}$, one has $V(\mathcal{U}^\lambda) V(\mathcal{U}) = V(\mathcal{U}) V((\mathcal{U}')^\lambda)$, and supp $((\mathcal{U}')^\lambda - 1) =$ supp $(\mathcal{U}^\lambda - 1)$, so that, by Lemma 2.5, $[V(\mathcal{U}), P(\mathcal{O})] = 0$ and eq. (8.2.30) follows.
ii) Since

$$V(\mathcal{U}) V(\mathcal{U}_n) P(\mathcal{O}) = V(\mathcal{U}'_n) V(\mathcal{U}) P(\mathcal{O}) = V(\mathcal{U}'_n) P(\mathcal{O}) V(\mathcal{U}),$$

and $V(\mathcal{U}'_n) P(\mathcal{O}) = T_n(\mathcal{O})$, one obtains $V(\mathcal{U}) T_n(\mathcal{O}) = T_n(\mathcal{O}) V(\mathcal{U})$; hence, $T_n(\mathcal{O})$ is gauge-invariant and therefore belongs to $\mathcal{A}(\mathcal{O})$. Actually it belongs to $\mathcal{Z}(\mathcal{O})$, since by gauge-invariance of the observables both $V(\mathcal{U}_n)$ and $P(\mathcal{O})$ commute with $\mathcal{A}(\mathcal{O})$, as well as with $\mathcal{A}$ (clearly, Lemma 2.5 applies to any observable A and implies $[A, P(\mathcal{O})] = 0$).
iii) Since $V(\mathcal{U}_n)$, $P(\mathcal{O})$ belong to $\mathcal{F}_W(\mathcal{O})$, the chiral transformations of $T_n(\mathcal{O})$ are given by the action of $V_R^5(\lambda)$, for R sufficiently large with respect to $\mathcal{O}$. Furthermore, $\forall \mathcal{U}^\lambda$, with supp $(\mathcal{U}^\lambda - 1) \subset \mathcal{O}_0$, by eq. (8.2.27), Lemma 2.5 applies to the $V_R^5(\lambda)$, for R sufficiently large and $[V_R^5(\lambda), P(\mathcal{O})] = 0$. Hence, by eq. (8.2.27),

$$\beta^\lambda(T_n(\mathcal{O})) = V_R^5(\lambda) V(\mathcal{U}_n) P(\mathcal{O}) V_R^5(\lambda)^{-1} = e^{-i2n\lambda} T_n(\mathcal{O}).$$

In conclusion, we have shown that the topology of the gauge group is displayed in the center $\mathcal{Z}$ of the algebra of observables by the *local operators $T_n(\mathcal{O})$*, which *provide a unitary representation of the gauge group modulo the subgroup of Gauss transformations*. Therefore, the spectrum of the $\mathcal{Z}$ reflects the non-trivial topology of the gauge group. This is the way the gauge group shows up at the observable level, through its topological invariant, the winding number. *Even if the gauge group acts trivially on the observables, its invariants classify the representations of $\mathcal{A}$.*

ii) Inevitable breaking of chiral symmetry

The above structure implies that the non-trivial topology of the gauge group and eq. (8.2.20) force the breaking of chiral symmetry.

Proposition 2.8 *The chiral symmetry β^λ, $\lambda \in \mathbf{R}$, is broken in any factorial (sub)representation π' of the (gauge-invariant) observable algebra $\mathcal{A}$ in the Gauss invariant subspace $\mathcal{H}'$.*

Proof. If the chiral symmetry β^λ is unbroken in π', then, in the corresponding representation space $\mathcal{H}_{\pi'}$, there is a one-parameter group of unitary operators $U^5(\lambda)$, $\lambda \in \mathbf{R}$, satisfying

$$\beta^\lambda(F) = U^5(\lambda)\, F\, U^5(-\lambda), \qquad (8.2.32)$$

for all local operators $F \in \mathcal{F}_W$ which leave $\mathcal{H}_{\pi'}$ invariant.

Since the $T_n(\mathcal{O})$ belong to the center of the local gauge-invariant algebras $\mathcal{A}(\mathcal{O})$, as well as to the center of $\mathcal{A}$, eqs. (8.2.31) and (8.2.32) are incompatible with the triviality of the center $\mathcal{Z}$ of $\mathcal{A}$ (which characterizes the factorial representations of $\mathcal{A}$). This implies the instability under chiral symmetry of any factorial subrepresentation of the observable algebra $\mathcal{A}$.

Remarks and comments

One might get the impression from the literature that there is a more direct way of obtaining the relation between the large gauge transformations and the labeling of the irreducible (or factorial) representations of the algebra of observables, in such a way to display the instability under chiral transformations.

By the proof of Proposition 2.5,

$$V(\mathcal{U}_n)\Psi = T_n S(\mathcal{U}_n)\Psi = T_n \Psi = T_n(\mathcal{O})\Psi, \quad \forall \Psi \in \mathcal{H}'. \qquad (8.2.33)$$

However, even if eq. (8.2.27) holds (if the $V(\mathcal{U}_n)$ are local implementers) one cannot deduce from it the non-trivial transformation of their restrictions T_n to $\mathcal{H}'$. Quite generally, the transformations of operators in $\mathcal{H}$ transfer to their restrictions to $\mathcal{H}'$ in the case of symmetries described by unitary operators which leave $\mathcal{H}'$ invariant, like the large gauge transformations and the space translations ($V(\mathcal{U}^\lambda)\,V(\mathcal{U})\,\Psi = V(\mathcal{U})\,V((\mathcal{U}')^\lambda)\,\Psi = V(\mathcal{U})\,\Psi,\ \forall \mathcal{U}^\lambda,\ \forall \Psi \in \mathcal{H}'$ implies $V(\mathcal{U})\,\mathcal{H}' \subset \mathcal{H}'$).

However, this is not the case of the chiral transformations β^λ, whose implementation by unitary operators in $\mathcal{H}$ or in $\mathcal{H}'$ is in question. Actually, the action of the unitary operators $V_R^5(\lambda)$, given by the last equality in eq. (8.2.27), does not admit a restriction to $\mathcal{H}'$, since, by Proposition 2.3, the restriction of $V_R^5(\lambda)$ to $\mathcal{H}'$ vanishes. In fact, given $V_R^5(\lambda)$, there are Gauss transformations and $\Psi \in \mathcal{H}'$ such that $V(\mathcal{U}^\lambda)\,V_R^5(\mu)\,\Psi \neq V_R^5(\mu)\,\Psi$ (see the proof of Proposition 2.3). Furthermore, if the representation of the field algebra $\mathcal{F}_W$ is irreducible, the T_n are multiples of the identity and cannot display a non-invariance under β^λ. If the $V(\mathcal{U}_n)$ are local, the operators T_n belong to the strong closure of $\mathcal{F}_W$ (since so do $S(\mathcal{U}_n)$ and $V(\mathcal{U}_n) \in \mathcal{F}_W$) and are gauge-invariant, but this does not mean that they belong to the center of $\mathcal{A}$.

By the same argument of Proposition 2.7, chiral symmetry is broken in any irreducible representation of $\mathcal{F}_W$ defined by a Gauss-invariant vacuum. In fact, if chiral symmetry is unbroken in $\mathcal{H}$, eq. (8.2.32) provides the unique strongly continuous extension of β^λ to the strong closure of $\mathcal{F}_W$. In particular,

$$\beta^\lambda(S(\mathcal{U}_n)) = \mathrm{s} - \lim_{|\alpha| \to \infty} \beta^\lambda(V^a) = S(\mathcal{U}_n), \qquad (8.2.34)$$

since, by eq. (8.2.19), all Gauss operators are invariant under chiral transformations. Then, by eqs. (8.2.27) and (8.2.32), one has

$$\beta^\lambda(T_n) = \beta^\lambda(V(\mathcal{U}_n))\beta^\lambda(S(\mathcal{U}_n)^{-1}) = e^{-i2n\lambda}\, V(\mathcal{U}_n)\, S(\mathcal{U}_n)^{-1} = e^{-i2n\lambda}\, T_n. \qquad (8.2.35)$$

This is incompatible with the irreducibility of the representation of $\mathcal{F}_W$, which implies $T_n = e^{i2n\theta}\mathbf{1}$.

iii) Topology and vacuum structure; θ sectors

We may now display the (observable) *role of the gauge group $\mathcal{G}$ in providing the labeling of the irreducible representation of the observable algebra*, not only through the invariants of the irreducible representations of $\mathcal{G}$, as discussed in Chapter 7, Sections 1 and 5.2, but also *through its topological invariants*.

Proposition 2.9 *The factorial subrepresentations, π_θ, of the observable algebra in $\mathcal{H}'$ are labeled by an angle θ (θ **sectors**):*

$$\pi_\theta(T_n) = e^{i\,2n\theta}\mathbf{1}, \quad \theta \in [0,\pi)$$

*(the corresponding ground states are called the θ **vacua**).*

Proof. The central variables $T_n(\mathcal{O})$ reduce to T_n in $\mathcal{H}'$, and irreducibility or factoriality requires that T_n are multiples of the identity; eq. (8.2.30) implies $T_n = T_1^n$, and therefore $\pi_\theta(T_n) = e^{i2n\theta}\,\mathbf{1}$. Clearly, different θ label inequivalent factorial representations of $\mathcal{A}$, since θ corresponds to a point of the spectrum of $T_n(\mathcal{O}) \in \mathcal{Z}$.

The link between the non-trivial topology of the gauge group and the labeling of the factorial representations of the local observable algebra (θ sectors) is clearly displayed in a (reducible) representation of the field algebra defined by a chirally invariant vacuum state. Such an invariance arises in an analysis based on the functional integral formulation and on semiclassical considerations,[32] as well as in rigorous treatments of soluble models (*in primis* the Schwinger model[33]). In general, one obtains chirally invariant correlation functions by using chirally invariant boundary conditions in the functional integral in finite volume.[34]

We shall therefore consider the case in which chiral symmetry is implemented in $\mathcal{H}$ by a one-parameter group of unitary operators $U^5(\lambda)$, i.e., the analog of eq. (8.2.32) holds in $\mathcal{H}$; in particular,

$$\beta^\lambda(T_n(\mathcal{O})) = U^5(\lambda)\, T_n(\mathcal{O})\, U^5(-\lambda) = e^{-i2n\,\lambda}\, T_n(\mathcal{O}). \qquad (8.2.36)$$

Furthermore, since $\beta^\alpha(V(\mathcal{U}^\lambda)) = V(\mathcal{U}^\lambda)$, $U^5(\alpha)$ commutes with the Gauss operators, and therefore leaves $\mathcal{H}'$ invariant.

[32] C. G. Callan, R. F. Dashen, and D. J. Gross, Phys. Lett. **63B**, 334 (1976); S. Coleman, *Aspects of symmetry*, Cambridge University Press 1985, Chap. 7.

[33] A. Z. Capri and R. Ferrari, Nuovo Cim. **A62**, 273 (1981); Jour. Math. Phys. **25**, 141 (1983); G. Morchio, D. Pierotti, and F. Strocchi, Ann. Phys. **188**, 217 (1988); F. Strocchi, *Selected Topics on the General Properties of Quantum Field Theory*, World Scientific 1993, Sect. 7.4.

[34] J. Löffelholz, G. Morchio, and F. Strocchi, Ann. Phys. **250**, 367 (1996); B. Booß-Bavnbeck, G. Morchio, F. Strocchi, and K. P. Wojciechowski, Jour. Geom. Phys., **22**, 219 (1997).

The Hilbert space $\mathcal{H}'$ has a central decomposition over the spectrum of T_1 and one has

$$\mathcal{H}' = \int_{\theta \in [0,\,\pi)} d\mu(\theta)\, \mathcal{H}_\theta, \quad T_n \mathcal{H}_\theta = e^{i2n\theta}\, \mathcal{H}_\theta,$$

since the spectral measure $d\mu(\theta)$ can be taken invariant under translations, by eq. (8.2.36), and by the chiral invariance of the vacuum.

By eq. (8.2.36), $U^5(\alpha)$ intertwines between the θ sectors

$$U^5(\alpha)\, \mathcal{H}_\theta = \mathcal{H}_{\theta - \alpha} \tag{8.2.37}$$

and satisfies $U^5(\pi) = (-1)^F$, with F the fermion number ($= 0$, by our definition of $\mathcal{F}_W$). This means that all θ occurs.[35]

Since the chiral symmetry commutes with time translations, the spectrum of the Hamiltonian is the same in all θ sectors, and "all the θ vacua have the same energy". The chiral symmetry intertwines between different θ vacua:

$$\omega_\theta(\beta^\lambda(A)) = \omega_{\theta - \lambda}(A), \quad \forall A \in \mathcal{A},$$

and therefore, as it happens for spontaneous breaking of internal symmetries, θ sectors define isomorphic representations of $\mathcal{A}$, physically indistinguishable from the $\theta = 0$ sector.

Furthermore, eq. (8.2.27) gives

$$U^5(\lambda)V(\mathcal{U}_n)U^5(-\lambda) = e^{-i2n\lambda}V(\mathcal{U}_n);$$

i.e., a correct derivation of the equation

$$[V(\mathcal{U}_n),\, Q^5] = 2\,n.$$

The standard derivation assumes the (problematic) existence of the axial charge density and that the gauge transformations of the axial charge density extend to its space integral, giving the transformation properties of the chiral charge Q^5 (the mathematical problems of such assumptions are discussed in Section 2.2). Our derivation of the relation between chiral symmetry and large gauge transformations, eq. (8.2.27), requires neither the (usually assumed) convergence of the space integral of J_0^5, nor that of its exponential, which are incompatible with Proposition 2.3.

It is worth to stressing that the breaking of chiral symmetry is governed by a quite different mechanism with respect to the Goldstone or the Higgs mechanism. In all the three cases the symmetry is a well-defined one-parameter group of automorphisms of the relevant algebra, which commutes with spacetime translations. In the Goldstone case, the symmetry breaking order parameter, typically an observable operator, has sufficiently strong localization properties (preserved under time evolution) and its transformations under the symmetry are generated by a local conserved current. In the Higgs case, in positive gauges such as the Coulomb gauge, the symmetry breaking order parameter is not an observable and it has a non-local time evolution, so that the (time-independent) symmetry is not generated at all times by the associated conserved local Noether current (see Chapter 7, Section 6.2).

---

[35] Such a picture is exactly the same as in the quantum mechanical model of QCD structures: J. Löffelholz, G. Morchio, and F. Strocchi, Ann. Phys. **250**, 367 (1996).

In the axial $U(1)$ case of QCD, contrary to statements appearing in the literature, the chiral transformations define a one-parameter group of time-independent automorphisms of the algebra of observables. The breaking of chiral symmetry does not require the occurrence of massless Goldstone bosons, because the correlation functions of the associated conserved Noether current J_μ^5, relative to a gauge-invariant vacuum, like a θ vacuum, do not exist (as we have seen, in the temporal gauge only the exponentials of J_μ^5 exist, the representation of the field algebra being non-regular), and one cannot write the symmetry breaking Ward identities required for the Goldstone theorem.

The chiral symmetry defines automorphisms of the algebra of observables $\mathcal{A}$, which may be locally generated by unitary operators, like the $V_R^5(\lambda)$, provided the representation of $\mathcal{A}$ is not factorial. In fact, as a consequence of the non-trivial topology of the gauge group, the local observable algebras have a center which is not left pointwise-invariant under the chiral symmetry, and such a non-trivial transformation cannot be obtained by the action of unitary operators if the elements of the center are multiples of the identity.

In conclusion, the temporal gauge and the associated Weyl quantization provides a non-perturbative derivation of chiral symmetry breaking, of the absence of Goldstone bosons, and of the θ vacuum structure as a consequence of the non-trivial topology of the gauge group. A careful treatment of the peculiar mathematical properties of such a gauge allows for an acceptable mathematical setting and proofs.

2.6   Regular temporal gauge

As discussed in the abelian case, one may look for an alternative realization of the temporal gauge, by weakening the condition of Gauss gauge invariance of the vacuum, so that the corresponding correlation functions of gauge-dependent fields, not only those of their exponentials, may be defined.

To be more precise, as before, one introduces a local field algebra $\mathcal{F}$, generated by $\mathbf{A}(\mathbf{f})$, $f_a^i \in \mathcal{S}(\mathbf{R}^4)$, by the fermion fields, by their gauge-invariant bilinears, by the axial current J_μ^5, by the exponentials $V_R^5(\lambda) = \exp(i\lambda J_0(f_R \alpha_R))$, and by local operators $V(\mathcal{U})$, which implement the time-independent gauge transformations $\alpha_\mathcal{U}$, eq. (8.2.8), represent the group $\mathcal{G}$, and satisfy for R large enough

$$V(\mathcal{U}_n)J_0^5(f_R\alpha_R)V(\mathcal{U}_n)^{-1} = J_0^5(f_R\alpha_R) + 2n. \tag{8.2.38}$$

We denote by $\mathcal{A}$ the gauge-invariant subalgebra of $\mathcal{F}$, and by V_G a generic monomial of the Gauss operators $V(\mathcal{U}^\lambda)$.

A **regular quantization of the temporal gauge** is defined by a (linear) vacuum functional ω on $\mathcal{F}$, which is invariant under spacetime translations and rotations, and such that its restriction to the observable algebra $\mathcal{A}$ satisfies positivity, Lorentz invariance, and the relativistic spectral condition.

From a constructive point of view, such a realization of the temporal gauge may be related to a functional integral quantization with a functional measure given by the Lagrangian of eq. (8.2.1) with the addition of the fermionic part (see Section 2.3). The invariance of the classical Lagrangian under the residual gauge group after the gauge fixing $A_0 = 0$, does not imply the corresponding residual gauge invariance of the correlation functions of $\mathcal{F}$, as discussed in the abelian case,[36] since an infrared regularization is needed which breaks the residual gauge invariance. Therefore, the Gauss constraint no longer holds.

The correlation functions of $\mathcal{F}$, given by a vacuum ω with the above properties, define a vector space $\mathcal{D} = \mathcal{F}\Psi_0$, with Ψ_0 the vector representing ω, and an inner product on it, $< ., . >$, which is assumed to be left invariant by the operators $V(\mathcal{U})$.

Instead of the strong Gauss invariance, ω is required to satisfy the following **weak Gauss invariance**:

$$\omega(A\, V_G) = \omega(A), \quad \forall A \in \mathcal{A}, \quad \forall V_G,$$

equivalently

$$< A\, \Psi_0, V_G\, \Psi_0 > = < A\, \Psi_0, \Psi_0 >, \quad \forall A \in \mathcal{A}, \quad \forall V_G. \tag{8.2.39}$$

It follows that the vectors Ψ of the subspace $\mathcal{D}_0' \equiv \mathcal{A}\,\Psi_0$ are weakly Gauss-invariant in the sense of eq. (8.2.39) with Ψ_0 replaced by Ψ, since by eq. (8.2.39), $\forall A, B \in \mathcal{A}$,

$$< BA\Psi_0, V_G\, A\Psi_0 > = < A^*BA\Psi_0, V_G\Psi_0 > = < BA\Psi_0,\, A\Psi_0 >.$$

Furthermore the spacetime translations $U(a)$ leave $\mathcal{D}_0'$ invariant. Thus, ω defines a vacuum representation of $\mathcal{A}$ in which the Gauss law holds.

The weak form of Gauss gauge invariance of the vacuum functional allows for the existence of the fields of $\mathcal{F}$ as operators on $\mathcal{D}$, but the inner product cannot be semidefinite on $\mathcal{D}$ (by the argument of Proposition 2.1). The subspace of vectors $\Psi \in \mathcal{D}_0'$ with null inner product, $< \Psi, \Psi > = 0$, is denoted by $\mathcal{D}_0''$.

Now, there is a substantial difference in the realization of the chiral symmetry, with respect to the representation defined by a Gauss-invariant vacuum. Due to the weak form of the Gauss gauge invariance, the (smeared) conserved current J_μ^5 may be defined as an operator in $\mathcal{D} = \mathcal{F}\Psi_0$ and the standard wisdom applies. In particular, for the infinitesimal variation $\delta^5 F$ of the fields under chiral transformations, following Bardeen, one has

$$\delta^5 F = i \lim_{R \to \infty} [\, J_0^5(f_R \alpha_R),\, F\,], \quad \forall F \in \mathcal{F}. \tag{8.2.40}$$

In general, the representation $\pi^{(0)}$ of the observable algebra defined by the vacuum vector Ψ_0 may not be irreducible, and therefore, in order to discuss the breaking of the chiral symmetry, one must decompose it into irreducible representations.

The above general structure is exactly reproduced in the regular temporal gauge realization of the Schwinger model; see Section 3 below.

Even if $\omega(\delta^5 F) = 0$, a symmetry breaking order parameter may appear in the irreducible components of $\pi^{(0)}$. Moreover, such a decomposition of the vacuum functional on $\mathcal{A}$ does not *a priori* extend to a decomposition of the expectations $< \Psi_0, J_0^5(f_R \alpha_R)\, A\, \Psi_0 >$, since $J_0^5(f_R \alpha_R)$ is not gauge-invariant. Thus, one of the basic

[36] J. Löffelholz, G. Morchio, and F. Strocchi, Jour. Math. Phys. **44**, 5095 (2003); J. Löffelholz, G. Morchio, and F. Strocchi, Groud state and functional integral representation of the CCR algebra with free evolution, arXiv math-ph/0212037.

assumptions of the Goldstone theorem may fail, and chiral symmetry breaking may not be accompanied by massless Goldstone bosons.

More definite statements can be made under the following reasonable assumption, hereafter referred to as the *existence of local implementers of the gauge transformations*:

i) the subspace $\mathcal{D}'$ generated by the vectors $V(\mathcal{U})\mathcal{D}'_0$, with $\mathcal{U}$ running over $\mathcal{G}$, satisfies the weak Gauss constraint and semidefiniteness of the inner product,

ii) if $\mathcal{U} - 1$ is localized in $\mathcal{O}$, then $V(\mathcal{U})$ can be obtained as a suitable weak limit of polynomials F_n of A_i^a and ψ localized in $\mathcal{O}$, in the following sense:

$$< \Psi, V(\mathcal{U})\Phi >= \lim_{n \to \infty} < \Psi, F_n \Phi >, \quad \forall \Psi, \Phi \in \mathcal{D}. \tag{8.2.41}$$

Property i) is supported by the fact that the states defined by the vectors $V(\mathcal{U})A\Psi_0$, $A \in \mathcal{A}$, are weakly Gauss-invariant and positive; in fact, $\forall A, B, C \in \mathcal{A}$, by eq. (8.2.39),

$$< A\,V(\mathcal{U})\,B\Psi_0,\, V_G\,V(\mathcal{U})C\Psi_0 >=< \alpha_\mathcal{U}^{-1}(A)\,B\Psi_0,\, V_G'C\Psi_0 >=$$

$$=< A\,B\,\Psi_0,\, C\Psi_0 >=< A\,V(\mathcal{U})\,B\,\Psi_0, V(\mathcal{U})\,C\,\Psi_0 >.$$

The stability of a weakly Gauss-invariant subspace, which includes $\mathcal{D}'_0$, is automatically satisfied if such a subspace is selected by a gauge-covariant subsidiary condition. Weak Gauss invariance of $\mathcal{D}'$ is also implied by the following stronger form of the weak Gauss invariance of the vacuum functional,

$$\omega(A\,V(\mathcal{U}^\lambda)\,V(\mathcal{U})) = \omega(A\,V(\mathcal{U})), \quad \forall A \in \mathcal{A}, \quad \forall \mathcal{U}^\lambda, \mathcal{U}. \tag{8.2.42}$$

Property ii) is supported by the localization of the gauge transformations, so that the $V(\mathcal{U})$ are local relative to the field algebra, with the localization region of $V(\mathcal{U})$ given by the support of $\mathcal{U} - 1$.

A field F which leaves $\mathcal{D}'$ invariant also leaves the subspace $\mathcal{D}''$ invariant, $\mathcal{D}'' \equiv \{\Psi \in \mathcal{D}'; < \Psi, \Psi >= 0\}$, and therefore defines a unique gauge-invariant operator $\hat{F}$ in the "physical" quotient space $\mathcal{D}_{phys} \equiv \mathcal{D}'/\mathcal{D}''$, which is the analog of the Gauss-invariant subspace $\mathcal{H}'$ of the non-regular realization of the temporal gauge.

Furthermore, one has

$$< \mathcal{D}', [\,F, V(\mathcal{U}_n)\,]\,\mathcal{D}' >= 0,$$

since $V(\mathcal{U}_n) = V(\mathcal{U}_n^a)V_G$, $[\,F, V(U_n^a)] = 0$, by locality for $|a|$ sufficiently large, and, by i) $V(\mathcal{U}_n^a)\mathcal{D}' \subset \mathcal{D}'$, $< \mathcal{D}', [F, V_G]\mathcal{D}' >= 0$. This means that F is *weakly gauge-invariant* and to all effects can be considered as an observable field; thus, in the following we shall take as *observable algebra localized in $\mathcal{O}$*, $\hat{A}(\mathcal{O})$, the algebra of operators in $\mathcal{D}_{phys}$ generated by fields localized in $\mathcal{O}$ which leave $\mathcal{D}'$ invariant and as observable algebra $\hat{A} \equiv \cup_\mathcal{O}\hat{A}(\mathcal{O})$.[37] In particular, the local operators $V(\mathcal{U}_n(\mathcal{O}))$ are weakly gauge-invariant, and therefore define unique operators $\hat{T}_{\mathcal{U}_n(\mathcal{O})} \in \hat{A}(\mathcal{O})$ in $\mathcal{D}_{phys}$.

---

[37] For discussion of weak gauge invariance and its characterization of the observables in indefinite quantum field theories, see F. Strocchi and A. S. Wightman, Jour. Math. Phys. **15**, 2198 (1974).

By the same arguments discussed previously, the $\hat{T}_{\mathcal{U}_n}$ depend only on n, are invariant under spacetime translations, and satisfy

$$\hat{T}_n \hat{T}_m = \hat{T}_{n+m}, \quad \hat{T}_0 = 1. \tag{8.2.43}$$

Moreover, for any F which leaves $\mathcal{D}'$ invariant, one has

$$< \mathcal{D}', V(\mathcal{U}_n)\, F\, \Psi_0 > = < \mathcal{D}', F\, V(\mathcal{U}_n)\Psi_0 >.$$

This implies that the $\hat{T}_n$ generate an abelian group $\mathcal{T}$ and $\forall \mathcal{O}$, $\hat{T}_n(\mathcal{O})$ belong to the center $\mathcal{Z}(\mathcal{O})$ of $\hat{A}(\mathcal{O})$, in $\mathcal{D}_{phys}$.

Furthermore, the local generation of the infinitesimal chiral transformations, eq. (8.2.40), implies weak continuity of the derivation δ^5 on the local field algebras $\mathcal{F}(\mathcal{O})$. By property ii), the chiral transformations properties of the local implementers $V(\mathcal{U}_n)$ of the large gauge transformations are determined by eq. (8.2.38); i.e.,

$$< \mathcal{D},\, \delta^5((V(\mathcal{U}_n))\, \mathcal{D} > = \lim_{m\to\infty} < \mathcal{D},\, \delta^5(F_m)\, D > =$$

$$\lim_{m\to\infty} i < \mathcal{D},\, [J_0^5(f_R \alpha_R),\, F_m]\, \mathcal{D} > = i < \mathcal{D},\, [J_0(f_R \alpha_R),\, V(\mathcal{U}_n)]\, \mathcal{D} > =$$

$$= i\,2n < \mathcal{D},\, V(\mathcal{U}_n)\, \mathcal{D} >, \tag{8.2.44}$$

for R sufficiently large, so that $f_R(\mathbf{x}) = 1$ on the localization region of $\mathcal{U}_n - 1$. Thus, one has

$$\delta^5(\hat{T}_n) = i\,2n\,\hat{T}_n, \tag{8.2.45}$$

and the observable algebra $\hat{A}$ (in $\mathcal{D}_{phys}$) has a non-trivial center $\mathcal{Z}$.

Summarizing, we have:

Proposition 2.10   *Under the above general assumptions, one has*
i) the non-trivial topology of the gauge group gives rise to a center of the observable algebra (in the physical space $\mathcal{D}_{phys}$), which is not left pointwise-invariant under the chiral symmetry,
ii) the chiral symmetry is broken in any factorial representation of the observable algebra,
iii) the decomposition of the physical Hilbert space $\mathcal{H}_{phys} \equiv \overline{\mathcal{D}_{phys}}$ over the spectrum of $\hat{T}_1$ defines representations of the observable algebra labeled by an angle $\theta \in [0, \pi)$, giving rise to the θ vacua structure,
iv) the expectations $\omega_\theta(J_0^5(f_R \alpha_R)\, A)$, $A \in \hat{A}$, cannot be defined, and a crucial condition of the Goldstone theorem fails.

Proof. Most of the arguments are essentially the same as in the non-regular realization. In particular, an unbroken chiral symmetry in a factorial representation of the algebra of observable is incompatible with the non-trivial chiral transformations of its center.

By eq.(8.2.43), the spectrum of $\hat{T}_1$ is $\{e^{i2\theta},\, \theta \in [0, \pi)\}$, and, even if J_0^5 is well defined as an operator in $\mathcal{D}$, the existence of the expectations $\omega_\theta(J_0^5(f_R \alpha_R)\, A)$ would lead to the same inconsistency as in eq. (8.2.22).

Thus, the mechanism of evasion of the Goldstone theorem in any irreducible representation of the observable algebra, the inevitable breaking of chiral symmetry, and the θ vacuum structure are reproduced along a similar pattern as in the non-regular realization of the temporal gauge.

3 A lesson from the Schwinger model

The general features discussed above are exactly reproduced by the Schwinger model (i.e. QED_{1+1}) in the temporal gauge, usually regarded as a prototype of the non-perturbative QCD structures.

The bosonized Schwinger model in the temporal gauge is formally described by the following Lagrangian density:

$$\mathcal{L} = \tfrac{1}{2}(\partial_0\varphi)^2 - \tfrac{1}{2}(\partial_1\varphi)^2 + \partial_0\varphi A_1 + \tfrac{1}{2}(\partial_0 A_1)^2, \tag{8.3.1}$$

where φ is the pseudoscalar field which bosonizes the fermion bilinears and therefore is an angular variable, and A_1 is the gauge vector potential. The time evolution is formally determined by the following canonical equations:

$$\pi = \partial_0\varphi + A_1, \quad \partial_0 A_1 = E, \quad \partial_0\varphi = \Delta\varphi, \quad \partial_0 E = \partial_0\varphi. \tag{8.3.2}$$

1) *Representation by a Gauss-invariant vacuum*

The exponential field algebra $\mathcal{F}_W$ is generated by the unitary operators

$$V_\varphi(f), \quad \int dx\, f = n, \quad V_A(h), \quad V_E(g), \quad V_\pi(g), \quad f, g, h, \in \mathcal{D}(\mathbf{R}), \tag{8.3.3}$$

formally corresponding to the exponentials $e^{i\varphi(f)}$, $e^{iA_1(h)}$, $e^{iE(g)}$, $e^{i\pi(g)}$, respectively, and satisfying the Weyl commutation relations, with the above restriction on f, required by the periodicity of φ.

The time-independent gauge transformations

$$\alpha_{\mathcal{U}}(V_A(h)) = V_A(h)\, e^{\mathcal{U}\partial_1\mathcal{U}^{-1}}, \quad \mathcal{U}(x_1) - 1 \in \mathcal{D}(\mathbf{R}),$$

$$\alpha_{\mathcal{U}}(V_\pi(g)) = V_\pi(g)\, e^{\mathcal{U}\partial_1\mathcal{U}^{-1}}, \tag{8.3.4}$$

φ and E being left invariant, are generated by the local operators

$$V(f) \equiv V_\varphi(f)\, V_E(-f) \equiv e^{i\sigma(f)}, \quad f = -i\mathcal{U}\partial_1\mathcal{U}^{-1}, \quad \int dx\, f = n. \tag{8.3.5}$$

The gauge functions f with $\int dx\, f = 0$, i.e., those of the form $f = \partial_1 g$, $g \in \mathcal{D}(\mathbf{R})$, define the Gauss transformations, and those f_n with $\int dx\, f_n = n$, $n \neq 0$, define the large gauge transformations.

Clearly, if $\mathcal{U} - 1$ is localized in $\mathcal{O}_0$, equivalently if in eq. (8.3.5) supp $f \subseteq \mathcal{O} = \mathcal{O}_0 \times (t_1, t_2)$, then $V(f) \in \mathcal{F}_W(\mathcal{O})$, i.e., *the implementation of the large gauge transformations by local operators is verified.* Then, the dynamics defined by eqs. (8.3.2) gives that $e^{i\sigma(f)}$ is independent of time.

The chiral transformations β^λ are defined by

$$\beta^\lambda(V_\varphi(f_n)) = e^{-i2n\lambda}V_\varphi(f_n), \tag{8.3.6}$$

all the other exponential fields being left invariant.

Thus, as argued in general, the anomaly of the gauge-invariant axial current $j^5_\mu = \partial_\mu\varphi$, $\partial^\mu j^5_\mu = \varepsilon_{\mu\nu}\partial^\mu A^\nu$ does not exclude the fact that *the chiral symmetry defines a one-parameter group of automorphisms of the exponential field algebra and of its gauge-invariant subalgebra* $\mathcal{A}$, locally generated by the unitary operators $V^5_R(\lambda) \equiv V_\pi(f_R\alpha_R)$.

The GNS representation of $\mathcal{F}_W$ by a Gauss-invariant state ω is characterized by a representative vector Ψ_0 which satisfies

$$V(\partial_1 g)\,\Psi_0 = \Psi_0, \quad \forall g \in \mathcal{D}(\mathbf{R}). \tag{8.3.7}$$

The Gauss invariance of the vacuum vector is independently required by the condition of positivity of the energy, using the positivity of the state ω and the invariance under space translations.[38]

By the same argument of Proposition 2.1, the Gauss invariance of the vacuum vector implies the vanishing of all the expectations $\omega(FV_A(h))$, with F any element of the gauge-invariant subalgebra $\mathcal{A}$ of $\mathcal{F}_W$, unless $h = 0$. Hence, we are left with the correlation functions of the gauge-invariant fields.

Since eqs. (8.3.2) imply $\Box E + E = \Delta\sigma$, and E is a pseudoscalar field, the vacuum correlation functions of E are those of a free pseudoscalar field of unit mass.

By the Gauss invariance of the vacuum, the vacuum expectations $\omega(V(f_n)\,\alpha_t(V_E(g)))$ depend on f_n only through the topological number n, and therefore, by the same argument of Proposition 2.7, define operators $T_n(\mathcal{O})$ in $\mathcal{H}$, and operators T_n on $\mathcal{H}'$, which are invariant under gauge transformations and under spacetime translations and satisfy

$$T_n\,T_m = T_{n+m}.$$

The residual arbitrariness is therefore that of the representation of the abelian algebra G_T, generated by the operators T_n in the subspace $\mathcal{A}\Psi_0$. The θ vacua are characterized by the expectations

$$\omega_\theta(V(f_n)\,A) = \omega_\theta(T_n A) = e^{i2n\theta}\omega_\theta(A), \quad \forall A \in \mathcal{A}. \tag{8.3.8}$$

On the other hand, the reducible representation defined by a chirally invariant vacuum is characterized by the expectations

$$\omega(V(f_n)\,A) = \omega(T_n\,A) = \delta_{n\,0}\,\omega(A), \quad \forall A \in \mathcal{A}, \text{ with } \beta^\lambda(A) = A$$

It is easy to check that all the general features of the QCD case discussed in the previous sections, in particular the evasion of the Goldstone theorem, the breaking of chiral symmetry in any irreducible or factorial representation of the observable algebra, as a consequence of the non-trivial topology of the gauge group and the θ vacua structure, are exactly reproduced.

[38] G. Morchio and F. Strocchi, Ann. Phys. **324**, 2236 (2009).

2) *Regular representation*

The local field algebra $\mathcal{F}$ is the canonical algebra generated by the fields $A_1(h)$, $E(g)$, $V_\varphi(f)$, with $\int dx\, f = n$, $\partial_0\varphi(f)$ with the following equal-time commutation relations:

$$[\,A_1(x_1,t),\, E(y_1,t)\,] = i\delta(x_1 - y_1),$$

$$[\,V_\varphi(f),\, (\partial_0\varphi + A_1)(g)\,] = i\int dx_1\, fg\, V_\varphi(f),$$

all the other commutators vanishing.

The Euclidean functional integral corresponding to the Lagrangian of eq. (8.3.1) yields well-defined correlation functions of E and $\partial_\mu\varphi$ satisfying the (weak) Gauss law constraint. As before, the two-point function $W_E(x)$ of E is that of the free pseudoscalar massive field. The correlation functions of $e^{i\varphi}$ involve a zero mode φ_0 and crucially depend on the boundary condition (in finite volume), which can be chosen so that positivity holds. Therefore, the (weak) Gauss law holds for the expectations of the gauge-invariant variables. Periodic boundary conditions in finite volume[39] give, for any polynomial function $\mathcal{P}$,

$$< e^{i\varphi_0}\mathcal{P}(\partial_1\varphi, E) > = \delta_{n\,0} < \mathcal{P}(\partial_1\varphi, E) >,$$

An infrared subtraction is needed for defining the correlation functions of A_1. The most general two-point function W_A of A_1 must satisfy $-(d^2/dt^2)W_A(x) = W_E(x)$ and therefore has the following form ($\omega(k_1) \equiv \sqrt{k_1^2 + 1}$):

$$\frac{1}{2\pi}\int dk_1\, \omega(k_1)^{-3}\, e^{i(\omega(k_1)x_0 - k_1 x_1)} - \tfrac{1}{2}i\,(\delta(x_1) - \tfrac{1}{2}e^{-|x_1|})x_0$$

$$+B(x_1)x_0 + C(x_1). \qquad (8.3.9)$$

Locality requires that $B(x_1) = -B(-x_1)$, $C(x_1) = C(-x_1)$. The term linear in x_0 violates positivity. As discussed in the QED case the function C can be removed by an operator time-independent gauge transformation and $B = 0$ if the vacuum functional w is invariant under the CP symmetry Γ: $\Gamma(\varphi(x_1)) = \varphi(-x_1)$, $\Gamma(A_1(x_1)) = A_1(-x_1)$.

The functional integral formulation based on the Lagrangian of eq. (8.3.1) suggests a Gaussian structure for the correlation functions involving the fields E, $\partial_\mu\varphi$ and A_1. Therefore we are left with the correlation functions involving the zero mode φ_0, i.e., the correlation functions involving $e^{i\sigma(f_n)}$, $n \neq 0$.

By the weak Gauss invariance, the one-point function $< e^{i\sigma(f_n)} > \equiv s_n$ depends only on the topological number and

$$s_n = \bar{s}_{-n}, \quad < e^{i\sigma(f_n)}\, e^{i\sigma(f_m)} > = < e^{i\sigma(f_n + f_m)} > = s_{n+m}.$$

[39] For a general discussion of the role of the boundary conditions in QCD and in related models see G. Morchio and F. Strocchi, Boundary terms, long range effects, and chiral symmetry breaking, Lectures at the XXIX Int. Universitätswochen Schladming, March 1990, in *Fields and Particles*, H. Mitter and W. Schweiger (eds.), Springer 1990, pp. 171–214; B. Booß-Bavnbeck, G. Morchio, F. Strocchi, and K. P. Wojciechowski, Jour. Geom. Phys., **22**, 219 (1997); J. Löffelholz, G. Morchio and F. Strocchi, Ann. Phys. **250**, 367 (1996).

Semidefiniteness of the subspace $\mathcal{A}\Psi_0 = \mathcal{D}'$ implies that the sequence $\{s_n\}$ is of positive type and (since $e^{i\sigma(f_n)}$ commutes with $\mathcal{A}$) the vacuum functional on $\mathcal{A}$ has the decomposition

$$w(A) = \int_0^\pi d\mu(\theta)\, \omega_\theta(A), \quad \forall A \in \mathcal{A},$$

$$\omega_\theta(e^{i\sigma(f_n)}\, A) = e^{i2n\theta}\, w(A), \quad \forall A = \mathcal{P}(\partial_1\varphi, E). \tag{8.3.10}$$

As expected, the vacuum expectations of the gauge-invariant fields are the same as in the non-regular realization. The (non-positive) extension to the gauge field algebra $\mathcal{F}$ is given by the correlation functions $< e^{i\sigma(f_n)} A_1(z_1) \ldots A_1(z_k) >$, $z_i = (x_1^{(i)}, x_0^{(i)})$. For simplicity, we consider the case of a chirally invariant vacuum functional w. Then, all such correlation functions for $n \neq 0$ vanish, and $s_n = \delta_{n0}$, corresponding to $d\mu(\theta) = d\theta/\pi$.

In agreement with the general analysis of Sections 2.3, 2.4 of this Chapter, the chiral symmetry cannot be locally generated in the physical space $\mathcal{D}_{phys} = \mathcal{D}'/\mathcal{D}''$. In particular, the density of the axial current $J_0^5 = \partial_0\varphi + A_1$, which generates the symmetry on $\mathcal{F}$, cannot be defined there, by the argument of Proposition 2.4. Thus, the breaking of the chiral symmetry in any factorial representation of the observable algebra does not require the existence of massless Goldstone bosons.

Bibliography

F. Acerbi, G. Morchio, and F. Strocchi, Lett. Math. Phys. **27**, 1 (1993)

F. Acerbi, G. Morchio, and F. Strocchi, Jour. Math. Phys. **34**, 899 (1993)

S. Adler and R. Dashen, *Current Algebra*, Benjamin 1968

S. Adler, Perturbation theory and anomalies, in *Lectures on Elementary Particles and Quantum Field theory*, 1970 Brandeis Summer Institute in Theoretical Physics, S. Deser *et al.* (eds.), MIT press 1970

Y. Aharonov and L. Susskind, Phys. Rev. **155**, 1428 (1967)

H. A. Al-Kuwari and M. O. Taha, *ibid.* **59**, 363 (1990)

P. W. Anderson, Phys. Rev. **130**, 439 (1964)

H. Araki, Ann. Phys. **11**, 260 (1960)

H. Araki, K. Hepp, and D. Ruelle, Helv. Phys. Acta **35**, 164 (1962)

J. Arsac, *Fourier Transforms and the Theory of Distributions*, Prentice-Hall 1961

W. A. Bardeen, Nucl. Phys. **B 75**, 246 (1974)

A. O. Barut and R. Rączka, *Theory of Group Representations and Applications*, World Scientific 1986

A. Bassetto, G. Nardelli, and R. Soldati, *Yang–Mills Theories in Algebraic Non-covariant Gauges*, World Scientific 1991

H. Bateman, *Tables of Integral Transforms*, McGraw-Hill 1954

K. Baumann, Jour. Math. Phys. **28**, 697 (1987)

K. Baumann, On irreducible quantum fields fulfilling CCR, Lecture at the Schladming School 1987, in *Recent Developments in Mathematical Physics*, H. Mitter and L. Pittner (eds.), Springer 1987

R. Beaume, J. Manuceau, A. Pellet, and M. Sirugue, Comm. Math. Phys. **38**, 29 (1974)

P. Blanchard, Comm. Math. Phys. **15**, 156 (1959)

J. Bognar, *Indefinite Inner Product Spaces*, Springer 1974

N. N. Bogoliubov and D. V. Shirkov, *Introduction to the Theory of Quantized Fields*, Interscience 1959

N. N. Bogoliubov, A. A. Logunov, A. I. Oksak, and I. T. Todorov, *General Principles of Quantum Field Theory*, Kluwer 1990

B. Booß-Bavnbeck, G. Morchio, F. Strocchi, and K. P. Wojciechowski, Jour. Geom. Phys. **22**, 219 (1997)

H. J. Borchers, R. Haag, and B. Schroer, Nuovo Cim. **29**, 148 (1963)

K. Brading and H. R. Brown, Noether's theorems and gauge symmetries, arXiv:hep-th/0009058 v1

O. Bratteli and D. W. Robinson, *Operator Algebras and Quantum Statistical Mechanics, Vol. 2*, 2nd ed., Springer 1996

D. Buchholz, Comm. Math. Phys. **52**, 147 (1977)

D. Buchholz, Comm. Math. Phys. **42**, 269 (1975)

D. Buchholz, Huyghens' principle in quantum field theory and its structural consequences, Lectures at the XVth Winter School of Theoretical Physics in Karpacz, 1978, Acta Universitatis Wratislaviensis, No. 519

D. Buchholz, Phys. Lett. **B174**, 31 (1986)

D. Buchholz, Phys. Lett. **B174**, 331 (1986)

D. Buchholz and S. J. Summers, Scattering in relativistic quantum field theory: Fundamental concepts and tools, in *Encyclopedia of Mathematical Physics*, J.-P. Françoise *et al.* (eds.), Elsevier 2006

D. Buchholz, S. Doplicher, G. Morchio. J. Roberts, and F. Strocchi, Ann. Phys. **209**, 53 (2001)

D. Buchholz, S. Doplicher, R. Longo, and J. E. Roberts, Rev. Math. Phys. **4**, 49 (1992)

D. Buchholz and K. Fredenhagen, Nucl, Phys. **B154**, 226 (1979)

D. Buchholz and K. Fredenhagen, Comm. Math. Phys. **84**, 1 (1982)

C. G. Callan, R. F. Dashen, and D. Gross, Phys. Lett. **B63**, 34 (1976)

A. Z. Capri and R. Ferrari, Nuovo Cim. **A62**, 273 (1981)

A. Z. Capri and R. Ferrari, Jour. Math. Phys. **25**, 141 (1983)

A. Casher, J. Kogut, and L. Susskind, Phys. Rev Lett. **31**, 792 (1973)

J. M. Chaiken, Comm. Math. Phys. **8**, 164 (1968)

Y. Choquet-Bruhat and C. DeWitt-Morette, *Analysis, Manifolds and Physics*, Part I, North-Holland 1991

V. Chung, Phys. Rev. **140B**, 1110 (1965)

S. Coleman, *Aspects of Symmetry: Selected Erice Lectures*, Cambridge University Press 1985

E. Corinaldesi and F. Strocchi, *Relativistic Wave Mechanics*, North-Holland 1963

R. J. Crewther, Chiral properties on quantum chromodynamics, in *Field Theoretical Methods in Particle Physics*, W. Rühl (ed.), Plenum Press 1980

D. G. Currie and T. F. Jordan, Interactions in relativistic classical particle mechanics, in (Boulder) Lectures in Theoretical Physics Vol. X-A, *Quantum Theory and Statistical Mechanics*, A. O. Barut and W. E. Brittin (eds.), Gordon and Breach 1968

D. G. Currie, T. F. Jordan, and E. C. G. Sudarshan, Rev. Mod. Phys. **35**, 350 (1963)

E. D'Emilio and M. Mintchev, Fortsch. Physik, **32**, 473 (1984)

G. F. De Angelis, D. De Falco, and F. Guerra, Phys. Rev. D **17**, 1624 (1978)

G. De Rham, *Differential Manifolds*, Springer 1984

G. F. Dell'Antonio and S. Doplicher, Jour. Math. Phys. **8**, 663 (1967)

G. F. Dell'Antonio, S. Doplicher, and D. Ruelle, Comm. Math. Phys. **2**, 223 (1966)

P. A. M. Dirac, Canad. J. Phys. **33**, 650 (1955)

P. A. M. Dirac, Proc. Roy. Soc. (London) **A112**, 661 (1926)

P. A. M. Dirac, Proc. Roy. Soc. (London) **A114**, 243 (1927)

P. A. M. Dirac, Rev. Mod. Phys. **21**, 392 (1949)

P. A. M. Dirac, Proc. Roy. Soc. (London) **A117**, 610 (1928)

P. A. M. Dirac, *The Principles of Quantum Mechanics*, Oxford University Press 1958

P. A. M. Dirac, V. A. Fock, and B. Podolsky, Z. Physik Sowjetunion **2**, 468 (1932)

J. Dollard, Jour. Math. Phys. **5**, 729 (1964)

S. Doplicher and J. E. Roberts, Comm. Math. Phys. **131**, 51 (1990)

S. Doplicher, in *Proceedings of the International Congress of Mathematicians*, Kyoto 1990, Springer 1991)

S. Doplicher, R. Haag, and J. Roberts, Comm. Math. Phys. **23**, 199 (1971)

K. Drühl, R. Haag, and J. Roberts, Comm. Math. Phys. **18**, 204 (1970)

I. Duck and E. C. G. Sudarshan, *Pauli and the Spin–Statistics Theorem*, World Scientific 1997

W. Dybalski, Lett. Math. Phys. **72**, 27 (2005)

F. Dyson, Phys. Rev. **85**, 631 (1952)

F. J. Dyson, Phys. Rev. **75**, 1736 (1949)

R. J. Eden, P. V. Landshoff, and D. I. Olive, *The Analytic S-matrix*, Cambridge University Press 1966

H. Ekstein, Comm. Math. Phys. **1**, 6 (1965)

S. Elitzur, Phys. Rev. D **2**, 3978 (1975)

F. Englert and R. Brout, Phys. Rev. Lett. **13**, 321 (1964)

H. Epstein and V. Glaser, Le rôle de la localité dans la renormalisation en théorie quantique des champs, in *Statistical Mechanics and Quantum Field Theory*, R. Stora and C. DeWitt (eds.), Gordon and Breach 1971

H. Epstein and V. Glaser, Adiabatic limit in perturbation theory, in *Renormalization Theory*, G. Velo and A. S. Wightman (eds.), Reidel 1976

L. D. Faddeev and A. A. Slavnov, *Gauge Fields: Introduction to Quantum Theory*, Addison-Wesley 1991

E. Fermi, Accad. Nazl. Lincei **2**, 881 (1920), reprinted in *Selected Papers on Quantum Electrodynamics*, J Schwinger (ed.), Dover 1958

E. Fermi, Rev. Mod. Phys. **4**, 87 (1932)

R. Fernandez, J. Fröhlich, and A. D. Sokal, *Random Walks, Critical Phenomena, and Triviality in Quantum Field Theory*, Springer 1992

R. Ferrari, L. E. Picasso, and F. Strocchi, Comm. Math. Phys. **35**, 25 (1974)

R. Ferrari, L. E. Picasso, and F. Strocchi, Nuovo Cim. **39A**, 1 (1977)

R. P. Feynman, Phys. Rev. **76**, 749 (1949)

R. P. Feynman, Phys. Rev. **76**, 769 (1949)

R. P. Feynman, Phys. Rev. **80**, 440 (1950)

R. P. Feynman, R. B. Leighton, and M. Sands, *The Feynman Lectures on Physics Quantum Mechanics*, Addison Wesley 1970

T. Frankel, *The Geometry of Physics. An Introduction*, 2nd ed. Cambridge University Press 2004

K. Friedrichs, *Mathematical Aspects of the Quantum Theory of Fields*, Interscience 1953

H. Fritzsch, M. Gell-Mann, and H. Leutwyler, Phys. Lett. **B 47**, 365 (1973)

J. Fröhlich, G. Morchio, and F. Strocchi, Ann. Phys. **119**, 241 (1979)

J. Fröhlich, G. Morchio, and F. Strocchi, Phys. Lett. **89**, 61 (1979)

K. Fujikawa, Phys. Rev. **D21**, 2848 (1980)

L. Gårding and J. L. Lions, Nuovo Cim. Supp. **14**, 45 (1959)

I. M. Gel'fand and N. Ya. Vilenkin, *Generalized Functions*, Vol. 4, Academic Press 1964

M. Gell-Mann and M. L. Goldberger, Phys. Rev. **91**, 398 (1953)

V. Glaser, Comm. Math. Phys. **37**, 257 (1974)

R. J. Glauber, Phys. Rev. Lett. **10**, 84 (1963)

R. J. Glauber, Phys. Rev. **131**, 2766 (1963)

J. Glimm and A. Jaffe, *Quantum Physics: A Functional Integral Point of View*, 2nd ed. Springer 1987

R. H. Good, Rev. Mod. Phys. **27**, 187 (1955)

T. Goto and I. Imamura, Prog. Theor. Phys. **14**, 196 (1955)

O. W. Greenberg, Phys. Rev. Lett. **13**, 598 (1964)

H. Grosse, *Models in Statistical Mechanics and Quantum Field Theory*, Springer 1988

H. Grundling and C. A. Hurst, Lett. Math. Phys. **15**, 205 (1988)

F. Guerra, Phys. Rev. Lett. **28**, 1213 (1972)

G. S. Guralnik, C. R. Hagen, and T. W. Kibble, Phys. Rev. Lett. **13**, 585 (1964)

G. S. Guralnik, C. R. Hagen, and T. W. Kibble, Broken symmetries and the Goldstone theorem, in *Advances in Particle Physics*, Vol. 2, R. L. Cool and R. E. Marshak (eds.), Interscience 1968

R. Haag, Dan. Mat. Fys. Medd. **29**, 12 (1955)

R. Haag, in *Colloque Int. sur les Problèmes Mathématiques de la Théorie Quantique des Champs*, Lille 1957, CNRS, Paris

R. Haag, Quantum theory of collision processes, in *Lectures in Theoretical Physics*, Vol. III, W. E. Brittin *et al.* (eds.), Interscience 1961, pp. 326–52

R. Haag, Phys. Rev. **112**, 669 (1958)

R. Haag, Nuovo Cim. **14**, 131 (1959)

R. Haag, *Local Quantum Physics*, Springer 1996

R. Hagedorn, *Introduction to Field Theory and Dispersion Relations*, Pergamon Press 1963

G. C. Hegerdeldt, K. Kraus, and E. P. Wigner, Jour. Math. Phys. **9**, 2029 (1968)

W. Heisenberg, *The Physical Principles of the Quantum Theory*, Dover 1930

W. Heisenberg, Z. Physik **90**, 209 (1934)

W. Heisenberg and W. Pauli, Z. Physik **56**, 1 (1929); **59**, 169 (1930)

K. Hepp, Helv. Phys. Acta **36**, 355 (1963)

K. Hepp, On the connection between the Wightman and the LSZ quantum field theory, in Brandeis Summer Inst. Theor. Phys. 1965, Vol. I, *Axiomatic Field Theory*, M. Chretien and S. Deser (eds.), Gordon and Breach 1966

P. W. Higgs, Phys. Lett. **12**, 132 (1964)

P. W. Higgs, Phys. Rev. Lett. **13**, 508 (1964)

P. W. Higgs, Phys. Rev. **145**, 1156 (1966)

P. W. Higgs, Spontaneous symmetry breaking, in *Phenomenology of Particle Physics at High Energy: Proc. 14th Scottish Univ. Summer School in physics 1973*, R. L. Crawford and R. Jennings (eds.), Academic Press 1974

G. 't Hooft, Phys. Rev. Lett. **37**, 8 (1976)

G. 't Hooft, Phys. Rev. D **14**, 3432 (1976)

C. Hurst, Dirac's theory of constraints, in *Recent Developments in Mathematical Physics*, H. Mitter *et al.* (eds.), Springer 1987, p. 18

R. Jackiw and C. Rebbi, Phys. Rev. Lett. **37**, 172 (1976)

R. Jackiw, Field theoretical investigations in current algebra, in S. B. Treiman, R. Jackiw, B. Zumino, and E. Witten, *Current Algebra and Anomalies*, World Scientific 1985

R. Jackiw, Topological investigations of quantized gauge theories, in S. B. Treiman, R. Jackiw, B. Zumino, and E. Witten, *Current Algebra and Anomalies*, World Scientific 1985

A. M. Jaffe, Ann. Phys. **32**, 127 (1965)

A. M. Jaffe, Phys. Rev. **158**, 1454 (1967)

A. Jaffe and E. Witten, Quantum Yang–Mills theory, in *The Millennium Prize Problems*, J. Carlson *et al.* (eds.), Amer. Math. Soc. 2006

J. M. Jauch and F. Rohrlich, *The Theory of Photons and Electrons*, 2nd exp. ed., 2nd corr. print, Springer 1980

R. Jost, *The General Theory of Quantized Fields*, Am. Math. Soc. 1965

G. Källen, Helv. Phys. Acta, **25**, 417 (1952)

G. Källen and W. Pauli, Dank. Mat. Fys. Fys. Medd. **30**, No. 7 (1955)

G. Källen, Consistency problems in quantum electrodynamics, CERN Report 57–43 (1957)

G. Källen, Renormalization theory, in Brandeis Summer Institute in Theoretical Physics 1961, *Lectures in Theoretical Physics*, Benjamin 1962

G. Källen, *Quantum Electrodynamics*, Springer 1972

D. L. Karatas and K. L. Kowalski, Am. J. Phys. **58**, 123 (1990)

D. Kershaw and C. H. Woo, Phys. Rev. Lett. **33**, 918 (1974)

T. W. Kibble, Broken symmetries, in *Proc. Oxford Int. Conf. on Elementary Particles*, 1965, Oxford University Press 1966, p. 19

T. W. Kibble, Phys. Rev. **155**, 1554 (1966)

T. W. Kibble, Phys. Rev. **173**, 1527 (1968)

T. W. Kibble, Phys. Rev. **174**, 1882 (1968)

T. W. Kibble, Phys. Rev. **175**, 1624 (1968)

T. W. B. Kibble, Coherent states and infrared divergences, in *Lectures in Theoretical Physics. Vol. XI-D. Mathematical Methods in Theoretical Physics*, K. T. Mahanthappa and W. E. Brittin (eds.), Gordon and Breach 1969

B. Klaiber, Nuovo Cim. **36**, 165 (1965)

P. P. Kulish and L. D. Faddeev, Theor. Math. Phys. **4**, 745 (1970)

L. L. Landau, On the quantum theory of fields, in *Niels Bohr and the Development of Physics*, W. Pauli *et al.* (eds.), McGraw-Hill, 1955

L. D. Landau and E. M. Lifshitz, *Quantum Mechanics*, Pergamon Press 1958

L. D. Landau and E. Lifshitz, *The Classical Theory of Fields*, Addison-Wesley 1962

J. Langerholc, DESY Report No. 66/24; quoted in R. A. Brandt, Phys. Rev. **166**, 1795 (1968)

J. W. Leech, *Classical Mechanics*, Methuen 1965

H. Lehmann, Nuovo Cim. **11**, 342 (1954)

H. Lehmann, K. Symanzik, and W. Zimmermann, Nuovo Cim. **1**, 1425 (1955)

H. Leutwyler, Nuovo Cim. **XXXVII**, 556 (1965)

B. A. Lippmann and J. Schwinger, Phys. Rev. **79**, 469 (1950)

J. Löffelholz, G. Morchio, and F. Strocchi, Ann. Phys. **250**, 367 (1996)

J. Löffelholz, G. Morchio, and F. Strocchi, Ground state and functional integral representation of the CCR algebra with free evolution, arXiv math-ph/0212037

J. Löffelholz, G. Morchio, and F. Strocchi, Jour. Math. Phys. **44**, 5095 (2003)

J. H. Lowenstein, BPHZ renormalization, in *Renormalization Theory*, G. Velo and A. S. Wightman (eds.), Plenum 1976

J. H. Lowenstein and J. A. Swieca, Ann. Phys. **68**, 172 (1971)

H. D. Maison, Nuovo Cim. **11A**, 389 (1972)

F. Mandl, *Introduction to Quantum Field Theory*, Interscience 1959

A. Martin, The rigorous analyticity–unitarity program and its success, in *Quantum Field Theory*, P. Breitenlohner and D. Maison (eds.), Springer 1999

P. T. Matthews and A. Salam, Nuovo Cim. **2**, 120 (1955)

R. Mirman, Phys. Rev. **186**, 1380 (1969)

V. V. Molotkov and I. T. Todorov, Comm. Math. Phys. **79**, 111 (1981)

G. Morchio and F. Strocchi, Ann. Inst. H. Poincaré, **A33**, 251 (1980)

G. Morchio and F. Strocchi, Nucl. Phys. **B211**, 471 (1984)

G. Morchio and F. Strocchi, Ann. Phys. **168**, 27 (1986)

G. Morchio and F. Strocchi, Infrared problem, Higgs phenomenon and long range interactions, Erice Lectures, in *Fundamental Problems of Gauge Field Theory*, G. Velo and A. S. Wightman eds., Plenum 1986

G. Morchio and F. Strocchi, Removal of the infrared cutoff, seizing of the vacuum and symmetry breaking in many body and in gauge theories, invited talk at the *IX Int. Cong. on Mathematical Physics*, B. Simon *et al.* (eds.), Hilger 1989

G. Morchio and F. Strocchi, Boundary terms, long range effects, and chiral symmetry breaking, Lectures at the XXIX Int. Universitätswochen Schladming, March 1990, in *Fields and Particles*, H. Mitter and W. Schweiger (eds.), Springer 1990

G. Morchio and F. Strocchi, Representations of *-algebras in indefinite inner product spaces, in *Stochastic Processes, Physics and Geometry: New Interplays. II*, Canad. Math. Soc. Conference Proceedings Vol. 29, 2000

G. Morchio and F. Strocchi, J. Math. Phys. **44**, 5569 (2003)

G. Morchio and F. Strocchi, Jour. Phys. A: Math. Theor. **40**, 3173 (2007)

G. Morchio and F. Strocchi, arXiv:0907. 2522v1 [hep-th], Ann. Phys. **324**, 2236 (2009)

G. Morchio, D. Pierotti, and F. Strocchi, Ann. Phys. **188**, 217 (1988)

U. Moschella and F. Strocchi, Lett. Math. Phys. **19**, 143 (1990)

U. Moschella and F. Strocchi, Lett. Math. Phys. **24**, 103 (1992)

N. Nakanishi and I. Ojima, *Covariant Operator Formalism of Gauge Theories and Quantum Gravity*, World Scientific 1990

T. Nakano, Prog. Theor. Phys. **21**, 241 (1959)

S. Narison, *QCD as a Theory of Hadrons: From Partons to Confinement*, Cambridge University Press 2004

E. Nelson, Jour. Funct. Analys. **12**, 97 (1973)

E. Nelson, A quartic interaction in two dimensions, in *Mathematical theory of Elementary Particles*, R. Goodman and I. Segal (eds.), MIT Press 1966

F. Nill, Int. J. Mod. Phys. **B 6**, 2159 (1992)

I. Ojima, Nucl. Phys. **B143**, 340 (1978)

A. I. Oksak and I. T. Todorov, Comm. Math. Phys. **14**, 271 (1969)

K. Osterwalder and R. Schrader, Comm. Math. Phys. **31**, 83 (1973)

K. Osterwalder and R. Schrader, Comm. Math. Phys. **42**, 281 (1975)

K. Osterwalder, Euclidean Green's functions and Wightman distributions, in *Constructive Quantum Field Theory*, G. Velo and A. S. Wightman (eds.), Lecture Notes in Physics, Vol. 25, Springer 1973

W. Pauli, Phys. Rev. **58**, 716 (1940)

P. A. Perry, *Scattering Theory by the Enss Method*, Harwood 1983

D. Pierotti, Jour. Math. Phys. **26**, 143 (1985)

R. T. Powers, Comm. Math. Phys. **4**, 145 (1967)

L. V. Prokhorov, Lett. Math. Phys. **19**, 245 (1990)

M. Reed and B. Simon, *Methods of Modern Mathematical Physics*, Vol. 1, Academic Press 1972

M. Reed and B. Simon, *Methods of Modern Mathematical Physics, Vol. 3, Scattering Theory*, Academic Press 1978

M. Requardt, Comm. Math. Phys. **50**, 259 (1976)

G. Rideau, Covariant quantizations of the Maxwell field, in *Field Theory, Quantization and Statistical Physics*, E. Tirapegui (ed.), D. Reidel 1981

D. Ruelle, Helv. Phys. Acta, **35**, 147 (1962)

D. Ruelle, *Statistical Mechanics*, Benjamin 1969

M. Salmhofer, *Renormalization: An Introduction*, Springer 1999

B. Schroer and P. Stichel, Comm. Math. Phys. **3**, 258 (1966)

S. S. Schweber, *An Introduction to Relativistic Quantum Field Theory*, Harper and Row 1961

J. Schwinger, Phys. Rev. **75**, 651 (1949)

J. Schwinger, Phys. Rev. **76**, 790 (1949)

J. Schwinger, Phys. Rev. **115**, 721 (1959)

J. Schwinger, Phys. Rev. Lett. **3**, 296 (1959)

J. Schwinger, Phys. Rev. **128**, 2425 (1962)

J. Schwinger, Trieste Lectures 1962, in *Theoretical Physics*, I. A. E. A. Vienna 1963

B. Simon, Ann. Phys. **58**, 76 (1970)

B. Simon, *The $P(\varphi)_2$ Euclidean (Quantum) Field Theory*, Princeton University Press 1974

B. Simon, An overview of rigorous scattering theory, in *Atomic Scattering Theory, Mathematical and Computational Aspects*, J. Nuttal (ed.), University of Western Ontario, 1978

B. Simon, *Functional Integration and Quantum Physics*, Academic Press 1979

B. Simon, Int. Jour. Quantum Chemistry, **XXI**, 3 (1982)

O. Steinmann, *Perturbative Quantum Electrodynamics and Axiomatic Field Theory*, Springer 2000

R. F. Streater and A. S. Wightman, *PCT, Spin and Statistics, and All That*, Benjamin 1964

F. Strocchi, Phys. Rev. **162**, 1429 (1967)

F. Strocchi, Phys. Rev. **D2**, 2334 (1970)

F. Strocchi, Comm. Math. Phys. **56**, 57 (1977)

F. Strocchi, Phys. Rev. **D17**, 2010 (1978)

F. Strocchi, Gauss law in local quantum field theory, in *Field Theory, Quantization and Statistical Physics*, D. Reidel 1981

F. Strocchi, *Elements of Quantum Mechanics of Infinite Systems*, World Scientific 1985

F. Strocchi, *Selected Topics on the General Properties of Quantum Field Theory*, World Scientific 1993

F. Strocchi, Foundations of Physics, **34**, 510 (2004)

F. Strocchi, *An Introduction to the Mathematical Structure of Quantum Mechanics*, 2nd ed., World Scientific 2008

F. Strocchi, *Symmetry Breaking*, 2nd ed., Springer 2008

F. Strocchi and A. S. Wightman, Jour. Math. Phys. **15**, 2198 (1974)

E. C. G. Sudarshan, Structure of Dynamical theories, in 1961 Brandeis Summer Institute *Lectures in Theoretical Physics*, Vol. 2, Benjamin 1962

J. A. Swieca, Comm. Math. Phys. **4**, 1 (1967)

J. Swieca, Goldstone's theorem and related topics, in *Cargèse Lectures in Physics*, Vol. 4, D. Kastler (ed.), Gordon and Breach 1970

J. Swieca, Phys. Rev. **D13**, 312 (1976)

K. Symanzik, Euclidean quantum field theory, in *Local Quantum Theory*, R. Jost (ed.), Academic Press 1969

K. Symanzik, Lectures on Lagrangian Field Theory, DESY report T-71/1

B. Thaller, *The Dirac Equation*, Springer 1992

S. B. Treiman, R. Jackiw, B. Zumino, and E. Witten, *Current Algebra and Anomalies*, World Scientific 1985

H. van Dam and E. P. Wigner, Phys. Rev. **138**, B1576 (1965)

H. van Dam and E. P. Wigner, Phys. Rev. **142**, 838 (1966)

B. L. van der Waerden, *Group Theory and Quantum Mechanics*, Springer 1980

G. Velo and A. S. Wightman eds., *Invariant Wave Equations*, Springer 1978

S. Weinberg, Phys. Rev. **B140**, 516 (1965)

S. Weinberg, Phys. Rev. Lett. **27**, 1688 (1971)

S. Weinberg, Rev. Mod. Phys. **46**, 255 (1974)

S. Weinberg, *The Quantum Theory of Fields*, Vol. I, Cambridge University Press 1995

S. Weinberg, *The Quantum Theory of Fields*, Vol. II, Cambridge University Press 1996

J. A. Wheeler and R. P. Feynman, Rev. Mod. Phys. **21**, 425 (1949)

G. C. Wick, Phys. Rev. **96**, 1124 (1954)

G. C. Wick, Phys. Rev. **80**, 268 (1950)

G. C. Wick, A. S. Wightman, and E. P. Wigner, Phys. Rev. **88**, 101 (1952)

G. C. Wick, A. S. Wightman, and E. P. Wigner, Phys. Rev. **D1**, 3267 (1970)

A. S. Wightman, Phys. Rev. **101**, 860 (1956)

A. S. Wightman, Nuovo Cim. (Suppl.) **1**, 81 (1959)

A. S. Wightman, L'invariance dans la mécanique quantique relativiste, in *Dispersion Relations and Elementary Particles*, Les Houches 1960, C. De Witt and R. Omnes (eds.), Wiley 1960

A. S. Wightman, Recent achievements of axiomatic field theory, in *Theoretical Physics*, Trieste 1962, IAEA 1963

A. S. Wightman, Annales Inst. H. Poincaré, **I**, 403 (1964)

A. S. Wightman, Introduction to some aspects of the relativistic dynamics of quantized fields, in *High Energy Electromagnetic Interactions and Field Theory*, M. Lèvy (ed.), Gordon and Breach 1967

A. S. Wightman, Orientation, in *Renormalization Theory* (Erice School in Mathematical Physics 1975) G. Velo and A. S. Wightman (eds.), Reidel 1976

A. S. Wightman, Invariant wave equations; general theory and applications to the external field problem, in *Invariant Wave Equations*, Erice 1977, G. Velo and A. S. Wightman (eds.), Springer 1978

A. S. Wightman, ϕ_ν^4 and generalized Borel summability, in *Mathematical Quantum Field Theory and Related Topics*, Montreal 1977, J. S. Feldman and L. M. Rosen (eds.), Am. Math. Soc. 1988

A. S. Wightman, Nuovo Cim. **110 B**, 751 (1995)

A. S. Wightman, Am. J. Phys. **67**, 742 (1999)

A. S. Wightman, Elect. J. Diff. Eqs., Conf. **04**, 207 (2000)

A. S. Wightman and L. Gårding, Arkiv f. Fysik. **28**, 129 (1964)

A. S. Wightman and S. Schweber, Phys. Rev. **98**, 812 (1955)

E. P. Wigner, Ann. Math. **40**, 149 (1939)

E. P. Wigner, Unitary representations of the inhomogeneous Lorentz Group including reflections, in *Group Theoretical Concepts and Methods in Elementary Particle Physics*, F. Gursey (ed.), Gordon and Breach 1964

E. P. Wigner, Relativistic interaction of classical particles, in *Fundamental Interactions at High Energy*, Coral Gables 1969, T. Gudehus *et al.* (eds.), Gordon and Breach 1969

K. G. Wilson, Quarks: from paradox to myth; Quarks and strings on a lattice, in *New Phenomena in Subnuclear Physics*, Proceedings of the 1975 Int. School of Subnuclear Physics, Erice, A. Zichichi (ed.), Plenum 1977

Z. Wizimirski, Bull. Acad. Polon. Sci. (Math., Astr. et Phys.) **14**, 91 (1966)

D. Yiennie, S. Frautschi, and H. Suura, Ann. Phys. **13**, 379 (1961)

H. Yukawa, Proc. Math. Soc. Japan, **17**, 48 (1935); reprinted in D. M. Brink, *Nuclear Forces*, Selected Readings in Physics, Pergamon Press 1965

W. Zimmermann, Comm. Math. Phys. **8**, 66 (1968)

W. Zimmermann, Brandeis Lectures, in *Lectures on Elementary Particles and Quantum Field Theory*, S. Deser *et al.* (eds.), MIT Press 1971

Yu. M. Zinoviev, Comm. Math. Phys. **174**, 1 (1995)

Yu. M. Zinoviev, Equivalence of the Euclidean and Wightman field theories, in *Constructive Physics*, V. Rivasseau (ed.), Springer 1995

D. Zwanziger, Phys. Rev. **D14**, 2570 (1976)

The manufacturer's authorised representative in the EU for product safety is Oxford
University Press España S.A. of El Parque Empresarial San Fernando de Henares,
Avenida de Castilla, 2 – 28830 Madrid (www.oup.es/en or product.safety@oup.com).
OUP España S.A. also acts as importer into Spain of products made by the manufacturer.

Printed in the USA/Agawam, MA
May 21, 2026